ARITHMETIC AND ALGEBRA ...AGAIN

Brita Immergut
Jean Burr Smith

Foreword by Sheila Tobias
author of *Overcoming Math Anxiety*

McGraw-Hill, Inc.

New York   St. Louis   San Francisco   Auckland   Bogotá
Caracas   Lisbon   London   Madrid   Mexico City
Milan   Montreal   New Delhi   San Juan
Singapore   Sydney   Tokyo   Toronto

Brita Immergut is professor of mathematics at Fiorello H. LaGuardia Community College, The City University of New York. After earning an M.S. from the University of Uppsala in her native Sweden, Professor Immergut did further graduate work in mathematics at New York University and received an Ed.D. degree in Mathematics Education from Teachers College, Columbia University, where her thesis dealt with basic skills in mathematics and attitudes toward math in working people. Her concern for and special understanding of those whose math skills do not come close to their skills in other areas make her the enthusiastic and effective teacher she is, whether in the classroom or in the workshops and courses for math-anxious people she has designed for numerous schools and organizations, including The New School for Social Research and the Port Authority of New York. Although Professor Immergut's career includes teaching and graduate research in chemistry, she finds her work in teaching basic skills to math-anxious people eminently more satisfying and challenging.

Jean Burr Smith taught mathematics for 22 years at Middlesex Community College (Conn.), where she brought excitement and inspiration to her classroom, replacing fear and failure with confidence and success. One of the originators of the concept of math anxiety, Professor Smith developed a national and international reputation in the field, serving as a consultant to colleges throughout the United States and in England, Australia, and Africa. She is the recipient of many awards, including the American Mathematical Association of Two-Year Colleges' Mathematical Excellence Award (1986), the first PIMMS (Project to Increase Mastery of Mathematics and Science) Vanguard Fellows' award (1988), and the Colby College Distinguished Educator award (1990). In 1992 the Jean Burr Smith Library at Middlesex Community College was dedicated in the honor of this respected and beloved teacher. Professor Smith received a Master of Arts in Teaching from Harvard University and a Certificate of Advanced Study from Wesleyan University. Before her career at Middlesex Community College she taught in secondary school and, for 3 years, at the University of Connecticut. She also served for 4 years as teacher coordinator at the Wesleyan Math Anxiety Clinic.

ARITHMETIC AND ALGEBRA...AGAIN

Copyright © 1994 by McGraw-Hill, Inc. All rights reserved. Printed in the United States of America. Except as permitted under the Copyright Act of 1976, no part of this publication may be reproduced or distributed in any form or by any means, or stored in a data base or retrieval system, without the prior written permission of the publisher.

3 4 5 6 7 8 9 10 11 12 13 14 15 16 17 18 19 20 SEM SEM 9 8 7 6 5

ISBN 0-07-031720-8

Sponsoring Editor: Jeanne Flagg
Production Supervisor: Leroy Young
Editing Supervisor: Patty Andrews
Book Designer: Rafael Hernandez
Cover design by Carla Bauer Design, Inc.

Library of Congress Cataloging-in-Publication Data

Immergut, Brita.
    Arithmetic and algebra—again / Brita Immergut, Jean Burr Smith; foreword by Sheila Tobias.
        p. cm.
    Includes index.
    ISBN 0-07-031720-8
    1. Mathematics. I. Smith, Jean Burr. II. Title.
QA39.2.I46 1994
513′.122—dc20   93-15956
                                                   CIP

CONTENTS

Chapter 3   Decimals and Percents   **63**

Chapter 4   Fractions   **93**

Chapter 5   Applications   **125**

PART THREE   PRACTICAL MATHEMATICS

Math Anxiety: An Update

By Sheila Tobias,

Math education activist and author[1]

Millions of people—and if you are using this book, you may be one of them—are blocked by fear of mathematics. "I can't do math," you say. "I never could do math." "There's no way I could pass that course," or learn algebra, or calculate mortgage payments for a client. It may surprise you to know this, but research shows that most of you are capable of doing much more mathematics than you think. You don't have a *mind* problem, you have a problem of *nerve*. But to get over your block, you're going to have to revisit mathematics in a new way.

The authors of this excellent new book know from their experience teaching mathematics to adults that you won't respond to a replay of the Run-Spot-Run mathematics you had in grade school. They want to give you a positive, reassuring *new* experience with math, one that erases painful early memories, one that makes you realize how much mathematics you have stored in your memory but have just forgotten how to use. Working your way through their text won't make you a mathematician, but it will give you the willingness and the capability *to use the math you know and learn the math you need as you need it*.

Fifteen years ago, when I began my work with math–anxious students and adults at a Connecticut university, I heard about a gifted teacher who was teaching

---

[1]Sheila Tobias, *Overcoming Math Anxiety* 1978, (new edition, in press); *Succeed with Math: Every Student's Guide to Conquering Math Anxiety* (1987).

a popular new kind of math course at a nearby community college. Jean Smith, who created that course, joined our Math Clinic staff and helped us to steer our students from "math anxiety reduction," to actual courses in mathematics. Hers were the first courses in the country to attract the "math anxious" back into math.

Now, after many more years of teaching and writing, Jean Smith has joined forces with Brita Immergut of LaGuardia Community College in New York, to give you a short course in everything you wanted to know about fractions, decimals, algebra, statistics, measurement, graphing and business math (to list just a few of their topics) but were until now afraid to ask.

This book is meant to give you back *control*, control over mathematics, over your life and work. Never again will you turn down a job opportunity, turn over a child's math homework problem to someone else in the family, refuse to "figure out the check." Here, finally, is the math book you always needed, the one that shows you how to apply what you're learning right away instead of at the end of the chapter; the one that talks your language and goes at a pace you can feel comfortable with; the one that makes you feel smart and capable and unafraid.

Good reading and good luck!

TO THE INSTRUCTOR

This arithmetic and beginning algebra text is written for students and anybody else whose experience with mathematics has been one of frustration but whose need for it in the working or the academic world has been constantly increasing. It is designed to help these nontraditional students gain confidence in their ability to cope with mathematics and to learn or relearn the basics of arithmetic and algebra necessary for success in their jobs or in other areas of study. Students who master this material will be ready to take a more rigorous algebra course or a college-level math course.

Although the book's primary audience is the student in a one-semester community or four-year college developmental course that covers the fundamental operations of arithmetic and elementary algebra, it can also be used for in-house courses and for teacher training. Part of the book is also suitable as a review course for high school students.

The order in which the topics are presented is life-experience oriented. Thus, integers, decimals, and percents are presented before fractions. This is a more appropriate order of presentation than one designed for elementary school students since adults can draw on their shopping, banking, and other real-world experiences. In working with nontraditional students, we have found that starting with material that is new to them, such as integers, is more successful than starting with material that has been troublesome, such as fractions.

The book covers operations with whole numbers, integers, decimals, fractions, percents, and the application of these skills in problem solving. Basic operations in algebra, solutions of first-degree equations, and graphing are included. The last two chapters deal with banking problems and with probability and statistics.

In general, the problems are geared to the adult's interests and day-to-day concerns. We have also included some problems from nineteenth century math books. Our students find them amusing and far removed from the grade-school variety they may associate with their earlier difficulties.

Since most adults use calculators at work, we encourage them to use calculators for problem solving but expect them to approximate their answers first. By showing students several ways of getting the answer (paper and pencil, calculator, and estimation), we feel that they will become more comfortable with math and that their math anxiety will gradually disappear.

Definitions and rules are highlighted throughout the text. At the end of each chapter is a summary with vocabulary and review problems, a check list to help students determine whether they have mastered all topics in the chapter, and a Readiness Check to indicate their readiness to go on to new material. Answers to all the exercises and the Readiness Checks will be found at the end of the book. A supplement with tests for each chapter is available for the instructor.

•

We wish to thank our families and friends for their patience and support during the preparation of this book. Our thanks and appreciation also go to Jeanne Flagg and Patty Andrews, editors at McGraw-Hill, for their encouragement and expert assistance in guiding the book toward publication.

TO THE STUDENT

This book is written for all of you whose worst memories of school are those having to do with numbers: the flash cards in elementary school when you hoped the teacher would finish before she ever got to you; the timed tests, where you not only had to be right, you had to be *fast*; the long division, where, after you had learned to go from right to left in addition, subtraction, and multiplication, you were told to go from left to right and *guess*; the *x* of algebra—why *x*?—; the proofs of geometry, which you memorized but never understood.

Perhaps worst of all was the fact that you could often get the right answer, but you didn't know *why* it was right. You lived in fear that sooner or later your bluff would be called and *everyone* would know how dumb you were.

It is not enough to know that much of the world feels as you do (although this is absolutely true). The fact that you are reading this book indicates that you need to learn to cope with numbers—whether you are looking for a job, want to advance in your present job, want to change jobs, or are making the decision to continue in or return to school. Wherever you are, you need to use numbers!

This book is intended to get you started. We'll try to get you over some hurdles. As a capable human being, you cope in all kinds of ways with all kinds of day-to-day problems. There are no studies that show that you cannot cope equally well with mathematics. You have likely had a great deal of experience with practical mathematics. You have paid for groceries with various bills and coins; you have paid bills by check, paid taxes, and figured out tips. Many people understand exactly how much to pay for clothing that is marked down 20% but are petrified when it comes to doing percent problems with paper and pencil in a school setting.

Why does mathematics affect people this way? There are some obvious reasons. From the earliest grades, the gold stars go to those who get the answers right and fast; feelings of helplessness and hopelessness with numbers overcome too many young children. Add to this the fact that mathematical knowledge is cumulative—that is, if you can't add, you can't subtract or multiply, and if you can't subtract or multiply, you can't divide, and without all four, fractions and decimals are impossible—and you get a feeling of how difficult and frustrating these first stages of mathematics can be.

Further and less obvious reasons are developed by Sheila Tobias in her book *Overcoming Math Anxiety*. Of these, four seem to us of particular importance.

The first is the "Sudden Death" experience. Students describe this experience in much the same way regardless of the mathematical level at which it occurred. Suddenly a concept was too hard to understand although it was simply the next step in a sequence of skills. It was as if a "curtain had been drawn." There was a "wall ahead" or a "dropoff" or a "steep cliff," and the student had the feeling that he or she would never be able to understand that particular concept and that it was useless to try. Along with this feeling of hopelessness is one almost of paranoia—that everyone will find out how little the student knows and how he or she never really understood. In an effort to cover up these feelings of failure, students at this point will not ask questions, so there is no way out of the dilemma. They feel guilty and ashamed and never consider the possibility that it may not be all their own fault.

Contributing to this feeling of helplessness is the common myth about mathematical ability—the mystique of "the mathematical mind." Certainly creative mathematicians, like creative people in all areas, have special talents. However, ordinary people who can do college-level work in English, psychology, or biology have the ability to do mathematics of the same level. No teacher of history ever told a student who handed in an inadequate term paper that he or she did not have a "historical mind." Yet with mathematics, people assume that having some sort of genetic predisposition is the only way to survive at any level.

A third factor contributing to feelings of frustration and hopelessness is that many mathematical operations involve ideas that are particularly frustrating to the verbally gifted student: "multiplying" fractions, which results in smaller numbers; "dividing," which gives you larger numbers; "adding" positive and negative numbers, when what you are actually doing is "subtracting." The only way to survive in such a world, anxious students say, is to memorize. The answers just don't seem logical.

The fourth factor is the inability to handle frustration—not frustration in general, but only as it occurs in dealing with mathematics. Students we've talked with about this agree that when they are writing an English paper, for example, they will put aside a paragraph if it is not going as they wished and take a fresh look at it later, but they believe that they shouldn't leave a math problem until they finish it.

As a start in dealing with these anxieties or frustrations, try keeping a log of good and bad experiences with numbers and, particularly, how you react to each experience. When you succeed, do you say, "That's just luck," and when you fail, "I knew I couldn't do it"? Try reversing these reactions. When you succeed, say, "I *knew* I could do it."

Because we feel that looking at your math history is important, we recommend that you answer the following questions for yourself.

1. The last math course I took was _____. I took it about _____ years ago.

2. An early experience I remember in math class was _____

   _____

   _____

   _____

3. One math teacher I remember is _____

   _____

   _____

4. I feel _____ was the hardest part of math for me to learn and _____ was the easiest.

5. I think I learned my present approach to doing math when_____

   _____

   _____

6. My goal in mathematics is _____

   _____

7. I expect this book to help in this way: _____

   _____

   _____

8. To improve my math attitude, I expect to do the following for myself: _____

_____

_____

To get the negatives out of the way, here is a check list of reasons people give for not being able to do mathematics and for avoiding it. Check the ones that apply, and fill in the blanks for 12 and 13 if you wish.

☐ 1. I never could do math.
☐ 2. Math just isn't my subject.
☐ 3. Everyone in my family has trouble with math.
☐ 4. Nobody is around who can help me.
☐ 5. Some people can do math, and some just can't.
☐ 6. My mind goes blank when I look at a math problem.
☐ 7. I'll never need it anyway.
☐ 8. It might ruin my grade-point average.
☐ 9. I can always hire someone to do it for me.
☐ 10. I never had a good math teacher.
☐ 11. When I got to geometry, it just didn't make sense.
☐ 12. _____

_____

☐ 13. _____

_____

When you have filled this in, you might talk about it with someone else. Share your feelings. You may find that each of you thinks that you are absolutely the worst!

Since too many people put all the blame for their mathematical difficulties on themselves, read the following and check the items that seem particularly helpful to you.

Math Anxiety Bill of Rights

I have the right to learn at my own pace and not feel put down or stupid if I'm slower than someone else.
I have the right to ask whatever questions I have.
I have the right to need extra help.
I have the right to ask a teacher for help.
I have the right to say I don't understand.
I have the right not to understand.
I have the right to feel good about myself regardless of my abilities in math.
I have the right not to base my self-worth on my math skills.
I have the right to view myself as capable of learning math.
I have the right to evaluate my math instructors and how they teach.
I have the right to relax.
I have the right to be treated as a competent adult.
I have the right to dislike math.
I have the right to define success in my own terms.

SANDRA DAVIS, *University of Minnesota*

We assume that you have found yourself in a position where you need to *use* mathematics. Therefore the emphasis in this book will be on applying the skills you have or will acquire. **Reading carefully is the first step.**

Here is a "reading" problem. Keep track of how you feel as you do it, and try to find someone to do it with you. It is helpful when "doing" mathematics to share the experience with someone else.

A fire fighter, standing on the middle rung of a ladder, goes up three rungs. (Rungs are the steps on a ladder.) After a burst of flames the fire fighter goes down five rungs and then up seven until the fire is out. Then he goes up seven more and into the building at the top rung. How many rungs are there in the ladder?

You've read the problem through once.
Read it again, slowly, and then draw some pictures.

Solution   The pictures above help you to visualize the problem. In this case you can "read" the answer.

From the middle to the top there are 12 rungs, so from the middle to the bottom there have to be 12 rungs, and then there's the middle rung. Answer: 25!

After you have read a problem carefully, the next step is to decide on an order of attack. That was easy in the ladder problem—the sequence was in the wording. Think about how you go about the process of putting on your coat. What if you tried to do it in this order:

1. Put one arm in one sleeve.
2. Button the buttons.
3. Pull coat up on your shoulders.
4. Put other arm in other sleeve.

Certainly the order matters. You couldn't very easily button the buttons with only one arm in a sleeve! So the answer is: 1, 4, 3, 2.

In mathematics an orderly approach is essential. But there is often more than one way to solve a problem. Never believe that there is only one sacred way to solve a math problem.

In this book we stress the applied mathematics rather than theory. So we have drawn upon experiences such as shopping and banking, emphasizing ways that mathematics (arithmetic and algebra) is used in daily life.

You most likely use a calculator at work, at home, or in other classes in school, and we encourage you to use a calculator as you work through examples and exercises in this book too. However, we suggest that you estimate the answers first, before you use the calculator. Keep in mind that the calculator does not always give the right answer. The battery may be wearing out, or you may not always push the correct key. If your estimate and the calculator answer are too different, use the calculator again.

Answers to all exercises will be found at the end of the book. To identify the answers to the exercises within the text, a triple numbering system is used—For example, Exercise 1.2.3 refers to chapter one, section two, exercise set number three.

WORDS OF ADVICE

1. Watch your frustration level. You handle frustrations in other areas. Try those methods here.

2. Put the math aside and do something else if you feel panic setting in.

3. Find others to work with.

4. Don't waste time worrying about whether or not you have a mathematical mind. If you do poorly on a biology test, do you think that you don't have a biological mind?

5. Remember, there is no one "best" way, and no extra credit for the "shortest" or "fastest."

6. Remember your "bill of rights."

PART

ARITHMETIC

THE ARITHMETIC OF WHOLE NUMBERS

Where it all began

During the late Stone Age, hunting and fishing societies developed in the Nile Valley and other parts of Africa. From the mathematical point of view, the most interesting find from these early societies is a carved bone, which was discovered at the fishing site at Ishango, on Lake Edward in Zaire. It is a bone tool handle with notches arranged in definite patterns, as shown in Figure 1.1.

Figure 1.1

There are three separate sets of markings on the bone. One set has four groups of 11, 13, 17, 19; another 11, 21, 19, 9; and the third, which is not shown, has seven groups of 3, 6, 4, 8, 10, 5, 5. This 8000-year-old tally stick has puzzled anthropologists for a long time. It is not known if the tally marks signify counting of perhaps animals or if they show relationships between numbers.

1.1 BASIC IDEAS

Vocabulary

There are four *basic operations* in mathematics: addition, subtraction, multiplication, and division.

Often when we talk about a collection of numbers, such as the numbers 1, 2, and 3, we use the word *set*. We could use set notation with braces, { }, to list the numbers: {1, 2, 3}. The set of *even numbers* could be written as

{2, 4, 6, 8, 10, . . .}, and the set of *odd numbers* as {1, 3, 5, 7, . . .}. (The three dots indicate that the numbers continue indefinitely. In any collection of numbers ending in dots, there is no largest number.)

In this chapter we deal with two sets of numbers: the *counting numbers* {1, 2, 3, 4, . . .} and the *whole numbers* {0, 1, 2, 3, . . .}. The whole numbers are just the counting numbers plus zero. When we count, we start with 1. When we answer the question "How many?" we need zero as a possible answer.

Symbols are necessary to make mathematical statements complete. For example, we use symbols for addition (+) and multiplication (×).

We use symbols to compare two numbers:

$=$ as in $8 + 3 = 11$ 8 plus 3 *equals* 11

$<$ as in $3 < 8$ 3 *is less than* 8

$>$ as in $8 > 3$ 8 *is greater than* 3

Notice that the symbols for less than and greater than are always open toward the larger number.

When statements are *not* true, we put a slash through the symbol:

$6 + 3 \neq 11$ $6 + 3$ does *not equal* 11

$5 \ngtr 7$ 5 is *not greater* than 7

$9 \nless 6$ 9 is *not less* than 6

Numerals are symbols for *numbers*, which are abstract ideas. For example, a fisherman 8000 years ago might record that he caught ||| fish. We would write 3 for the amount. ||| and 3 are symbols for the same number. Our number symbols are called arabic numerals. For the most part, in this book we say "number" instead of "numeral."

Digits are the number symbols (numerals) 0, 1, 2, 3, 4, 5, 6, 7, 8, and 9 in our number system. Numbers are written as combinations of any of these ten digits.

A whole number is written as a string of digits. 7 is a one-digit number; 32 is a two-digit number with 3 as the first digit and 2 as the second digit. 487 is a three-digit number with 4 as the first digit, 8 as the second digit, and 7 as the third digit.

Depending on its place in the number, a digit will take on different values. For example, we read the number 67 as sixty-seven, that is, 60 plus 7. 6 is in the tens place and is worth 6 tens, or 60, while 7 is in the ones or units place and is therefore worth 7.

We read 128 as one hundred twenty-eight: $100 + 20 + 8$. 1 is in the hundreds place, 2 is in the tens place, and 8 is in the units place.

2346 is read as two thousand three hundred forty-six: $2000 + 300 + 40 + 6$. 2 is in the thousands place, 3 is in the hundreds place, 4 is in the tens place, and 6 is in the units place.

The names of the *place values* are often illustrated in a chart such as Figure 1.2.

Figure 1.2

EXERCISE 1.1.1

1. Use set notation to indicate:
   (a) the counting numbers from two through seven.
   (b) the counting numbers larger than five.
   (c) the first three whole numbers.
   (d) the first five odd whole numbers.

2. Find the value of the digit 5 in the following numbers.
   (a) 50
   (b) 125
   (c) 5326
   (d) 5

3. Translate into mathematical symbols.
   (a) Four hundred five
   (b) Six hundred fifty
   (c) Three thousand fifty-six
   (d) Six thousand four hundred

4. Write the following in words.
   (a) 5236
   (b) 8204
   (c) 7029
   (d) 1002

1.2 OPERATIONS

The Number Line

Numbers can be shown on a *number line*, a line divided into equal sections and numbered:

We can *graph* a number on the number line by putting a circle on the mark that indicates the number. For example, here's a graph of 0, 2, 4, and 5:

We can show distances as well as points on number lines. To show the distance 2, we draw an arrow above two points that are 2 spaces apart on the number line. For the distance 4, the arrow is 4 spaces long.

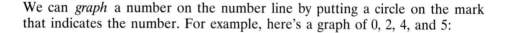

Addition

Numbers that are added or subtracted are called *terms*; the answer in addition of terms is the *sum*. For example, in $2 + 1 = 3$, 2 and 1 are terms and 3 is the sum.

When you are asked "What is $8 + 5$?" can you give the correct answer quickly? How about $6 + 7$? Most of us are not completely comfortable with certain number pairs. One way to identify these is to make a table. To shorten the time you need to fill in the table, look at these examples.

EXAMPLE

Find the sums.
(a) three and two; two and three
(b) four and three; three and four

Solution
(a) $3 + 2 = 5$; $2 + 3 = 5$
(b) $4 + 3 = 7$; $3 + 4 = 7$

The additions in part (a) of the example are shown by the two pairs of arrows over the number line in Figure 1.3. In each pair, the tail of the second arrow is placed at the head of the first. Their combined length is the sum.

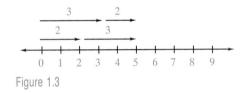

Figure 1.3

We say that addition is *commutative*, which means that the order in which you add the numbers does not matter.

EXERCISE 1.2.1

1. Fill in the Addition Facts table.

ADDITION FACTS										
+	0	1	2	3	4	5	6	7	8	9
0	—	—	—	—	—	—	—	—	—	—
1	—	—	—	—	—	—	—	—	—	—
2	—	—	—	—	—	—	—	—	—	—
3	—	—	—	—	—	—	—	—	—	—
4	—	—	—	—	—	9	—	—	—	—
5	—	—	—	—	9	—	—	—	—	—
6	—	—	—	—	—	—	—	—	—	—
7	—	—	—	—	—	—	—	—	—	—
8	—	—	—	—	—	—	—	—	—	—
9	—	—	—	—	—	—	—	—	—	—

$4 + 5 = 9$, so we write 9 across from the 4 and down from the 5.
$5 + 4 = 9$, so we write 9 across from the 5 and down from the 4.

After completing the table, circle any of the addition facts you are not comfortable with. Then write them on a 3×5 card, so you can refer to the card whenever you need to.

Some number pairs might be more comfortable for you than others, and you might be able to change an uncomfortable pair into a comfortable pair. For example, if $7 + 5$ is hard to remember, but $5 + 5$ is easy, split 7 into $2 + 5$. Then you have $2 + 5 + 5 = 2 + 10 = 12$.

When you have to add larger numbers, by all means use a calculator. But if you want to practice on a few examples, we have worked one out in detail for you.

EXAMPLE

Add   247
           +896

Solution   Start at the right-hand column. Add $7 + 6$.

$$\begin{matrix} \overset{1}{2}47 \\ +896 \\ \hline 3 \end{matrix}$$ $7 + 6 = 13$ Write down 3 in the units column and *carry* 1 into the 10s column (put 1 over 4).

Add second column plus carry: $1 + 4 + 9$.

$$\begin{matrix} \overset{1\,1}{2}47 \\ +896 \\ \hline 43 \end{matrix}$$ $1 + 4 + 9 = 14$ Write 4 in the 10s column and *carry* 1 into the 100s column.

Add $1 + 2 + 8$.

$$\begin{matrix} \overset{1\,1}{2}47 \\ +896 \\ \hline 1143 \end{matrix}$$ $1 + 2 + 8 = 11$ Write 1 in the 100s column and *carry* 1 into the 1000s column.

EXAMPLE

Add $2 + 356 + 45 + 193$

Solution   Write the numbers in columns with the digits in the ones place below each other. Then proceed with the addition.

$$\begin{matrix} \overset{1\,1}{}2 \\ 356 \\ 45 \\ +193 \\ \hline 596 \end{matrix}$$

EXERCISE 1.2.2

1. Graph the following on number lines.
    (a) 1
    (b) 3
    (c) the distance 3
    (d) the distance 5
    (e) $5 + 3$

2. Write the numbers in columns and add.
   (a) 39 + 146
   (b) 398 + 469
   (c) 457 + 872
   (d) 1045 + 307
   (e) 236 + 421 + 132
   (f) 609 + 373 + 290
   (g) 123 + 7 + 15
   (h) 311 + 238 + 5114 + 3625
   (i) 6732 + 2 + 107 + 642 + 28
   (j) 1000 + 2010 + 1926 + 17

Multiplication

We multiply *factors* and get the *product* as the answer. In $2 \times 3 = 6$, 2 and 3 are factors and 6 is the product.

A number can often be *factored*—expressed as a product of its factors. For example, 12 can be written as 2×6, 6×2, 4×3, 3×4, 12×1, or 1×12.

12 is a *composite* number because it can be factored into factors other than itself and 1. The factors of 12 are 1, 2, 3, 4, 6, 12. Some other composite numbers are 4, 6, 8, 9, 10, 12, 14.

A number that cannot be factored by anything other than 1 and itself is called a *prime number*. Examples of prime numbers are 2, 3, 5, 7, 11, 13, 17, and 19.

When a number is multiplied by $1, 2, 3, \ldots$, we get *multiples* of the number. For example, 5, 10, 15, and 20 are multiples of 5. 8, 16, 24, and 32 are multiples of 8.

EXERCISE 1.2.3

1. Write all possible factors of
   (a) 30
   (b) 42
   (c) 49
   (d) 53

2. Write all prime numbers between 50 and 60.

3. Which prime number comes after 31?

4. Write the composite numbers larger than 20 but smaller than 30.

Multiplication is a kind of shorthand for addition. For example,

$$3 \times 2 = 3 \text{ twos} = 2 + 2 + 2 = 6$$
$$2 \times 3 = 2 \text{ threes} = 3 + 3 = 6$$

From this example we see that multiplication is also commutative. The order does not matter: $3 \times 2 = 2 \times 3 = 6$. This property is shown with arrows on the number line in Figure 1.4.

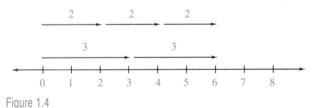

Figure 1.4

EXERCISE 1.2.4

1. Fill in the Multiplication Facts table below.

MULTIPLICATION FACTS										
×	0	1	2	3	4	5	6	7	8	9
0	—	—	—	—	—	—	—	—	—	—
1	—	—	—	—	—	—	—	—	—	—
2	—	—	—	—	—	—	—	—	—	—
3	—	—	—	—	—	—	18	—	—	—
4	—	—	—	—	—	—	—	—	—	—
5	—	—	—	—	—	—	—	—	—	—
6	—	—	—	18	—	—	—	—	—	—
7	—	—	—	—	—	—	—	—	—	—
8	—	—	—	—	—	—	—	—	—	—
9	—	—	—	—	—	—	—	—	—	—

$3 \times 6 = 18$, so we write 18 across from 3 and down from 6. $6 \times 3 = 18$, so we write 18 across from 6 and down from 3.

Again, *circle* any facts that you are not comfortable with in the table. Write them on a 3×5 file card, and refer to it when you need to.

Multiplication is usually harder than addition, particularly once you get beyond the fives. Sixes, sevens, and eights are the hard ones for many people. If they are for you, you are in good company!

Here too you can change uncomfortable pairs to comfortable ones. For example, if 7×8 is hard for you, you can change 8 to 4×2 and get $7 \times 8 = 7 \times 4 \times 2 = 28 \times 2 = 56$. You could also split 8 into $7 + 1$ and get $7(7 + 1)$ or split 8 into $5 + 3$ and get $7(5 + 3)$. In both cases use the *distributive* law, which distributes the multiplication over the addition:

$$7(7 + 1) = 7 \times 7 + 7 \times 1 = 49 + 7 = 56$$

or

$$7(5 + 3) = 7 \times 5 + 7 \times 3 = 35 + 21 = 56$$

It is important to remember that zero times any number is always zero.

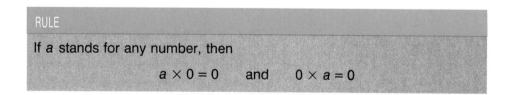

RULE

If *a* stands for any number, then

$$a \times 0 = 0 \quad \text{and} \quad 0 \times a = 0$$

The symbol for addition is always the plus sign $(+)$, but multiplication can be written in several different ways. For example,

"two times three" can be written

$$2 \times 3, \quad 2 \cdot 3, \quad 2(3), \quad (2)3, \quad (2)(3), \quad or \quad 2*3 \text{ (used on computers)}$$

In algebra the most common way to write multiplication of two numbers is to use one or two pairs of parentheses, such as 9(3), or (9)(3).

When you have to multiply large numbers, use the calculator, but practice now and then with paper and pencil. Here are examples you can try by yourself without a calculator.

EXAMPLE

Multiply 83 × 4.

> **Solution**   First write the problem as
>
> $$\begin{array}{r} 83 \\ \times\ 4 \\ \hline \end{array}$$

Then begin with the ones column.

$$\begin{array}{r} \overset{1}{8}3 \\ \times\ 4 \\ \hline 332 \end{array}$$

$$4 \times 3 = 12 \quad \text{Write 2 and carry 1.}$$
$$4 \times 8 = 32 \quad 32 + 1 = 33$$

EXAMPLE

Multiply 45 × 27.

> **Solution**   One way to solve this problem is to think of it as
> 45 × (20 + 7) = (45 × 20) + (45 × 7). Start by multiplying 45 × 7:

$$7 \times 5 = 35 \qquad \text{Write down 5 and } \textit{carry} \text{ 3.} \qquad \begin{array}{r} \overset{3}{4}5 \\ \times\ 7 \\ \hline 5 \end{array}$$

You do not have to write down what you carry if you can keep it in your head. But if it helps, write it down somewhere.

$$\begin{array}{r} \overset{3}{4}5 \\ \times\ 7 \\ \hline 315 \end{array} \qquad \text{Multiply } 7 \times 4 = 28. \text{ Add the 3 you carried.}$$
$$28 + 3 = 31$$

Now multiply 45 by 20.

$$\begin{array}{r} 45 \\ \times 20 \\ \hline 00 \\ 90 \\ \hline 900 \end{array}$$

0 × 45 = 0; it isn't necessary to write that down, but leave the zeros if you feel more secure. Now multiply 2 × 45.   2 × 5 = 10. Since 2 is in the tens place, we move the answer one step to the left. Write down 0 and carry 1. 2 × 4 = 8, and 8 + 1 = 9.

We must recognize the *place value* when we multiply.

We now know that 7 × 45 = 315 and that 20 × 45 = 900.

Add: 900 + 315 = 1215

In summary, we have

$$\begin{array}{r} 45 \\ \times 27 \\ \hline 315 \\ +\,900 \\ \hline 1215 \end{array}$$

EXAMPLE

Multiply 276 × 34.

Solution   First multiply 276 by 4.

$$
\begin{array}{r}
\overset{\scriptstyle 3\,2}{276} \\
\times\ \ \ 4 \\
\hline
1104
\end{array}
$$

4 × 6 = 24
Write down 4 and carry 2.
4 × 7 = 28, and 28 + 2 = 30
Write down 0 and carry 3.
4 × 2 = 8, and 8 + 3 = 11
Write down 11.

Now multiply by 3. The answer will be in the 10s place because we are actually multiplying by 30.

$$
\begin{array}{r}
\overset{\scriptstyle 2\,1}{276} \\
\times\ \ 30 \\
\hline
8280
\end{array}
$$

3 × 6 = 18. Write down 8 and carry 1.
3 × 7 = 21. Add the 1.
Write down 2 and carry 2.
3 × 2 = 6, and 6 + 2 = 8
Write down 8.
Add: 1104 + 8280 = 9384

In summary,

$$
\begin{array}{r}
276 \\
\times\ \ 34 \\
\hline
1104 \\
828\ \ \\
\hline
9384
\end{array}
$$

EXAMPLE

Find 102 × 204.

Solution

$$
\begin{array}{r}
204 \\
\times\,102 \\
\hline
408 \\
000\ \ \\
+\,204\ \ \ \ \\
\hline
20808
\end{array}
$$

You don't have to write the row 000, but make sure you move the next row to the proper position. You could also write down only the first zero, and the 4 would automatically be in the hundreds place. The answer is 20,808.

EXERCISE 1.2.5

Multiply.

1. 6 × 39
2. 8 × 75
3. 34 × 34
4. 35 × 83
5. 68 × 143

6. 86 × 103
7. 272 × 727
8. 204 × 903
9. 1246 × 25
10. 36 × 2405

MULTIPLICATION BY 9 AND BY 99

The nines have an interesting characteristic. Look at these answers:

$$9 \times 2 = 18$$
$$9 \times 3 = 27$$
$$9 \times 4 = 36$$
$$9 \times 5 = 45$$
$$9 \times 6 = 54$$
$$9 \times 7 = 63$$
$$9 \times 8 = 72$$
$$9 \times 9 = 81$$

The tens digit of the answer is one less than the number you are multiplying 9 by, and the sum of the two digits is always 9! For example, $4 = 5 - 1$, and $4 + 5 = 9$. This is true only for multiplication of 9 by a one-digit number.

Now multiply some two-digit numbers by 99:

$$99 \times 12 = 1188$$
$$99 \times 23 = 2277$$
$$99 \times 46 = 4554$$
$$99 \times 87 = 8613$$

Can you see the pattern? This pattern will be true only for multiplication of 99 and a two-digit number.

The first two digits of the answer are one less than the number you are multiplying by, and the sum of the two pairs is 99! For example, 45 is one less than 46, and $45 + 54 = 99$.

Properties

The Commutative Property We have already seen that the order of addition or multiplication of two numbers does not matter. In other words, $2 + 3 = 3 + 2$ and $5(6) = 6(5)$. This is the *commutative* property.

The Associative Property When we have three numbers, such as $3 + 4 + 5$, it doesn't matter if we first add 3 and 4 to get 7 and then add 5 to get $7 + 5 = 12$ or if we first add $4 + 5$ to get 9 and then add 3 to get $9 + 3 = 12$. We still get 12. In short, $(3 + 4) + 5 = 3 + (4 + 5)$.

Looking at it on the number line, we have

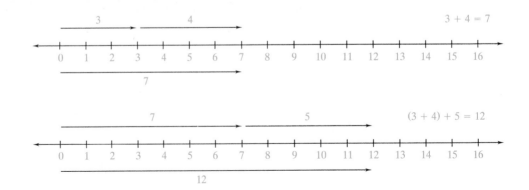

or

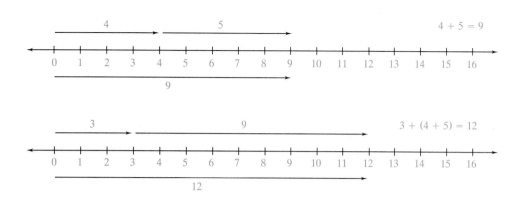

Again: $(3 + 4) + 5 = 3 + (4 + 5)$. This is the *associative* property of addition. Multiplication is also associative:

$$(2 \cdot 3)4 = 2(3 \cdot 4)$$

$$(6)4 = 2(12)$$

$$24 = 24$$

This can also be shown on the number line:

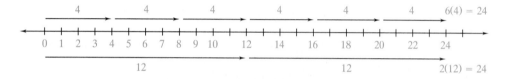

THE COMMUTATIVE PROPERTY

If *a* and *b* stand for any numbers, then

$$a + b = b + a$$

and

$$a(b) = b(a)$$

THE ASSOCIATIVE PROPERTY

If *a*, *b*, and *c* stand for any numbers, then

$$(a + b) + c = a + (b + c)$$

and

$$(a \cdot b)c = a(b \cdot c)$$

The Distributive Property Multiplication can be distributed over addition (or subtraction). For example,

$$5 \times (2 + 3) = 5 \times 2 + 5 \times 3 = 10 + 15 = 25.$$

Identity Properties

$$7 + 0 = 7, \quad 19 + 0 = 19, \quad 231 + 0 = 231$$

When we add 0 to any number, the number does not change. We say that 0 is the *identity element* for addition.

$$7 \times 1 = 7, \quad 19 \times 1 = 19, \quad 231 \times 1 = 231$$

When we multiply any number by 1, the number does not change. We say that 1 is the *identity element* for multiplication.

THE DISTRIBUTIVE PROPERTY

If *a*, *b*, and *c* stand for any numbers, then

$$a(b + c) = a \cdot b + a \cdot c$$

IDENTITY PROPERTIES

Let *a* be any number.

$$a + 0 = a$$

0 is the identity element in addition.

$$a \times 1 = a$$

1 is the identity element in multiplication.

EXERCISE 1.2.6

1. Which property or properties are illustrated in each of the following?
   (a) $5 + 0 = 5$
   (b) $6 + 2 = 2 + 6$
   (c) $3(1) = 3$
   (d) $(5 \cdot 6)7 = 5(6 \cdot 7)$
   (e) $2(5) = 5(2)$
   (f) $(5 \cdot 7)2 = (2 \cdot 5)7$
   (g) $4 \cdot 3 \cdot 11 = 3 \cdot 4 \cdot 11$
   (h) $(3 + 2) + 5 = 5 + (3 + 2)$
   (i) $3(2 + 7) = 3 \cdot 2 + 3 \cdot 7$

Subtraction

The answer in subtraction is called the *difference*. In $4 - 3 = 1$, 4 and 3 are terms and 1 is the difference.

We have not shown tables for subtraction or division because the tables you have for addition and multiplication can be used backwards for these operations. Make sure your tables are correct before you start to use them for subtraction or division.

Let's look at the Addition Facts table in Exercise 1.2.1 again. For example, $4 + 5 = 9$. When we subtract $9 - 4$, we are really saying $4 + ? = 9$.

Subtraction is the reverse of addition. $11 - 7 = ?$ becomes $7 + ? = 11$. From the row that starts with 7 in the Addition Facts table, move over to 11 and read the number at the top of the column. It's 4. $7 + 4 = 11$, so $11 - 7 = 4$.

EXAMPLE

Find the difference 25 − 13.

Solution   These numbers are not in the table, but we subtract one place at a time.

$$5 - 3 = 2$$
$$20 - 10 = 10$$
$$10 + 2 = 12$$
$$25 - 13 = 12$$

EXAMPLE

Subtract 394 from 489.

Solution   When we have to subtract large numbers, we set them up the same way as in addition. 489 − 394 is rewritten in column form:

$$\begin{array}{r} 489 \\ -394 \\ \hline \end{array}$$

We subtract 4 from 9 and get 5. Subtracting 9 from 8 doesn't work, so we have to "borrow" 1 from 4. This "1" becomes "10" when it moves to the place value to the right. $10 + 8 = 18$ and 9 subtracted from 18 is 9:

$$\begin{array}{r} {}^{3\,1}\!\!\!\!\!\not{4}89 \\ -\ 394 \\ \hline 95 \end{array}$$

We subtract 3 from 3 and get 0. Since this 0 comes at the beginning of the number, we do not write it.

EXAMPLE

Subtract 904 from 1000.

Solution   1000 − 904 becomes

$$\begin{array}{r} 1000 \\ -\ 904 \\ \hline \end{array}$$

Here we have to borrow from the 1 and then borrow again twice; we subtract 4 from 10, 0 from 9, and 9 from 9:

$$\begin{array}{r} {}^{9\ 9\ 1}\!\!\not{1}\,\not{0}\,\not{0}\,0 \\ -9\,0\,4 \\ \hline 96 \end{array}$$

Make sure you think this over in your head. $904 is $900 + $4. $900 from $1000 is $100, and $4 from $100 leaves you with $96. You can check your subtraction by going backwards: $96 + 904 = 1000$.

EXERCISE 1.2.7

1. 761 − 482

2. 1236 − 408

3. 2512 − 1387

4. 3702 − 2985

5. 4108 − 287

6. 3002 − 407

Division

Division is the reverse of multiplication:

$$24 \div 3 \qquad \text{becomes} \qquad 3 \times ? = 24$$

In the Multiplication Facts table you prepared in Exercise 1.2.4, start in the row that begins with 3 and move to 24. The number at the top of the column is 8. $3 \times 8 = 24$, so $24 \div 3 = 8$.

$$\text{divisor} \overline{)\,\text{dividend}}^{\text{quotient}}$$

In *division* the answer is called the *quotient*. In $21 \div 3 = 7$, 21 is the *dividend*, 3 is the *divisor*, and 7 is the *quotient*.

Look at the Multiplication Facts table and review the facts you had difficulty with.

The symbols for division are

For example, 63 divided by 9 can be written as

$$63 \div 9 \quad \text{or} \quad 9\overline{)63} \quad \text{or} \quad 9 \lfloor 63 \quad \text{or} \quad 63/9 \quad \text{or} \quad \frac{63}{9}$$

EXAMPLE

What is $63 \div 9$?

> **Solution**   Start with the row that begins with 9. The answer 63 is in the column below 7. Thus $63 \div 9 = 7$.

We saw that both addition and multiplication are commutative operations. Subtraction and division are *not* commutative. If the temperature in the morning is 10 degrees and drops 6 degrees, it is then 4 degrees. But if the morning temperature is 6 degrees and it drops 10 degrees, we would then have a very cold temperature of −4 degrees.

If you divide 12 apples among 6 people, each person gets $12 \div 6 = 2$ apples. But if you have only 6 apples for 12 people, each person gets only $6 \div 12$ or half an apple.

Division also presents some particular problems. What is the difference between

12 divided by 6   and   12 goes into 6?

12 divided by 6 is 12 ÷ 6   or   6)‾12‾, which is 2.

12 into 6 is 12)‾6‾, or 6 ÷ 12, which is not a whole number.

Twelve divided by six is not the same as six divided by twelve. Division is *not* commutative.

3 divided by 2 does not come out even. We would have 1 whole and 1 left over. We say "1 remaining" or "1 as a remainder." For example,

$$\begin{array}{r} 1 \\ \overline{2)\,3} \end{array} + \text{remainder}$$

In this section we will only do examples where the quotient is a whole number.

If you have difficulty with "by" and "into," rewrite the division problem as *divided by* before you solve it. On the calculator you can only "divide by."

Division also presents problems when we try to divide by zero. When we multiply by zero we have

$$1 \times 0 = 0, \quad 2 \times 0 = 0, \quad 3 \times 0 = 0, \quad \text{and so on}$$

When 0 is a factor, the products are all zero. So, since division is the reverse of multiplication, *zero divided by any number is zero*. The only exception to this rule is 0 ÷ 0.

RULE

If *a* stands for any number but 0, then

$$0 \div a = 0$$

If you divide a large number by a very small number, you get a large quotient. But when you divide by zero, you don't know what you get. We say that the quotient is *undefined*. We can find how many times a number goes into zero (always 0 times), but we *cannot* divide *by* zero!

EXAMPLE

Solve (a) 5 ÷ 0,   (b) 0 ÷ 10,   (c) 176 ÷ 0.

   Solution   (a) and (c) undefined; (b) 0

RULE

Division by zero is undefined.

In longer division problems such as 728 ÷ 8 we set up a *long division* example. We always start by filling in the "box" first with the number being divided (the dividend), in this case 728.

$$)\,\overline{728}$$

Then we write the divisor, in this case 8, to the left of the box:

$$8\overline{)728}$$ This is read "8 into 728."

Then work in steps: 8 does not go into 7, so 8 into 7 is 0 (we don't have to write the 0 since this is the beginning of a whole number).

$$\begin{array}{r} 9 \\ 8\overline{)728} \end{array}$$ 8 into 72 is 9. Write 9 above the 2.

$$\begin{array}{r} 91 \\ 8\overline{)728} \\ -72 \\ \hline 8 \\ -8 \\ \hline 0 \end{array}$$

$9 \times 8 = 72$
Subtract, and bring down the 8.
8 goes into 8 one time. $1 \times 8 = 8$
$8 - 8 = 0$, so there is no remainder.

Check: $8 \times 91 = 728$

EXAMPLE

Solve $153 \div 9$.

Solution
$$\begin{array}{r} 17 \\ 9\overline{)153} \\ -9 \\ \hline 63 \\ -63 \\ \hline 0 \end{array}$$

Check: $17 \times 9 = 153$

EXAMPLE

Solve $7208 \div 8$.

Solution
$$\begin{array}{r} 901 \\ 8\overline{)7208} \\ -72 \\ \hline 08 \\ -8 \\ \hline \end{array}$$

Check: $8 \times 901 = 7208$

EXERCISE 1.2.8

1. 12 divided by 3
2. 18 divided by 9
3. 7 into 21
4. 11 into 121
5. 63 divided by 7
6. 7 into 49
7. $546 \div 13$

8. 1545 ÷ 15

9. 2781 ÷ 103

10. 9282 ÷ 238

11. 72 into 4032

12. 2337 divided by 57

13. (a) 11(0)   (b) 0(55)

14. (a) 0 ÷ 5   (b) 5 ÷ 0

15. (a) (0)(0)   (b) 0 ÷ 0

16. 7777 ÷ 77 (Just because the sevens are usually the most uncomfortable numbers!)

1.3 ORDER OF OPERATIONS

We have learned that neither order nor grouping matter in addition or multiplication.

$$3 + 2 = 2 + 3 \qquad (3 + 2) + 1 = 3 + (2 + 1)$$
$$3 \cdot 2 = 2 \cdot 3 \qquad (3 \cdot 2) \cdot 1 = 3 \cdot (2 \cdot 1)$$

Addition and multiplication are commutative and associative.

Now we will see what happens when two or more operations occur in the same example.

In $3 + 2 \cdot 4$ we can apparently get two different answers.

1. If we add and then multiply, we get $3 + 2 \cdot 4 = 5 \cdot 4 = 20$.

2. If we multiply and then add, we get $3 + 2 \cdot 4 = 3 + 8 = 11$.

Both answers seem reasonable. Because we read from left to right, you might prefer the first. However, at some point in history the decision was made that multiplication comes before addition or subtraction. Division also comes before addition or subtraction.

EXAMPLE

Find the value of:
(a) $2 \cdot 4 + 6$
(b) $9 - 2 \cdot 3$
(c) $12 \div 2 + 6$
(d) $5 - 8 \div 4$

Solution
(a) $2 \cdot 4 + 6 = 8 + 6 = 14$
(b) $9 - 2 \cdot 3 = 9 - 6 = 3$
(c) $12 \div 2 + 6 = 6 + 6 = 12$
(d) $5 - 8 \div 4 = 5 - 2 = 3$

When both multiplication and division appear in the same example, carry out these operations in order from left to right.

EXAMPLE

Find the value of:
(a) $8 \div 2 \cdot 3$
(b) $3 \cdot 2 + 6 - 10 \div 5$
(c) $19 - 8 \div 2 \cdot 3$
(d) $4 \cdot 6 \div 2 \cdot 3$

Solution
(a) $8 \div 2 \cdot 3 = 4 \cdot 3 = 12$
(b) $3 \cdot 2 + 6 - 10 \div 5 = 6 + 6 - 2 = 12 - 2 = 10$
(c) $19 - 8 \div 2 \cdot 3 = 19 - 4 \cdot 3 = 19 - 12 = 7$
(d) $4 \cdot 6 \div 2 \cdot 3 = 24 \div 2 \cdot 3 = 12 \cdot 3 = 36$

When grouping symbols, ( ), [ ], or { }, appear in an example, simplify inside the grouping symbols first.

EXAMPLE

Find the value of:
(a) $3 + 2(5 - 4)$
(b) $9 - 4 \div (3 + 1)$
(c) $2\{8 + 1\} - 4[5 - 2]$

Solution
(a) $3 + 2(5 - 4) = 3 + 2(1) = 3 + 2 = 5$
(b) $9 - 4 \div (3 + 1) = 9 - 4 \div 4 = 9 - 1 = 8$
(c) $2\{8 + 1\} - 4[5 - 2] = 2\{9\} - 4[3] = 18 - 12 = 6$

RULE

When two or more operations appear in the same problem, solve it in this sequence:
1. Simplify inside grouping symbols.
2. Do multiplication and/or division in order from left to right.
3. Do the addition and/or subtraction.

EXAMPLE

Find the value of:
(a) $3 + (7 - 5)3 - 8 \div (4 - 2)$
(b) $[8 + 2(4 - 3) - 6] - 2$

Solution
(a) $3 + (7 - 5)3 - 8 \div (4 - 2)$
    Simplify: $3 + (2)3 - 8 \div (2)$
    Multiply and divide: $3 + 6 - 4$
    Add and subtract: $9 - 4 = 5$

(b) This time we start at the innermost grouping symbol $(4 - 3) = (1)$.
Simplify: $[8 + 2(4 - 3) - 6] - 2 = [8 + 2(1) - 6] - 2$
Multiply: $[8 + 2(1) - 6] - 2 = [8 + 2 - 6] - 2$
Simplify inside the grouping symbols: $[8 + 2 - 6] - 2 = [4] - 2$
Subtract: $4 - 2 = 2$

Brackets and braces are not usually used until parentheses have already been used. The accepted hierarchy among the grouping symbols is parentheses within brackets and brackets within braces.

$$\{ \, [\, (\quad) \,] \, \}$$

EXAMPLE

Find the value of:
(a) $2[3(7 + 4) + 9]$
(b) $50 - 2\{3[2(5 - 4) + 6]\}$

Solution
(a) $2[3(7 + 4) + 9] = 2[3(11) + 9]$
$= 2[33 + 9]$
$= 2[42] = 84$
(b) $50 - 2\{3[2(5 - 4) + 6]\} = 50 - 2\{3[2(1) + 6]\}$
$= 50 - 2\{3[2 + 6]\}$
$= 50 - 2\{3[8]\}$
$= 50 - 2\{24\}$
$= 50 - 48 = 2$

EXERCISE 1.3.1

Find the value of each expression.

1. $2 \times 8 - 7$
2. $7 + 12 - 10$
3. $15 - 6 + 18$
4. $24 - 2(10 - 2)$
5. $5 + 3 \times 9 - 2$
6. $12 - 24 \div 3$
7. $40 \div 8 + 14$
8. $54 \div 9 + 72 \div 8$
9. $60 \div 6 \times 5$
10. $8 \times 6 \div (1 + 2)$
11. $24 \div 12 \div 2$
12. $24 \div (12 \div 2)$
13. $12 \div (2 \times 6)$
14. $12 \div 2 \times 6$

15. $12 \div 6 \times 2$

16. $40 - 4(2 + 3 \times 2)$

17. $10 - 2\{4[3 - (2 - 1)] - 2(2 + 1)\}$

18. $2\{3 + [4 - (5 - 3)]\}$

The problem $2 + 4 \times 7$ becomes $2 + 28 = 30$, but with the parentheses in $(2 + 4) \times 7$, we have $6 \times 7 = 42$.

In $18 \div 9 \times 2$ we get $2 \times 2 = 4$, but with parentheses, $18 \div (9 \times 2) = 18 \div 18 = 1$. You can see that grouping symbols is very important.

In the following exercises, we give you the answers and you insert the parentheses.

EXAMPLE

Insert parentheses to give a true statement.
(a) $10 \times 2 + 3 = 50$
(b) $3 + 9 \div 3 = 4$

Solution
(a) The problem must be rewritten as $10 \times (2 + 3)$
(b) $3 + 9 \div 3$ must be rewritten as $(3 + 9) \div 3 = 4$.

EXERCISE 1.3.2

Insert parentheses to give a true statement.

1. $7 + 4 \times 5 = 55$

2. $2 \times 6 - 3 = 6$

3. $3 \times 4 + 2 \times 1 = 18$

4. $18 \div 5 - 3 = 9$

5. $16 - 8 \div 4 = 2$

6. $8 \times 7 \div 4 = 14$

7. $24 \div 3 \times 2 = 4$

8. $8 \div 2 \div 2 = 8$

The following problems are missing both operations and parentheses.
    $6 \quad 6 \quad 6 \quad 6 = 5$ has to be changed to $(6 \times 6 - 6) \div 6$
which is the same as $(36 - 6) \div 6 = 30 \div 6 = 5$.

EXERCISE 1.3.3

Insert operations and parentheses to give a true statement.

1. $6 \quad 6 \quad 6 \quad 6 = 8$

2. $6 \quad 6 \quad 6 \quad 6 = 13$

3. $4 \quad 4 \quad 4 \quad 4 = 2$

4. $4 \quad 4 \quad 4 \quad 4 = 17$

5. $1 \quad 1 \quad 1 \quad 1 = 2$

6. $1 \quad 1 \quad 1 \quad 1 = 3$

CHECKPOINT

This first chapter is rather long, so we break here for a summary.

SUMMARY

Properties

The commutative property: If a and b stand for any numbers, then
$$a + b = b + a \qquad \text{and} \qquad a(b) = b(a)$$
The associative property: If a, b, and c stand for any numbers, then
$$(a + b) + c = a + (b + c) \qquad \text{and} \qquad (a \cdot b)c = a(b \cdot c)$$
The distributive property: If a, b, and c stand for any number, then
$$a(b + c) = ab + ac$$
Identity properties: Let a be any number.

Addition: $\qquad\qquad a + 0 = a \qquad$ 0 is the identity element

Multiplication: $\qquad a \times 1 = a \qquad$ 1 is the identity element

Rules

Multiplication by zero: If a stands for any number, then
$$a \times 0 = 0 \qquad \text{and} \qquad 0 \times a = 0$$

Division: $0 \div a = 0 \qquad$ (here a stands for any number but 0)

Division by zero is undefined.

Order of operations:

1. Simplify inside grouping symbols.

2. Do multiplication and/or division in order from left to right.

3. Do addition and/or subtraction.

Place Value

Hundred thousands | Ten thousands | Thousands | Hundreds | Tens | Ones

VOCABULARY

Addition: Combining numbers into one sum.

Basic operations: Addition, subtraction, multiplication, division.

Composite number: A number that can be factored into factors other than itself and 1.

Counting numbers: {1, 2, 3, 4, 5, . . .}

Difference: The answer in subtraction.

Digits: The symbols 0, 1, 2, 3, 4, 5, 6, 7, 8, 9

Dividend: The number that will be divided.

Division: Repeated subtraction; the reverse of multiplication.

Divisor: The number to divide by.

Even number: {0, 2, 4, 6, . . .}

Factor: A number that divides a whole number evenly.

Identity element: 0 for addition; 1 for multiplication.

Multiple: The product of a certain number and any other number; for example, 5, 10, 15, . . . , 55, . . . are multiples of 5.

Multiplication: Repeated addition.

Number line: A line with numbers arranged in consecutive order and with unit spacing.

Numeral: Symbol for a number.

Odd numbers: {1, 3, 5, 7, . . .}

Place value: Position in a number.

Prime number: A number that has only 1 and itself as factors.

Product: The answer in multiplication.

Quotient: The answer in division.

Set: A collection of things.

Subtraction: The reverse of addition.

Sum: The answer in addition.

Symbols: $+, -, \div, \times, >, =, <$, etc.

Terms: The building blocks of addition.

Whole numbers: {0, 1, 2, 3, 4, 5, . . .}

CHECK LIST

Check the box for each topic you feel you have mastered. If you are unsure, go back and review.

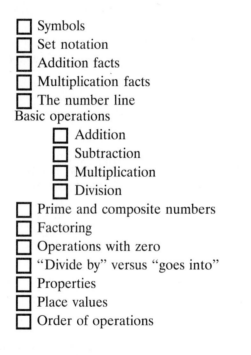

- ☐ Symbols
- ☐ Set notation
- ☐ Addition facts
- ☐ Multiplication facts
- ☐ The number line

Basic operations
- ☐ Addition
- ☐ Subtraction
- ☐ Multiplication
- ☐ Division
- ☐ Prime and composite numbers
- ☐ Factoring
- ☐ Operations with zero
- ☐ "Divide by" versus "goes into"
- ☐ Properties
- ☐ Place values
- ☐ Order of operations

READINESS CHECK

Solve these problems to satisfy yourself that you have mastered the first three sections of Chapter 1.

1. Write in set notation: the set of counting numbers that are multiples of 3.
2. Divide 3000 by 0.
3. Translate five thousand twenty into mathematical symbols.
4. List the factors of 36.
5. Is 51 prime or composite? Explain why.
6. List all multiples of 7 between 80 and 100.
7. Insert the correct symbol $<$ or $>$ between 75 and 31.
8. Evaluate: $10 - 2[2 - (5 - 4) + 8 \div 4 \times 2]$.

9. Tell which properties are illustrated in

$$(7 + 2) + 3 = (7 + 3) + 2$$

10. A sum of four thousand eighty dollars is divided equally among four people. How much money does each person get?

1.4 USING A CALCULATOR

We recommend strongly that you use a calculator in your math work. It is important that you know how to do mental arithmetic and that you know all your addition and multiplication facts, but it is not important to do long sums, complicated multiplications, or long divisions by hand.

Using a calculator does *not* mean that you are lazy or taking the easy way out. You still have to decide which operation to use—when to add, subtract, multiply, or divide. These are the real mathematical decisions. The calculator will simply relieve you of the drudgery. Remember, however, that the calculator always follows your directions. In general, enter an example into your calculator from left to right of the problem. When you press = the answer appears. Always do the problem twice. If the answer changes, do the problem over again.

Become familiar with your own calculator by trying to do the examples below. Also do some examples from the instruction booklet that came with your calculator.

EXAMPLE

Solve 25 + 16.

> **Solution**   Press the ON key (solar calculators are ready to go without that key) followed by **2**, **5**, **+**, **1**, **6**, **=** . Your answer window should show **41**.

EXAMPLE

Solve 78 − 46.

> **Solution**   Did you get 32? Be sure to use the **-** key and not the **+/-** key for subtraction. We need the **+/-** key for negative numbers, which will be discussed in Chapter 2.

Multiplication is also straightforward. For example, $23 \times 96 = 2208$.

Division can create problems. "56 divided by 7" is $56 \div 7$ and works fine; the answer is 8. But "7 into 56" must be rewritten as "56 divided by 7" to work on the calculator. All calculators are programmed to do *divide by* only.

EXERCISE 1.4.1

Do each problem twice: first on paper without the calculator and then with the calculator. If your answers do not come out the same, do the problem again on the calculator; we often make mistakes when we use the calculator keys. If your answers still disagree, rework the problem on paper.

1. 1263 + 522

2. 4817 + 987

3. 967 + 284 + 315

4. 4124 + 3590 + 206

5. 723 + 38 + 5094 + 844

6. 691 + 485 + 384 + 567 + 209

7. 71 − 47

8. 753 − 86

9. 619 − 284

10. 257 − 198

11. 300 − 173

12. 1240 − 987

13. 55 × 52

14. 41 × 39

15. 83 × 41

16. 68 × 73

17. 203 × 57

18. 69 × 482

19. 408 ÷ 8

20. 6272 ÷ 7

21. 408 ÷ 12

22. 644 ÷ 14

23. 5824 ÷ 56

24. 9775 ÷ 23

1.5 PLACE VALUE AND NAMES FOR LARGER NUMBERS

As we saw in Figure 1.2, each digit in a number has a name for its place. Starting from the rightmost digit in a whole number, we have the ones or units place. One step to the left takes us to the tens place, and the next position is the hundreds place. Continuing to move to the left, we enter the set of "thousands"—thousands, ten thousands, and hundred thousands. Then we have the set of millions—millions, ten millions, and hundred millions. To the left of millions are billions, and to their left, trillions.

These place values are illustrated below:

Billions			Millions			Thousands					
100	10	1	100	10	1	100	10	1	100s	10s	1s

EXAMPLE

Read 2385280.

Solution The number 2,385,280 is read: two million three hundred eighty-five thousand two hundred eighty.

EXAMPLE

Give the names for the places of the digits in the number
(a) 386,294   (b) 2,503,807

   Solution
   (a) 386,294 has 3 in the hundred thousands place, 8 in the ten thousands
       place, 6 in the thousands place, 2 in the hundreds place, 9 in the tens
       place, and 4 in the ones place. We read the number as three hundred
       eighty-six thousand two hundred ninety-four.
   (b) 2,503,807 has 2 in the millions place, 5 in the hundred thousands place,
       0 in the ten thousands place, 3 in the thousands place, 8 in the
       hundreds place, 0 in the tens place, and 7 in the units place. We read
       the number as two million five hundred three thousand eight hundred
       seven.

   In order to read 12859641327, we divide it into groups of three digits by
commas, starting from the right:
                     12,859,641,327
   Each group of three numbers is read as if it were alone: twelve, eight
hundred fifty-nine, six hundred forty-one, and three hundred twenty-seven.
   But 641 represents thousands, 859 represents millions, and 12 represents
billions, so the number is read: twelve billion eight hundred fifty-nine million
six hundred forty-one thousand three hundred twenty-seven.

EXERCISE 1.5.1

1. What is the name of the place occupied by the digit 5 in each of the
   following?
   (a) 50
   (b) 124,005
   (c) 5,002,398
   (d) 245,003
   (e) 342,517
   (f) 657,900
   (g) 53,789,000
   (h) 125,436,789,000

2. Put in the commas and write the number.
   (a) 4389
   (b) 247932
   (c) 9999999
   (d) 45592640
   (e) 20681030
   (f) 367006973800

1.6 ROUNDING AND APPROXIMATING NUMBERS

Rounding

Many times you do not need an exact number. For example, knowing that you
have a little less than $100 in your checking account is usually as good as
knowing that you have $97.89, and it's much easier to remember. (Of course,
you want the exact number in your checkbook and your bank statement!)
Similarly, knowing that it is approximately 200 miles from Boston to New York
is often as good as remembering that it is exactly 229 miles. When we
approximate a number, we round it off to a certain place.

There are rules for rounding numbers. For example, if you want to round the number 4683 to the nearest thousands place, you find the digit in the thousands place. In this case the digit is 4. The digit to the right of 4 is 6, and the rule we follow states that since 6 is greater than 5 we must add 1 to the 4 in the thousands place. In other words, 4683 rounded off to the thousands place is 5000. It is clear that 4683 is closer to 5000 than 4000.

RULES FOR ROUNDING

If the digit to the right of the rounded digit is less than 5, leave the digit the same and replace the digits to the right by zeros.
   If the digit to the right of the rounded digit is 5 or more, increase the rounded digit by one and replace the digits to the right by zeros.

EXAMPLE

Round 312 to the nearest hundred.

   Solution   The digit 3 is in the hundreds place, and the digit to the right of 3 is 1. 1 is less than 5, so 312 becomes 300. The zeros are used to keep the place value of the 3.

EXAMPLE

Round 3761 to the nearest thousand.

   Solution   3 is in the thousands place, and the digit to the right is 7. Since 7 is more than 5, we add 1 to 3, and 3761 becomes 4000. 7, 6, and 1 are replaced by zeros.

EXAMPLE

Round 1431 to the nearest hundred.

   Solution   The digit in the hundreds place is 4, and 3 is less than 5. Keep the 4, and make the digits to the right of it zeros. In other words, 1431 is closer to 1400 than it is to 1500.

EXAMPLE

Round 75,000 to the ten thousands place.

   Solution   The digit in the ten thousands place is 7, and the following digit is 5. Our rule states that we add 1 to 7. 75,000 becomes 80,000.

EXAMPLE

Round 37,952 to the nearest hundred.

   Solution   The digit in the hundreds place is 9, and the following digit is 5. Add 1 to 9 and get 38,000.

EXERCISE 1.6.1

Round the number to the indicated place.

  1. 2453;   tens

  2. 196;   tens

  3. 1431;   hundreds

  4. 2576;   hundreds

  5. 18,227;   thousands

  6. 10,750;   hundreds

  7. 102,876;   thousands

  8. 191,482;   hundred thousands

  9. 8,935,249;   millions

10. 7,894,291;   ten thousands

Approximating the Answer

When solving math problems, approximating the answer first can serve as a check as to whether your actual answer is correct or not. *Make a habit* of approximating the answer first when you use your calculator.

In approximating, it is customary to write the number with only one nonzero digit. That is, we usually want to see numbers such as 8000 and not 7500. For example, 311 is approximately 300 (written $311 \approx 300$), and 2976 is approximately 3000 ($2976 \approx 3000$).

EXAMPLE

Approximate 12,976.

    Solution   12,976 is closer to 10,000 than to 20,000, so the approximation is 10,000. However, you might decide that for your purposes 13,000 is a better approximation. The rules are not strict here. Do what is convenient.

When we need to approximate the answers in an operation, such as in the sum $212 + 596$, we approximate each term.

$$212 + 596 \approx 200 + 600 = 800$$
$$212 + 596 = 808$$

EXAMPLE

Approximate the product 112×98.

    Solution $112 \times 98 \approx 100 \times 100 = 10,000$
                   $112 \times 98 = 10,976$

EXAMPLE

Approximate the quotient $5940 \div 198$.

    Solution $5940 \div 198 \approx 6000 \div 200 = 30$
                   $5940 \div 198 = 30$

As numbers get larger you might prefer to use two digits in your approximations. For example, 12,976 × 12,976 could be approximated either as

$$10,000 \times 10,000 = 100,000,000$$

or as

$$13,000 \times 13,000 = 169,000,000 \approx 200,000,000$$

The exact answer is 12,976 × 12,976 = 168,376,576.

EXERCISE 1.6.2

First find the approximate answer; then solve with the calculator. Compare the answer with your approximation.

1. 81 × 27
2. 1021 + 2788
3. 225 − 183
4. 105 ÷ 7
5. 25 × 42
6. 3299 + 78
7. 5133 − 1295
8. 114 ÷ 38
9. 385 − 97
10. 819 ÷ 7
11. 39 + 101
12. 19,328 × 52
13. 45 × 810
14. 1184 ÷ 16
15. 781 + 2193 − 832
16. 32,532 − 19,865 + 10,534

1.7 APPLICATIONS

Nobody is reading this text to spend his or her time sitting in a corner adding, subtracting, multiplying, or dividing. You need these operations for use in problem solving! For most people this is the nightmare of mathematics. It is true that many math problems are not well written, interesting, or practical. However, it is hard to find or create good problems that illustrate each step in the learning process of mathematics. Some problems will have to be artificial in order to show a particular point. If you are going to master mathematics in a practical way, you need to understand many smaller steps.

Below are some problems from an 1843 arithmetic book, *Practical and Mental Arithmetic on a New Plan*. The numbers are small, so the arithmetic will not be complicated. Try them this way:

1. Read the problem through.
2. Decide your strategy: whether to add, subtract, multiply, or divide.
3. Get an answer.
4. Check whether the answer "makes sense."

EXAMPLE

How many hats, at 4 dollars apiece, can be bought with 20 dollars?

> **Solution**   Strategy: divide. Here we divide 20 by 4 and find that 5 hats can be bought.

EXAMPLE

A man was 45 years old and had been married 19 years. How old was he when he married?

> **Solution**   Strategy: subtract. $45 - 19 = 26$ gives us the solution that he was 26 years old when he got married.

EXAMPLE

If I pay $141 for a bolt of cloth containing 47 yards, for how much must I sell it in order to make $1 a yard from the sale?

> **Solution**   Think before you start to do too many calculations on this one. How much money did you expect to earn? One dollar per yard would give you $47, and since you spent $141 for the whole piece, you must sell it for $141 + $47 = $188.

EXERCISE 1.7.1

1. If you have 4 apples in one pocket and 2 in the other, how many apples do you have?

2. If you give 4 cents for a yard of tape, how many cents will you need to buy 3 yards? 5 yards? 7 yards? 11 yards?

3. A man had to travel 24 miles and has traveled all but 4 miles. How many miles has he traveled?

4. How many legs have 2 cats and a bird?

5. If a man earns $7 in one week, how many dollars will he earn in 5 weeks?

6. Twelve men are to be given 96 dollars for performing a piece of work. How much will each man receive?

7. One bushel of clover seed costs $12. How much will 5 bushels cost?

8. I sold 8 pencils for 80 cents. How much is this for each pencil?

Now we will try a set of problems with larger numbers. The following slightly different five-step strategy may help.

STRATEGY FOR SOLVING PROBLEMS

1. Read the problem through first without worrying about exact numbers. Decide what you are looking for. (If it will help, draw pictures.) Guess an answer, if you can.
2. Decide how you are going to get the answer (whether you are going to add, subtract, multiply, or divide).
3. Approximate an answer.
4. Find the exact answer.
5. Check with your approximation.

EXAMPLE

It is 795 miles from New York City to Chicago, and then 410 miles from Chicago to Minneapolis. How long a drive will it be from New York to Minneapolis by way of Chicago?

Solution

1. We need to find the driving distance between New York and Minneapolis by way of Chicago. Draw a quick sketch of our route (Figure 1.5).

2. The total driving distance is the distance from New York to Chicago *plus* the distance from Chicago to Minneapolis. So we will add.

3. $795 + 410 \approx 800 + 400 = 1200$. The distance is approximately 1200 miles.

4. $795 + 410 = 1205$

5. $1205 \approx 1200$. The distance from New York City to Minneapolis is 1205 miles.

Figure 1.5

EXAMPLE

Harvey drives 23 miles round trip to and from school each day. How many miles does he drive during the fall semester (76 days)?

Solution

1. Draw a picture. Harvey drives 23 miles 76 times.

2. We could add 23 miles 76 times, but it will be quicker to multiply.

3. $23 \times 76 \approx 20 \times 80 = 1600$

4. $23 \times 76 = 1748$

5. $1748 \approx 1600$. Harvey drives 1748 miles during the fall semester.

EXAMPLE

Pat bought a boat for $329, patched it, and painted it. She then sold the boat for $450. How much profit did she make on the sale?

Solution

1.   Pat bought something and sold it for more.
2.   To find the profit, subtract the original cost from the selling price.
3. $450 - 329 \approx 500 - 300 = 200$
4. $450 - 329 = 121$
5. $121 \approx 200$. Pat earned $121 profit.

In this case you could have made both 450 and 329 smaller, that is, $450 \approx 400$ and $329 \approx 300$. The approximation ($100) would then be closer to $121.

EXAMPLE

In 1850 there were 213 libraries in the United States, and they contained 942,312 books. What was the average number of books in each library?

Solution

1.   A certain number of libraries contained a certain number of books. The average is the number of books per library.
2.   To find the average, divide the total number of books by the total number of libraries.
3. $942,312 \div 213 \approx 900,000 \div 200 = 4500$
4. $942,312 \div 213 = 4424$
5. $4424 \approx 4500$. There were an average of 4424 books in each library.

EXAMPLE

The U.S. total budget receipts were $1031 billion in 1990. Of that amount, $466,884 million came from individual income taxes. How much came from other sources?

Solution $1031 billion equals $1,031,000,000,000. This number minus $466,884 million, which is $466,844,000,000, would be $564,116,000,000.
It is more convenient in this case to work with millions of dollars. 1031 billion = 1,031,000 million. Now subtract 466,884 million from that. The answer is $564,116 million.

If your calculator cannot handle the largest numbers, you can either solve the problem by hand or round the numbers you work with to the same place. For example, use 250 million instead of 250,000,000.

EXERCISE 1.7.2

1. In New Zealand, people of two very different cultures live side by side: 280,000 Maoris (the original settlers) and about 3 million other New Zealanders of various national origins. How many people live in New Zealand?

2. The federal budget deficit for 1987 was $150 billion. Legislation was proposed that would have reduced it to $99 billion by 1989. However, the legislation was defeated, and the actual deficit was $153 billion in 1989.
   (a) By how much would the deficit have been reduced?
   (b) By how much was it actually increased?
   (c) How much more was the budget deficit than it would have been if the proposed legislation had taken effect?

3. Jorge can buy a stereo either for $285 cash or for $40 down and $25 per month for a year. How much will he save by paying cash?

4. Margot earns $47,947 a year. This is $1438 less than she had expected. What had she expected?

5. The current cost of the military retirement program is $18 billion. This is four times as much as it was 20 years ago. How much was it then?

6. According to a recent estimate, people in the United States used approximately 45,000,000,000 cans in a year.
   (a) If the population of the United States for that year was 250,000,000, what would be the number of cans per person?
   (b) If the average lifespan for a man is 72 years, how many cans would a man use in his lifetime?
   (c) If the average lifespan for a woman is 78, how many cans would a woman use in her lifetime?

7. The average household of four people uses 72 gallons of water per day.
   (a) How much water is used per person per day?
   (b) How much water is used per person per year?

8. If the U.S. public debt was approximately $3224 billion dollars and the U.S. population 248,000,000, what was the debt per person?

9. 20,000 Italian lire equals approximately $10.
   (a) How much in dollars is an 80,000-lire leather handbag?
   (b) What is the cost in dollars of a 40,000-lire pair of gloves?
   (c) About how many lire does a $20 sweater cost?

10.

			Africa	
Year	World Production	South Africa	Ghana	Zaire
1972	44,843,374	29,245,273	724,051	140,724
1989	63,497,633	19,531,550	429,469	112,528

WORLD GOLD PRODUCTION
(Troy Ounces)

   (a) From the table, determine how much of the world production of gold did *not* come from Africa in 1972.
   (b) How much more gold was produced in 1989 than in 1972 in the world? Round the answers to the nearest hundred thousand.

HOW BIG IS A MILLION?

To make a million marks with a pencil at one mark a second, it would take you 278 hours of nonstop writing, or 11 days and 14 hours!

Or look at it this way. A common housefly is only three-tenths of an inch long. If it were a million times bigger, it would be 25,000 feet—almost 5 miles—long!

A human hair multiplied by a million would be wider than a city block.

A 6-foot man whose height was multiplied by a million would be approximately 1136 miles tall. (Lying down, he'd stretch from Chicago to Galveston, Texas.)

If you spend $1000 a day, it would take you $2\frac{3}{4}$ years to spend a million dollars.

SUMMARY

Place Value

	Billions			Millions			Thousands				
100	10	1	100	10	1	100	10	1	100s	10s	1s

Rules for Rounding

If the digit to the right of the rounded digit is less than 5, leave the digit the same.

If the digit to the right of the rounded digit is 5 or more, increase the first digit by one.

Strategy for Problem Solving:

1. Read the problem through.
2. Try to estimate the answer.
3. Decide your strategy: whether to add, subtract, multiply, or divide.
4. Get an answer.
5. Check whether the answer "makes sense."

VOCABULARY

Approximate: Write the number with one or two nonzero digits by rounding it.

Estimate: Approximate or guess the final answer.

Rounding: Using only a required number of digits and replacing the rest by zeros.

CHECK LIST

Check the box for each topic you feel you have mastered. If you are unsure, go back and review.

☐ Place values for larger numbers
☐ Rounding numbers

☐ Approximating numbers
☐ Estimating answers
☐ Using the calculator
☐ Applications

REVIEW EXERCISES

1. Try this crossnumber puzzle. Keep track of your feelings as you do it. Are there some problems that are harder than others? Check any that are, and practice on them.

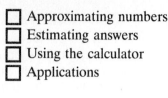

	Across		Down
1.	38 + 45	1.	112 − 28
3.	72 ÷ 9	2.	48 + 135 + 72 + 50
4.	23 × 18	3.	24 × 34
7.	480 ÷ 12	5.	10 + 4
8.	7 × ___ = 84	6.	242 × 20
10.	2 × 2 × 2 × 2 × 3	9.	715 − 511
11.	526 × 10	12.	112 + 80 + 17 + 13
13.	93 × 50	14.	5 × 8 × 16
15.	675 + 428 + 325	15.	700 + 336
17.	726 + 483 + 198	16.	1088 − 276
20.	2 × 2 × 2 × 2 × 2	18.	8 × 6 × 10
22.	1008 ÷ 36	19.	23 × 32
23.	13 × 3	21.	7 + 7 + 7 + 7
25.	4092 ÷ 6	24.	The largest number
26.	3 × 23		named by two digits

2. Draw a number line for each of the following.
   (a) Show that 3 + 5 = 5 + 3. What property does this illustrate?
   (b) Show that (4 × 2) × 3 = 4 × (2 × 3). What property does this illustrate?

(c) Plot the prime numbers between 25 and 38.
(d) Plot the composite numbers between 15 and 25.

3. What is the place value of 7 in each of the following numbers?
   (a) 297
   (b) 370
   (c) 10,742
   (d) 7243

4. Round the number to the indicated place.
   (a) 8296 to the hundreds place.
   (b) 7406 to the tens place.
   (c) 9516 to the thousands place.

5. Approximate the answer, and check with your calculator.
   (a) 79 + 213
   (b) 923 − 872
   (c) 983 × 189
   (d) 2098 ÷ 1049

6. Find the answers.
   (a) 3 + 4 · 5
   (b) (3 + 4)5
   (c) 18 ÷ 3 + 6
   (d) 18 ÷ (3 + 6)
   (e) 5 + 2(3 + 4 · 6)
   (f) 8 ÷ 2 + 6 × 3
   (g) [8 ÷ (2 + 6)]3 − 1

7. Replace the squares with digits to make the addition correct.

$$
\begin{array}{r}
7\ \square\ 2\ \square \\
+\ \square\ 4\ \square\ 2 \\
\hline
1\ 3\ ,\ 1\ 9\ 1
\end{array}
$$

8. The output of raw steel in the United States was 98,906,000 metric tons in 1990 and 66,982,686 metric tons in 1940. Find the increase.

9. Australia produced 6,751,647 troy ounces of gold in 1990. If gold sold for $350 per troy ounce, what was the total dollar value of the Australian gold in billions of dollars?

10. One U.S. dollar is worth approximately 600 Colombian pesos. How many dollars do you need to exchange to get 33,000 pesos?

READINESS CHECK

Solve the problems to satisfy yourself that you have mastered Chapter 1.

1. What is the value of the digit 2 in 320,465,894?
2. Read 3,056,070 in words.
3. A number has a 3 in both the hundreds and the hundred thousands places, a 7 in the ten thousands place as well as in the ones place, a 6 in the thousands place, and 0 everywhere else. What is the number?
4. List the prime numbers between 100 and 120.
5. Approximate the product 45×789 to one nonzero digit followed by the proper number of zeros.
6. Round 49,883 to the thousands place.
7. Multiply the two numbers 17 and 23. Is the product prime? Explain.
8. Add 345 million and 4,687,349.
9. What is the difference between $5 million and $350,000?
10. Brita spent $21 for a pin in Mexico. If the exchange rate is $1 to 3000 Mexican pesos, how many pesos did she pay?

INTEGERS

The other side of zero

Babylonians calculated with large numbers from early historical times. From before 2000 B.C., they used two wedge-shaped symbols, ∇ (for 1) and $\triangleleft$ (for 10), pressed on clay tablets to make numerals as shown in Figure 2.1.

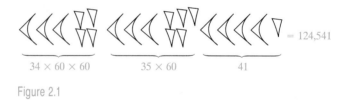

$$\underbrace{}_{34 \times 60 \times 60} \quad \underbrace{}_{35 \times 60} \quad \underbrace{}_{41} = 124{,}541$$

Figure 2.1

    The Babylonians used place values similar to our units and tens, but while we use only powers of 10, they used both 10 and 60. In Figure 2.1 there are three sets of tens and ones. The set to the right is worth $41 \times 1 = 41$; the set in the middle is worth $35 \times 60 = 2100$, and the set to the left is worth $34 \times 60 \times 60 = 122{,}400$. $41 + 2100 + 122{,}400 = 124{,}541$.

    At about the same time, the Egyptians had a much more elaborate system. They wrote with ink on papyrus, and their number symbols were much more sophisticated. (See Figure 2.2.) Figure 2.3 illustrates 124,541 in Egyptian.

| 1 | 10 | 100 | 1000 | 10,000 | 100,000 | 1,000,000 |

Figure 2.2

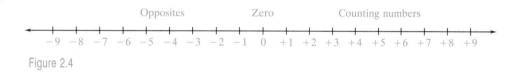

Figure 2.3

The Egyptians developed geometry and astronomy and a calendar of 365 days. Neither of these ancient civilizations, however, recognized the existence of zero as a number or, therefore, the existence of negative numbers.

It was the Hindus in the period A.D. 200–1200 who first recognized zero as a number. They made use of the idea of negative numbers to signify debts and positive numbers to signify assets, and they even recorded debts in red. These new ideas and kinds of notation came from the Hindus to the Arabs, who brought them to Spain and eventually, by the 14th century, into Italy. Reactions to this new notation were similar to ours regarding the metric system. There was great controversy about its acceptance. As late as 1498 there was a law on the books of Venice making it a crime punishable by death to use the new notation when keeping the books of the city. In the early 1500s, Robert Recorde of England wrote a book called *Grownde of Artes* that promised to teach the new numerals to "servants, children, women and other intellectual inferiors"!

Once zero was accepted, however, symbols were needed for numbers on the *other* side of zero, the opposites of the familiar counting numbers.

There are many practical examples of these opposite numbers:

Money: I *have* $150, and I *owe* $320.

Temperature: It is 17° *above* zero today. Yesterday it was 12° *below.*

2.1 ABSOLUTE VALUE

The Number Line for the Integers

In each of the examples above we have a starting point for counting: the zero. Our measurements then can go in both positive and negative directions.

One way to portray this concept is through the number line:

Opposites	Zero	Counting numbers

$$-9 \quad -8 \quad -7 \quad -6 \quad -5 \quad -4 \quad -3 \quad -2 \quad -1 \quad 0 \quad +1 \quad +2 \quad +3 \quad +4 \quad +5 \quad +6 \quad +7 \quad +8 \quad +9$$

Figure 2.4

We indicate the counting numbers as positive ($+$) and their opposites as negative ($-$). Together, positive numbers, negative numbers, and zero are called *integers*. If a number has no sign, the number is considered positive. Whole numbers are part of the integers. The counting numbers are also called *positive integers*. The number zero together with the positive integers form the set of *whole numbers*.

The numbers on the number line are arranged in order from the smallest to the largest. For example, 2 is smaller than 4, $(2 < 4)$, and -4 is smaller than -2, $(-4 < -2)$.

EXAMPLE

Write the numbers $6, -5, -2$ in order from the smallest to the largest.

Solution   On the number line the order would be

EXAMPLE

Write $-7, 0, 3, 9, -4$ from the largest to the smallest.

Solution   Graph the numbers.
The order from the largest to the smallest is $9, 3, 0, -4, -7$.

Absolute Value

Chicago is 963 miles west of Boston, and Boston is 963 miles east of Chicago. The distance *between* the two cities is 963 miles.

If we want to point out that Chicago is 963 miles west of Boston, we can say that the distance from Boston to Chicago is -963 miles and that from Chicago to Boston the distance is $+963$ miles. Here we are talking about a *directed* distance.

The distance *between* zero and $+1$ is the same as the distance *between* zero and -1; the distance *between* zero and $+2$ is the same as the distance *between* zero and -2 and so on.

Whether you receive \$7 or give away \$7 will make a difference to you, but the value of the \$7 is still \$7, whether you receive it or give it. Similarly, $+7°$ is as far from $0°$ as $-7°$. When we talk about *magnitude* of a number or *distance* between zero and the number, we are talking about *absolute value*.

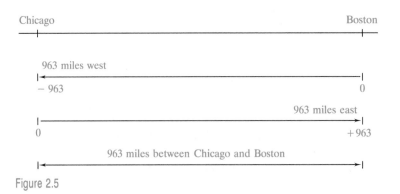

Figure 2.5

The way we write absolute value, the mathematical notation, is to use two vertical bars | |. For example, |7| means the absolute value of seven, which is seven. | − 7| means the absolute value of negative seven, which also is seven. | − 7| = |7| = 7.

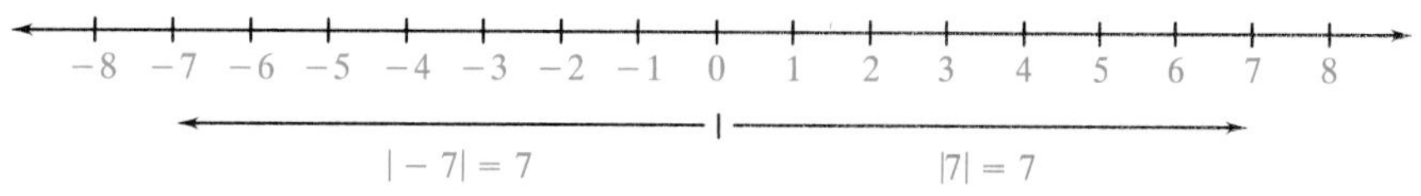

Figure 2.6

EXAMPLE

Find the absolute value of (a) |5|,   (b) | − 5|,   (c) | − 6| − | − 3|, (d) 2| − 25|.

Solution
(a) |5| = 5
(b) | − 5| = 5
(c) | − 6| − | − 3| = (6) − (3) = 6 − 3 = 3
(d) 2| − 25| = 2(25) = 50.

Part (d) of the solution demonstrates that a number followed by another number inside absolute value bars implies multiplication.

EXERCISE 2.1.1

1. Locate the following on a number line (place 0 anywhere you like): 0, 8, −3, −9, 1, 4, −7

2. Arrange the numbers in Problem 1 in order from the smallest to the largest.

3. Give the absolute values of the numbers in Problem 1.

4. Find:
   (a) | − 4| + |2|
   (b) | − 6| − | − 4|
   (c) 4 − | − 3|
   (d) |−5| − | − 1|
   (e) 4| − 3|
   (f) | − 4|| − 3|
   (g) 6| − 5|
   (h) | − 3|| − 5|

2.2 ADDITION AND SUBTRACTION OF INTEGERS

Addition

An understanding of absolute values will help us add two or more integers, even if they have different signs. Let's look at four examples using the number line.

EXAMPLE

(a) $(+4) + (+3)$
(b) $(-4) + (-3)$
(c) $(-4) + (+3)$
(d) $(+4) + (-3)$

Solution

(a) To add $(+4) + (+3)$ on the number line, we start at zero. We move right four units. Then, to add $+3$, we move right three more units.

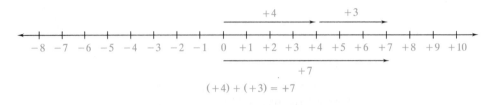

$$(+4) + (+3) = +7$$

(b) To add $(-4) + (-3)$, we first move left four units from zero to locate -4. Since the number we are adding is negative, we move left three units to add -3.

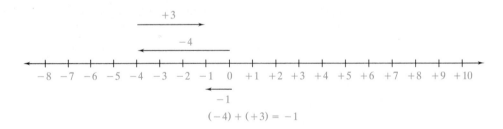

$$(-4) + (-3) = -7$$

(c) To add $(-4) + (+3)$, we first locate -4 to the left of 0. The number we are adding is positive, so we then move to the right, in this case three units, to add $+3$.

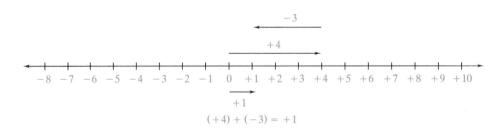

$$(-4) + (+3) = -1$$

(d) To add $(+4) + (-3)$, we first locate $+4$ to the right of zero. Since we are adding a negative 3, we then move to the left three units.

$$(+4) + (-3) = +1$$

> **RULE**
>
> **Addition of Integers**
>
> **1.** When signs are alike, add the absolute values. The sign of the sum is the sign of the numbers added.
> **2.** When the signs are different, find the difference between the absolute values. The sign of the answer is the sign of the number with the larger absolute value.

EXAMPLE

Add: $(+11) + (+3)$

> **Solution** Add the absolute values: $11 + 3 = 14$. The sign of both numbers is positive, so the sum is $+$.
> $$(+11) + (+3) = +14$$

EXAMPLE

Add: $(-7) + (-5)$

> **Solution** Add the absolute values: $7 + 5 = 12$. The sign of both numbers is negative, so the sum is $-$.
> $$(-7) + (-5) = -12$$

EXAMPLE

Add: $(+5) + (-3)$

> **Solution** The absolute values are 5 and 3. The difference is $5 - 3 = 2$. The sign of the number with the largest absolute value is positive, so the answer is positive.
> $$(+5) + (-3) = +2$$

EXAMPLE

Add: $(-5) + (+3)$

> **Solution** The absolute values are 5 and 3. The difference is $5 - 3 = 2$. The sign of the number with the largest absolute value is negative (-5). So the answer is negative.
> $$(-5) + (+3) = -2$$

EXERCISE 2.2.1

Find the sum without using your calculator.

1. $(-3) + (+5)$
2. $(+3) + (-5)$
3. $(-7) + (-4)$

4. $(+7) + (+4)$

5. $(-11) + (-11)$

6. $(+24) + (-20)$

7. $(+45) + (-60)$

8. $(-76) + (-16)$

9. $(-102) + (-88)$

10. $(+85) + (-115)$

Applications

Go back to Chapter 1 and review the steps for solving word problems. In these examples, write the problem as an addition example, recall the rules for signs, and then solve.

EXAMPLE

The temperature was $-2°$ and went up $10°$. What is the temperature now?

    Solution $-2 + 10 = 8$. The temperature is $8°$.

EXAMPLE

Ellen paid off $250 on her $10,420 student loan. How much is the loan now?

    Solution $-10,420 + 250 = -10,170$. The loan is $10,170.

EXAMPLE

Jean travels 110 miles west and then 200 miles further west. How far is she from her starting point?

    Solution If we consider west negative as on the number line, we get $-110 + (-200) = -310$. Jean is 310 miles west of her starting point.

EXERCISE 2.2.2

1. At 6 P.M. the temperature was $62°$. It dropped $25°$ overnight. What is the temperature in the morning?

2. Suppose the temperature was $5°$ below zero and then dropped $15°$. How cold was it then?

3. Jason has $200 in his checking account. In April he writes checks for $91, $85, and $25, deposits $50, and writes a check for $38. What is his balance? Is he overdrawn (is his balance negative)? Was he overdrawn at any time during April?

4. Michelle owes $295 and pays 6 monthly installments of $40. What is her balance?

5. The first elevator in the copper mine takes Craig down 600 feet. The next elevator takes him down another 300 feet. How far below ground is he?

6. Tanya leaves her home and drives north 45 miles. She goes back 20 miles for gas and then north 50 miles to stop for lunch. How far is Tanya from home?

7. Billy drives east 15 miles before he realizes he is going in the wrong direction. He turns around and drives west 85 miles. How far is Billy from his starting point?

8. A plane was flying at 51,000 feet. It descended 37,000 feet as it approached the Pacific coast. How high is it now? (*Hint*: Let up be the positive direction.)

9. Death Valley is 282 feet below sea level at its lowest point. At this point a 272-foot well is drilled. Express as a negative number the depth of the bottom of the well below sea level.

10. Rachel owes $350 on her car and just charged $69 for new tires. What does she owe now?

Subtraction

As we have seen in Chapter 1, subtraction is the reverse of addition. $5 - 2 = 3$ and $3 + 2 = 5$.

(a)	$6 - 1 = 5$	and	$5 + 1 = 6$
(b)	$10 - 2 = 8$	and	$8 + 2 = 10$
(c)	$17 - 5 = 12$	and	$12 + 5 = 17$
(d)	$-12 - 3 = -15$	and	$-15 + 3 = -12$
(e)	$-2 - 5 = -7$	and	$-7 + 5 = -2$

Let's look at the subtraction examples again.

(a)	$6 - 1 = 5$	and	$6 + (-1) = 5$
(b)	$10 - 2 = 8$	and	$10 + (-2) = +8$
(c)	$17 - 5 = 12$	and	$17 + (-5) = +12$
(d)	$-12 - 3 = -15$	and	$-12 + (-3) = -15$
(e)	$-2 - 5 = -7$	and	$-2 + (-5) = -7$

Subtracting a positive number gives the same result as adding a negative number.

Now, what happens when we subtract a negative number?

First look at this subtraction pattern:

$$10 - 3 = 7$$
$$10 - 2 = 8$$
$$10 - 1 = 9$$
$$10 - 0 = 10$$
$$10 - (-1) = ?$$

According to this pattern, $10 - (-1)$ should equal 11. From the reasoning earlier,

$$10 - (-1) = ? \quad \text{can be rewritten as} \quad ? + (-1) = 10$$

We also know that $11 + (-1) = 10$, so $10 - (-1) = 11$. But $10 + 1 = 11$, so here we can also change the subtraction to addition:

$$10 - (-1) = 10 + 1 = 11$$

RULE

Subtraction of Integers

1. Change the subtraction to addition.

2. Change the sign of the second number.

3. Add the two numbers. (Follow the rules for addition.)

EXAMPLE

Subtract: (a) $5 - 11$
(b) $-5 - 11$
(c) $5 - (-11)$
(d) $-5 - (-11)$

Solution
(a) $5 - 11 = 5 + (-11) = -6$
(b) $-5 - 11 = -5 + (-11) = -16$
(c) $5 - (-11) = 5 + 11 = 16$
(d) $-5 - (-11) = -5 + 11 = 6$

EXERCISE 2.2.3

Solve.

1. $3 - 8$
2. $-3 - 8$
3. $10 - 11$
4. $-10 - 11$
5. $44 - 44$
6. $-44 - 44$
7. $27 - 19$
8. $-27 - 19$
9. $36 - 85$
10. $-36 - 85$
11. $2 - (-3)$
12. $-2 - (-3)$
13. $5 - (-1)$
14. $-5 - (-1)$
15. $6 - (-4)$
16. $-6 - (-4)$
17. $1 - (-1)$
18. $-1 - (-1)$
19. $-35 - (-25)$
20. $35 - (-25)$

We mentioned earlier that $-12 - 3 = -12 + (-3) = -15$. Can you find the answer for $-12 - 3$ without changing the subtraction to addition of the opposite?

EXERCISE 2.2.4

Here are Problems 1–8 of Exercise 2.2.1 in simplified form. Solve them and see if you get the same answers as before.

1. $-3 + 5$
2. $3 - 5$
3. $-7 - 4$
4. $7 + 4$
5. $-11 - 11$
6. $24 - 20$
7. $45 - 60$
8. $-76 - 16$

Applications

In the following examples, write the problem as a subtraction example, recalling the rules for the signs, and then solve. When we use the words "from ... to," we go in reverse order. For example, if the temperature changed from 12° to 16°, the change is $16 - 12 = 4$, which shows that the temperature went up 4°. If we say the change was from 16° to 12°, the change is $12 - 16 = -4$. The temperature went down 4°. As you saw earlier, the word "between" implies absolute value.

EXAMPLE

Yesterday the temperature was 94°. Today it is 72°. What is the change from yesterday to today?

   Solution $(+72) - (+94) = -22°$

EXAMPLE

Dave is scuba diving 15 feet below the surface of the lake, and Lloyd is on a cliff 37 feet above the lake. What is the vertical distance from Dave to Lloyd?

   Solution $37 - (-15) = 37 + 15 = 52$

EXERCISE 2.2.5

In each problem, write a subtraction example, remember the rules for signs, then solve.

1. (a) The temperature rises from 72° to 81°. What is the change?
    (b) What is the change if the temperature falls from 81° to 72°?

2. Arthur's test score is 23 points above average, and Carl's is 4 points below average. What is the difference between the scores?

3. The city of Katmandu is located at 4700 feet above sea level. The Gilgit Pass is at 10,000 feet. What is the difference in altitude between Katmandu and the pass?

4. Carla and Luis each borrowed $500 for expenses in this college year. Carla paid back $231, and Luis paid $309. Who owes more? How much more than the other does that person owe?

5. Death Valley is 282 feet below sea level, and Mt. Everest is 29,000 feet above sea level. What is the difference in altitude between the top of Mt. Everest and the bottom of Death Valley?

6. Starting at a small town in the desert, Ben walks east 11 miles and Susan walks west 16 miles. What is the distance from Susan to Ben?

7. A helicopter 200 feet above sea level observes a submarine 50 feet below sea level. How far apart are the helicopter and the submarine?

8. A scuba diver dives to 78 feet below sea level, then rises 52 feet. What is his position with relation to sea level?

Combining More Than Two Integers

Go back to Chapter 1 and review the order-of-operations rule:

1. Simplify inside the grouping symbols.
2. Perform multiplication and division from left to right.
3. Lastly, do addition and subtraction.

EXAMPLE

Solve: $-3 - (-2) + 6 + (-4)$

Solution   Rewrite the problem without parentheses:

$$-3 + 2 + 6 - 4$$

Evaluate by doing addition and subtraction in order from left to right:

$$-3 + 2 + 6 - 4 = -1 + 6 - 4 = 5 - 4 = 1$$

Thus, $-3 - (-2) + 6 + (-4) = 1$.

Alternative Solution   Rewrite the problem as before:

$$-3 + 2 + 6 - 4$$

Now, collect terms with the same sign:

$$-3 - 4 = -7 \quad \text{and} \quad 2 + 6 = 8; \quad 8 - 7 = 1$$

EXAMPLE

Solve: $-3 + (-5) - (8) + 11$

Solution $-3 + (-5) - (8) + 11 = -3 - 5 - 8 + 11$

$$-3 - 5 - 8 + 11 = -8 - 8 + 11 = -16 + 11 = -5$$

Thus, $-3 + (-5) - (8) + 11 = -5$.

Alternative Solution $-3 - 5 - 8 = -16$, and $-16 + 11 = 5$.

EXERCISE 2.2.6

Solve.

1. $-2 + (-3) + 4$
2. $3 - (-2) - 6$
3. $1 - (-2) - 5 + (-6)$
4. $3 - (3) + (-4) - 5$
5. $-1 + (-3) - 5 - (-2)$
6. $4 - (-5) + (-11) - (-9)$
7. $-15 - (-27) + (-13) - (-1)$
8. $12 + (-4) - 9 - (-16) + (-8)$

Positive and Negative Numbers on the Calculator

These examples can be done by calculator if you have a **+/-** key. Most calculators show **-5** if you press **5 +/-** on the calculator. If your calculator has a different system, read the instruction booklet. Sometimes you get a wrong answer if you use the minus sign instead of the **+/-** sign.

EXAMPLE

Use your calculator to solve $-3 + (-5)$.

> **Solution**   **3 +/- + 5 +/- = -8**. Now try **- 3 + - 5 =**. If you get **-8**, it's fine.

EXAMPLE

$2 - (-8)$

> **Solution**   **2 - 8 +/- = 10**

Now try **2 - - 8 =**. Is the answer still 10? If not, read the instructions for your calculator.

EXERCISE 2.2.7

Solve using your calculator. Check by hand.

1. $(7) + (-5)$
2. $(-11) + (-2)$
3. $(-13) + (+15)$
4. $(+10) + (+17)$
5. $(+96) - (-87)$
6. $(-66) - (9)$
7. $(38) - (-47)$
8. $(-44) - (-88)$

2.3 MULTIPLICATION AND DIVISION OF INTEGERS

Multiplication

Multiplication is a short form of addition.

$$3 \times 7 = 7 + 7 + 7 = 21 \quad \text{(seven, three times)}$$

$$7 \times 2 = 2 + 2 + 2 + 2 + 2 + 2 + 2 = 14 \quad \text{(two, seven times)}$$

$$5 \times 0 = 0 + 0 + 0 + 0 + 0 = 0 \quad \text{(zero, five times)}$$

Remember also that multiplication can be written 3×5, $3 \cdot 5$, $3(5)$, $(3)5$, $(3)(5)$, or $3*5$.

When we multiply negative numbers, we use parentheses, for example, $3(-4)$.

$$3(-4) = (-4) + (-4) + (-4) = -12 \quad \text{(negative four, three times)}$$

What about $-4(3)$? Multiplication is a commutative operation, so when two numbers are multiplied, the *order* does *not* matter. $3 \times 5 = 5 \times 3$, for example. Therefore, $-2(7)$ is the same as $7(-2)$.

$$7(-2) = -14, \quad \text{so} \quad -2(7) = -14$$

The concept of multiplication of numbers with signs can also be explained the following way:

-2 and 2 are opposite numbers.

$-2(7)$ is the opposite of $2(7)$. Therefore, the answer to $-2(7)$ is the opposite of 14 [the answer to $2(7)$], or -14.

Similarly, $-2(-7)$ is the opposite of $2(-7) = -14$, so $-2(-7) = 14$.

EXAMPLE

Multiply: $-3(-5)$

> **Solution** $-3(-5)$ is the opposite of $3(-5) = -15$. The opposite of -15 is 15, so $-3(-5) = 15$.

RULE

Multiplication of Two Integers

1. Multiply the absolute values.

2. (a) If the signs are alike, the sign of the product is +.
   (b) If the signs are different, the sign of the product is −.

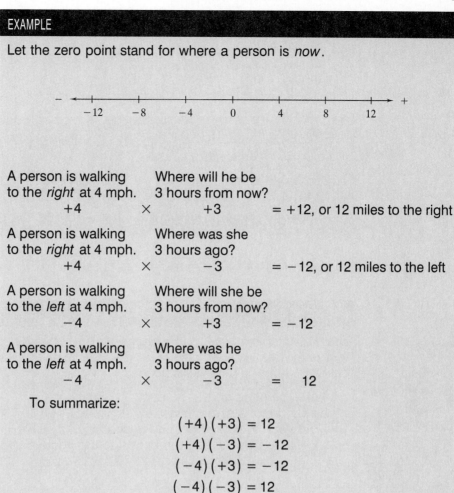

EXAMPLE

Let the zero point stand for where a person is *now*.

A person is walking   Where will he be
to the *right* at 4 mph.   3 hours from now?
     +4   ×   +3   = +12, or 12 miles to the right

A person is walking   Where was she
to the *right* at 4 mph.   3 hours ago?
     +4   ×   −3   = −12, or 12 miles to the left

A person is walking   Where will she be
to the *left* at 4 mph.   3 hours from now?
     −4   ×   +3   = −12

A person is walking   Where was he
to the *left* at 4 mph.   3 hours ago?
     −4   ×   −3   =   12

To summarize:

$$(+4)(+3) = 12$$
$$(+4)(−3) = −12$$
$$(−4)(+3) = −12$$
$$(−4)(−3) = 12$$

EXERCISE 2.3.1

First determine the sign of the answer. Then calculate the answer.

1. 5(3)

2. 5(−3)

3. −5(3)

4. −5(−3)

5. 10(−100)

6. (−10)(−100)

7. −13(−11)

8. 7(−21)

9. (−25)16

10. (−102)(−23)

Division

Division is the reverse of multiplication. $12 \div 2 = 6$ and $6 \times 2 = 12$.
   $−2(6) = −12$. Therefore, $−12 \div (−2) = 6$ and $−12 \div 6 = −2$
   $−2(−6) = 12$. Therefore, $12 \div (−2) = −6$ and $12 \div (−6) = −2$

EXAMPLE

(a) $-72 \div (-8)$ (b) $81 \div (-3)$ (c) $2 \cdot 3 \div 6$ (d) $12 \div 4(-3)$

Solution   (a) 9   (b) -27 (c) 1   (d) -9

RULE

Division of Two Integers

1. Divide the absolute values of the two numbers.

2. (a) If the signs are alike, the sign of the quotient is $+$.
    (b) If the signs are different, the sign of the quotient is $-$.

MEMORY TIP

Lest you despair of ever making sense of the rules of signs for multiplication or division, Mary Dolciani, a very fine mathematician and teacher, developed this help for her students:

Let $+$ be good and $-$ be bad.
To do good to something good is good $(+)(+) = (+)$
To do bad to something bad is good $(-)(-) = (+)$
BUT To do good to the bad or bad to the good is bad.
$$(+)(-) = (-), \qquad (-)(+) = (-)$$

EXERCISE 2.3.2

First determine the sign of your answer. Then calculate the answer.

1. $8 \div (-2)$

2. $-32 \div (-16)$

3. $-18 \div 3$

4. $56 \div (-7)$

5. $72 \div 12$

6. $-63 \div 9$

7. $-12 \div (-6)$

8. $28 \div (-4)$

9. $-589 \div (-19)$

10. $-48 \div 16$

Multiplication and Division with More Than Two Numbers

When more than two numbers are multiplied or divided, we find the sign of the answer by counting the minus signs:

$$(-)(+)(-) = + \qquad\qquad (-)(+)(-)(+) = +$$
$$(+)(-)(+) = - \qquad\qquad (+)(-)(-)(-) = -$$
$$(-) \div (+) \div (-) = + \qquad (-) \div (+) \div (-) \div (+) = +$$
$$(+) \div (-) \div (+) = - \qquad (+) \div (-) \div (-) \div (-) = -$$

From the above we can conclude that

One negative gives a negative answer.

Two negatives give a positive answer.

Three negatives give a negative answer.

Four negatives give a positive answer.

> **RULE**
>
> **Multiplication and Division of Signed Numbers**
> An even number of negative signs gives a positive answer.
> An odd number of negative signs gives a negative answer.

EXAMPLE

Multiply: $(-1)(-2)(-3)$

Solution Three negatives; the product is negative.
$$1(2)(3) = 6, \quad \text{so} \quad (-1)(-2)(-3) = -6$$

EXAMPLE

Divide: $-8 \div 2 \div (-4)$

Solution Two negatives; the quotient is positive.
$$8 \div 2 = 4 \quad \text{and} \quad 4 \div 4 = 1, \quad \text{so} \quad -8 \div 2 \div (-4) = 1$$

EXAMPLE

Solve: $-10 \div 5(-3)(-2)$

Solution Three negatives; the answer is negative.
$$10 \div 5(3)(2) = 2(3)(2) = 12, \quad \text{so} \quad -10 \div 5(-3)(-2) = -12$$

EXERCISE 2.3.3

1. $(-1)(2)(-3)$
2. $(-3)(-4)(-3)$
3. $(-6)(0)(-5)$
4. $-2(-3)(-4)(1)$
5. $-7(-1)(3)(5)$
6. $(-3)(-1)(-6)(-5)$
7. $(-1)(-1)(-1)(-1)(-1)$
8. $(-2)(-1)(-3)(-1)(-2)(-1)$

   9. $-7(-2) \div (2)(3)$

   10. $12 \div 6 \cdot (-2)$

   11. $12 \div 2 \cdot 6$

   12. $(-12) \div (-6) \div 2$

   13. $(-12) \div (-2) \div (-6)$

   14. $-20 \div 4(-4)$

Applications

Review the rules for problem solving.

Write each of the following first as a multiplication or division problem.

EXAMPLE

The temperature dropped 3 degrees per hour. Now it is $-12°$.

(a) What was it 7 hours ago?

(b) If it continues to drop at $3°$ per hour, what will the temperature be 3 hours from now?

   Solution   Both a "drop" and "hours ago" are negative.

   (a) $-12 + (-7)(-3) = -12 + 21 = 9$

   Check: Is it true that if the temperature was $9°$ and dropped $3°$ per hour, after 7 hours it was down to $-12°$? $9 + 7(-3) = -12$

   (b) "Hours from now" is positive, so

   $$-12 + (3)(-3) = -12 - 9 = -21.$$

   The temperature will be $-21°$.

EXERCISE 2.3.4

1. The temperature dropped 2 degrees a day for 8 days. How much did it drop?

2. After taking off from JFK airport, a plane rose 2500 feet per minute for 9 minutes. Later in the flight the plane descended 2000 feet a minute for 3 minutes. What was its altitude then?

3. Liz is walking east at 3 miles per hour (mph).
   (a) Where will she be in 5 hours?
   (b) Where was she 3 hours ago if she was walking at 4 mph?
   (c) Suppose she's walking west at 3 mph. Where will she be in 2 hours? Where was she 2 hours ago?

4. Jeff's checking account shows that he is overdrawn by 3 times as much as Laura is. Laura's balance is $-\$23$. What is Jeff's balance?

5. Five friends drove to Florida for spring break. They each contributed $50 toward expenses. The cost of the trip was $285. Did they each get money back, or did they have to pay more? How much?

6. Pat borrowed $3258 from her parents to buy a car. She made 9 payments of $325 each. How much does she still owe?

7. On a true-false test (scored rights minus wrongs) Joe's scores were $17, -3, -9, 0, 10, -9$. What was his average score? The average is the sum of the scores divided by the number of scores (here 6).

8. The lowest point in Death Valley is 279 ft below sea level (-279). The Dead Sea on the border between Jordan and Israel is 1395 ft below sea level (-1395). How many times lower than Death Valley is the lowest point in the Dead Sea? (Use negative numbers.)

2.4 EXPONENTS AND RADICALS

Exponential Notation

Suppose we have $2 \times 2 \times 2 \times 2 \times 2 \times 2 \times 2$. This equals 128, but it is a tedious example to write. Instead of writing the repeated multiplication we can use a shorthand notation called *exponential notation*. We write $2 \times 2 \times 2 \times 2 \times 2 \times 2 \times 2$ as 2^7, meaning seven 2s multiplied together. The 2 is called the *base*, and the 7 is the *exponent*. When we read 2^7, we say "2 raised to the seventh power" or simply "2 to the seventh power" or "2 to the seventh." 2^7 is an *exponential expression*. When we evaluate this expression, we get 128.

$2^3 = 2 \times 2 \times 2 = 8$; 2 is the base and 3 is the exponent.

The exponential expression 2^3 equals 8; we might also say that 8 is "the third power of 2." The exponential expression 3^2, where 3 is the base and 2 is the exponent, is evaluated as 3×3 or 9.

DEFINITION

a^n is an exponential expression.

a is the base.

n is the exponent.

$a^n = a \times a \times \cdots \times a$

There are n factors.

EXAMPLE

Find the value of (a) 3^4,   (b) 4^1,   (c) $(-2)^2$,   (d) -2^2.

Solution
(a) $3^4 = 3 \times 3 \times 3 \times 3 = 81$
(b) $4^1 = 4$
(c) $(-2)^2 = (-2)(-2) = 4$
(d) $-2^2 = -(2)^2 = -(2)(2) = -4$

It is important to see the distinction between examples (c) and (d). In (c) the minus sign is inside the parentheses and thus a negative number has to be multiplied by itself. However, in (d) the exponent belongs only to the base 2. Here we have only one minus sign.

EXAMPLE

(a) $(-5)^2$ (b) $-(5)^4$ (c) $(-1)^5$ (d) $(-1)^{10}$

Solution

(a) $(-5)^2 = (-5)(-5) = 25$

(b) $-(5)^4 = -(5)(5)(5)(5) = -625$

(c) $(-1)^5 = (-1)(-1)(-1)(-1)(-1) = -1$

(d) $(-1)^{10} = (-1)(-1)(-1)(-1)(-1)(-1)(-1)(-1)(-1)(-1) = 1$

RULE

Powers of Negative Numbers

If a negative number is raised to an even-numbered power, the result is a positive number. If it is raised to an odd-numbered power, the result is a negative number.

EXERCISE 2.4.1

Evaluate.

1. 1^4

2. 2^5

3. 5^2

4. 5^3

5. 4^3

6. $(-1)^8$

7. $(-2)^6$

8. $(-7)^2$

9. -3^2

10. $(-3)^2$

11. -2^3

12. $(-2)^4$

13. -5^3

14. -6^2

Radicals

Subtraction is the inverse of addition.

$$8 - 2 = 6 \qquad \text{and} \qquad 8 = 6 + 2$$

Division is the inverse of multiplication.

$$8 \div 2 = 4 \qquad \text{and} \qquad 8 = 4 \times 2$$

Raising to an exponent and finding a root are similarly related: Three to the second power is nine, or three squared, and the *square root* of nine is three. In symbols:

$$3^2 = 9 \qquad \text{and} \qquad \sqrt[2]{9} = 3, \text{ or simply } \sqrt{9} = 3$$

The symbol $\sqrt{}$ is called a *radical sign*.

EXAMPLE

Find the square root of (a) 4,   (b) 16,   (c) 25,   (d) 81.

> **Solution**
> (a) $2^2 = 4$, so $\sqrt{4} = 2$
> (b) $4^2 = 16$, so $\sqrt{16} = 4$
> (c) $5^2 = 25$, so $\sqrt{25} = 5$
> (d) $9^2 = 81$, so $\sqrt{81} = 9$

The square root of a negative number does not exist as a real number, since the product of two equal numbers is always positive. Also $(-2)^2 = 4$ but $\sqrt{4} = 2$. The symbol $\sqrt{4}$ implies a positive number.

Two cubed (2 raised to the third power, or 2^3) equals 8, and the cube root of 8 is 2. In symbols this is

$$2^3 = 8 \quad \text{and} \quad \sqrt[3]{8} = 2$$

Negative two cubed is negative eight, and the cube root of negative eight is negative two. In symbols:

$$(-2)^3 = -8 \quad \text{and} \quad \sqrt[3]{-8} = -2.$$

EXAMPLE

Find the fourth root of 81.

> **Solution** $81 = 3 \times 3 \times 3 \times 3 = 3^4,$ so $\sqrt[4]{81} = 3$

DEFINITION

Let n be a positive integer. Then for any a if n is odd, and for a positive a if n is even,

$$\sqrt[n]{a^n} = a$$

That is, the nth root of a^n is a.
   n is called the *index*.

One, zero, and negative numbers are not used as a root index.

EXERCISE 2.4.2

1. $\sqrt{100}$

2. $\sqrt{144}$

3. $\sqrt{0}$

4. $\sqrt[3]{-27}$

5. $\sqrt[4]{-16}$

6. $-\sqrt[4]{16}$

7. $\sqrt{4 \cdot 16}$

8. $\sqrt{4} \cdot \sqrt{16}$

9. $\sqrt[5]{-32}$

10. $\sqrt[6]{-64}$

As we mentioned in Chapter 1, the symbols ( ), [ ], { } are used to group numbers. Radical signs are also considered grouping symbols. For example,

$$\sqrt{4 + 12} = \sqrt{16} = 4$$

Exponential terms and radicals are evaluated before multiplication and division. The rules for *order of operations* can now be rewritten in complete form.

RULE

Order of Operations

1. Simplify inside any grouping symbols.

2. Solve in this order:
   (a) Evaluate exponential and radical expressions.
   (b) Multiply and divide in order from left to right.
   (c) Add and subtract.

EXAMPLE

Solve: $(2 + 3)^2$

Solution   First simplify inside the parentheses:

$$2 + 3 = 5$$

Next, evaluate the exponential expression:

$$(2 + 3)^2 = (5)^2 = 25$$

EXAMPLE

Solve: $3(2 + 3)^2$

Solution
1. Simplify inside the parentheses to get $3(5)^2$.
2. Evaluate the exponential expression: $3(5)^2 = 3(25)$
3. Multiply: $3 \times 25 = 75$

EXAMPLE

Solve: $3(2 + 3^2)$

Solution   This time the exponent is inside the parentheses, so to simplify inside the parentheses we need to evaluate 3^2 as a *first* step.

$$3(2 + 3^2) = 3(2 + 9)$$

Then we carry out the addition inside the parentheses and go on to solve the problem.

$$3(2 + 9) = 3(11)$$
$$3(11) = 33.$$

EXAMPLE

Evaluate: (a) $\sqrt{3^2 + 4^2}$ (b) $(3 + \sqrt{16})^2$

Solution
(a) $\sqrt{3^2 + 4^2} = \sqrt{9 + 16} = \sqrt{25} = 5$
(b) $(3 + \sqrt{16})^2 = (3 + 4)^2 = 7^2 = 49$

EXAMPLE

Evaluate:
(a) $2[3 + 5(4 - 1) + 3] + 1$
(b) $10 - 3\{4 - 2[7 - (9 - 6) - 2] + 1\}$

Solution
(a) Start inside the brackets and simplify until there is only one number left inside the brackets:

$$[3 + 5(4 - 1) + 3] = [3 + 5(3) + 3] = [3 + 15 + 3] = [21]$$

We now have $2[21] + 1 = 42 + 1 = 43$.

(b) $10 - 3\{4 - 2[7 - (9 - 6) - 2] + 1\} = 10 - 3\{4 - 2[7 - (3) - 2] + 1\}$
$= 10 - 3\{4 - 2[2] + 1\}$
$= 10 - 3\{4 - 4 + 1\}$
$= 10 - 3\{1\}$
$= 10 - 3 = 7$

EXERCISE 2.5.1

Evaluate.

1. 2×3^2

2. $(2 \times 3)^2$

3. $(2 + 3)^2$

4. $4 \times \sqrt{25}$

5. $(4 \times \sqrt{25})^2$

6. $(4 + \sqrt{25})^2$

7. $(4 - 5^2)$

8. $(4 - 5^2)^2$

9. $6(2 + 3)^2$

10. $6(2)(3)^2$

11. $[(18 - 6) \times 2] - 4$

12. $[(5 - 12) \times 7] + 1$

13. $\sqrt{100} - [500 \div (4 - 6)^2]$

14. $2 \cdot 5^2 - \{[2(10 - 7)]^3 \div 3^2\}$

15. $(18 - 3) \div [3 + (4 - 2)]$

16. $10 - 5[2(6 - 3) - 3(7 - 5)]$

Definitions

a^n is an exponential expression, where a is the base and n is the exponent. a is a factor n times.

$\sqrt[n]{a^n} = a$ where n is a positive integer and a is a positive number if n is even. n is called the index.

Rules

Addition of Two Integers

1. When signs are alike, add the absolute values. The sign of the sum is the sign of the numbers added.
2. When the signs are different, find the difference between the absolute values. The sign of the answer is the sign of the number with the largest absolute value.

Subtraction of Two Integers

1. Change the subtraction to addition.
2. Change the sign of the second number and add.

Multiplication of Two Integers

1. Multiply the absolute values.
2. (a) If the signs are alike, the sign of the product is $+$.
    (b) If the signs are different, the sign of the product is $-$.

Division of Two Integers

1. Divide the absolute values of the two numbers.
2. (a) If the signs are alike, the sign of the quotient is $+$.
    (b) If the signs are different, the sign of the quotient is $-$.

Multiplication and Division of Signed Numbers

An even number of negative signs gives a positive answer.

An odd number of negative signs gives a negative answer.

If a negative number is raised to an even-numbered power, the result is a positive number. If it is raised to an odd-numbered power, the result is a negative number.

Order of Operations

1. Simplify inside any grouping symbols.
2. Solve in this order:
    (a) Evaluate exponential and radical expressions.
    (b) Multiply and divide in order from left to right.
    (c) Add and subtract.

VOCABULARY

Absolute value: Distance between zero and a number on the number line. It is always positive. Symbols: $|+2| = 2$ and $|-2| = 2$. The absolute value symbol also acts as parentheses. $|-7 + 5| = |-2| = 2$.

Base: The number that is the factor in exponential notation; in a^n, a is the base.

Exponent: The n in a^n; the number of a factors whose product is a^n.

Exponential notation: a^n; a is the base and n is the exponent.

Integers: $\{\ldots, -3, -2, -1, 0, 1, 2, \ldots\}$

Inverse operations: Operations that cancel each other out. For example, addition/subtraction, multiplication/division, raising to a power/taking the root.

Number line: A line that shows the relationship of numbers.

Radical: The inverse of an exponential expression. $\sqrt[n]{a^n} = a$, n positive.

Root: Same as radical. The nth root of $a = \sqrt[n]{a}$.

Signed numbers: Numbers preceded by a positive or negative sign.

CHECK LIST

Check the box for each topic you feel you have mastered. If you are unsure, go back and review.

- ☐ Absolute value
- ☐ Applications
- ☐ Basic operations with integers
- ☐ Exponential notation
- ☐ Radicals
- ☐ Positive and negative numbers on the calculator

REVIEW EXERCISES

1. Locate the following on a number line:
   $-5, 2, -1, -8, 0, 3, -9$
2. Give the absolute values of the numbers in Review Exercise 1.
3. Evaluate:
   (a) $3|-2|$
   (b) $-|-2|$
   (c) $|-3||-2|$
   (d) $3 - |-5|$
   (e) $|-5| - |-8|$
   (f) $|-7 + 2|$
   (g) $|-7| + |2|$
   (h) $|-3||-2| |-1|$
4. Show on the number line:
   (a) $3 + 2 + 5$
   (b) $2(5)$
   (c) $3(-2)$
   (d) $4 - 5$
5. Simplify:
   (a) $-2 + 3$

(b) $(-6) + (-3)$
(c) $4 + 2 - 7$
(d) $(-3) + (-5) + 8$
(e) $(17) + (-32) + (-11)$

6. The temperature was $17°$ above zero and dropped 28 degrees. What is the temperature now?
7. In Bethel, Alaska, on a certain day, the low temperature was $-47°$, and the high temperature was $-18°$. What was the difference between the high and low temperatures?
8. Lee had $300 in his checking account. He wrote checks for $71, $92, $15, $75, and $82. Is he overdrawn? (Is his balance negative?)
9. Simplify:
   (a) $(-2) - (3)$
   (b) $(-6) - (-8)$
   (c) $3 + 2 - (6)$
   (d) $-7 - (-3) - (-4)$
   (e) $0 - (5) - (11)$

Write Review Exercises 10 and 11 as subtraction examples and solve.

10. Yesterday the temperature was 63°, and today it is 41°. What is the change from yesterday to today?

11. A scuba diver dives to 57 feet below sea level, then rises 29 feet. What is his position in relation to sea level?

12. Simplify:
    (a) $(-2)(7)$
    (b) $(-3)(-4)$
    (c) $(-25)(2)$
    (d) $(-3)(-2)(-1)$
    (e) $(-1)(3)(0)(-5)$

13. Simplify:
    (a) $(8) \div (-2)$
    (b) $(-16) \div (-4)$
    (c) $(-8) \div 0$
    (d) $(72) \div 6 \div (-4)$
    (e) $(-15) \div (-3) \div (-1)$

14. Mrs. Murphy's stocks have been losing $5 every year for 7 years. How much were they worth 4 years ago compared to now?

15. The temperature dropped 20 degrees in 5 hours. How many degrees did it change per hour?

16. Mary lost 2 lb every month for 6 months. What was her change in weight?

17. The U.S. trade balance with Japan was close to $-\$42$ billion in 1990. What was the average trade balance per month in millions of dollars?

18. Evaluate:
    (a) 3^2
    (b) 2^3
    (c) 7^1
    (d) $(-2)^4$
    (e) -2^4
    (f) $(-1)^7$

19. Evaluate:
    (a) $\sqrt{64}$
    (b) $\sqrt{1}$
    (c) $\sqrt[3]{27}$
    (d) $\sqrt[5]{32}$
    (e) $\sqrt[3]{-64}$
    (f) $\sqrt{9} \cdot \sqrt{4}$

20. Simplify:
    (a) $19 - 11 + 14 - 16$
    (b) $(-6)^3 + 0(-2)^4$
    (c) $24 \div 8 \cdot 3(-1)$
    (d) $5^2\sqrt{16} - 48 \div (-12) \div 2$
    (e) $(\sqrt{5} + 4)(3^1)(7 - 4)$
    (f) $7 - [3 - 2(3 + 6)] + (2^3)$

READINESS CHECK

Solve these problems to satisfy yourself that you have mastered Chapter 2.

1. Solve: $|-2||-4| + |-1||5|$
2. Evaluate: $19 + (-11)$
3. Evaluate: $-11 - (-17)$
4. Evaluate: $-(-2)(-3)(-6)$
5. Evaluate: $25 \div (-5)$
6. Evaluate: $-5 - (-7) + (-3) - (2)(-6) - 4$
7. Evaluate: $-3^2 + (-3)^2$
8. Solve: $\sqrt{5^2 + 12^2}$
9. Evaluate: $15 \div 5(-3)$
10. The temperature dropped from $-2°$ to $-11°$. How many degrees was the drop?

CHAPTER 3

DECIMALS AND PERCENTS

Ordering by tens

More than incidental use of decimals can be found in the mathematics of ancient China and medieval Arabia. However, it was in the *Compendio de lo abaco*, published in the year Columbus journeyed to America, that the use of the decimal point appeared to denote the division of an integer by powers of 10. The earliest systematic treatment of decimals was given by the Flemish mathematician Simon Stevin in his arithmetic text of 1595 entitled *La Disme*. He developed a notation quite similar to ours; each digit is followed by a circled number that indicates the number of places that the digit is to the right of the units place. For example, 3.14 would be written as 3 ⓪ 1 ① 4 ②.

In this chapter we discuss numbers that can be written with decimal points or as percents. Our money is expressed in decimals, and interest is expressed in percents. Everyone has experience with money and with loans. You will find that you are already familiar with many of the operations involving decimals and percents.

3.1 DECIMALS

Place Values

Our number system is called a *decimal* system. The prefix *deci* means ten. In the decimal system, we use ten symbols—0, 1, 2, 3, 4, 5, 6, 7, 8, and 9—and we count in groups of ten: 10, 20, 30, 40, and so on. (For the Laplanders, with one hand hidden in a mitten, 5 was the central number or base of their number system. The Mayans, in Mexico, used 20, possibly because they were barefoot and could use both fingers and toes as aids to counting.) In computer science, the base 2 is used because in an off/on system there are only two possibilities: a switch is either on or off, there is or is not an electric signal, etc.

As an example of place values in the decimal system, look at the number 222. Starting at the left, the first 2 has a value of 200 (2 × 100), the second has a value of 20 (2 × 10), and the last 2 has a value of 2 (2 × 1). In each place we can have any of our ten digits (0, 1, 2, 3, 4, 5, 6, 7, 8, and 9). From each digit's place in the number we are considering, we know its value (e.g., in 34, 3 represents 30 and 4 represents 4).

Consider the number 5. If you move the 5 one place to the left, you get 50; you have multiplied 5 by 10. Move 5 two places to the left and you get 500; you have multiplied 5 by 100 or 10 × 10. (This is, by the way, how you enter numbers when you use a calculator or when you deposit money into a cash machine at the bank: you press one number key at a time, and the digits you have already keyed in move one place to the left each time.)

Now, instead of moving 5 to the left, move it one place to the right. What happens?

5   becomes   .5

The rightmost place value for a whole number is the ones (units) place. We can mark that with a period called the *decimal point* and then continue with new places to the right of the decimal point. Thus the whole number 5 could be written as 5., and when we move the digit 5 one place to the right we have .5 as the new number.

EXAMPLE

Line 1 indicates how one dollar ($1.00) would be entered in this table.

Ones Place

```
1.  __  __  __   1.   0   0
2.  __  __  __   __   __  __
3.  __  __  __   __   __  __
4.  __  __  __   __   __  __
5.  __  __  __   __   __  __
```

(a) Show where 10 dollars belongs in line 2.
(b) Show where 100 cents belongs in line 3.
(c) Show where 10 cents belongs in line 4.
(d) Show where 1 cent belongs in line 5.

Solution   Since 100 cents = $1.00
10 cents = $0.10
1 cent = $0.01
we get the following table:

	hundreds	tens	ONES		tenths	hundredths
Line 1			1	.	0	0
Line 2		1	0	.	0	0
Line 3			1	.	0	0
Line 4			0	.	1	0
Line 5			0	.	0	1

The decimal point shows where the ones (or units) place is. Notice that the decimal numbers are symmetrical around the ones place: one step to the left of 1 gives us the place value of ten, while one step to the right of 1 has the value of one tenth. Two steps to the left of 1 is the hundreds place; two steps to the right is the hundredths place.

Earlier, when we moved 5 one place to the right, 5 moved to the tenths place. We call its new value "five tenths." (In daily language we say "point five.")

To go from 100 to 10, we divide by 10.

To go from 10 to 1, we divide by 10.

To go from 1 to the tenths place, we also divide by 10. That place's value is $1/10$, or one tenth. We can write one tenth as .1 or 0.1.

DEFINITION

Place values:

Ones.Tenths Hundredths Thousandths Ten-thousandths . . .
(see Figure 3.1)

Reading Decimals

The number 1.234 has 1 in the units place, 2 in the tenths place, 3 in the hundredths place, and 4 in the thousandths place. We read the number as "one and two hundred thirty-four thousandths." Notice that we say "and" to indicate the decimal point. We always use the last place as the name for the decimal part. In daily life, however, we usually say, "one point two three four."

12.0578 is read "twelve and five hundred seventy-eight ten-thousandths." 0.1 and .1 are both read as "one tenth"; in such a case we ignore the digit zero in the ones place. However, we might also say "zero point one."

EXAMPLE

Read: (a) 0.678,   (b) 600.078,   (c) 60.0078,   (d) 60007.8

Solution
(a) six hundred seventy-eight thousandths
(b) six hundred and seventy-eight thousandths
(c) sixty and seventy-eight ten-thousandths
(d) sixty thousand seven and eight tenths
Remember to read the decimal point as "and"—(a) and (b) illustrate what a difference that makes.

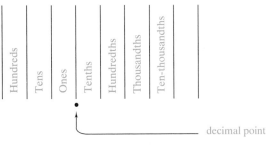

Figure 3.1 Place values.

EXAMPLE

Write the number five thousand and two hundred fifty-three ten-thousandths in mathematical symbols.

    Solution   The number is 5000.0253. We must fill the empty places with zeros.

RULE

READING DECIMALS

The decimal point is read as "and." The decimals are read as whole numbers followed by the name of the rightmost place.

EXERCISE 3.1.1

1. In which place is the digit 5?
   (a) 3.056
   (b) 45.37
   (c) 10.2358
   (d) 1,053,698.23
   (e) 203.561
   (f) 1.236952

2. Write out the number as you would read it.
   (a) 98.6
   (b) 45.34
   (c) 0.678
   (d) 7.190
   (e) 10.06
   (f) 15.3829

3. Write in decimal notation.
   (a) Six and five tenths
   (b) Seven and three hundredths
   (c) Twenty-two and fifteen hundredths
   (d) Thirty and one hundred two thousandths
   (e) One hundred and six hundredths
   (f) Thirty-six ten-thousandths

Ordering Decimals

You can always tell which one of two numbers is the larger, so you can arrange any group of numbers in order. To order whole numbers or decimals we use our knowledge of place value. 20 is greater than 18 (20 > 18) because the digit in the largest place in 20 (the 2 in the tens place) is larger than the digit in the corresponding place in 18 (2 > 1).

EXAMPLE

Which is larger, 0.3 or 0.098?

Solution   Look at corresponding place values.

tenths
↓
0.3
0.098

   The largest place is the tenths, and 3 > 0. Therefore 0.3 is larger than 0.098. We don't have to pay attention to the hundredths or the thousandths place in this example.

EXAMPLE

Which is larger, 0.245 or 0.268?

   Solution   If we compare 0.245 and 0.268, we see that in both numbers there is a 2 in the tenths place. Therefore we go to the next place to the right, the hundredths, and compare the digits. 6 is larger than 4 (6 > 4), so 0.268 is larger than 0.245. Two hundred sixty-eight thousandths is larger than two hundred forty-five thousandths.

EXAMPLE

Find the smallest number:
(a) 0.095, 0.1, 0.16
(b) 3, 0.333, 0.3
(c) 0.85, 0.9, 0.099

   Solution
   (a) 0.095
       0.1
       0.16
       The digits in the tens place are 0 and 1. 0 < 1, so 0.095 is the smallest number.
   (b) 3.000
       0.333
       0.300
       If we fill in the empty spaces with zeros, we see clearly that 0.3 is the smallest number.
   (c) 0.850
       0.900
       0.099
       The smallest number is 0.099, because it has zero in the tens place.

RULE

ORDERING DECIMALS

Arrange the numbers in a column with the decimal points below each other. Compare the place values going from left to right.

EXERCISE 3.1.2

1. Find the largest number. (Remember that if there is no decimal point visible, it is to the right of the number: 3 is 3., 29 is 29., and so on.)
   (a) 2, 0.1, 0.9
   (b) 0.035, 0.0099, 0.2
   (c) 0.006, 0.060, 0.02
   (d) 0.95, 0.099, 0.9
   (e) 1.03, 1.1, 1.008, 1, 1.0
   (f) 2.05, 2.049, 2, 2.1, 2.13

2. Arrange in order from the smallest to the largest.
   (a) 2, 0.020, 0.2
   (b) 0.95, 0.09, 0.9
   (c) 1.2, 0.120, 12
   (d) 0.0005, 0.05, 0.005, 0.5
   (e) 0.089, 0.091, 0.19, 0.91, 0.10
   (f) 0.014, 0.019, 0.020, 0.010, 0.050

Rounding Decimals

The procedure for rounding decimals is the same as for rounding whole numbers, except that we don't add zeros at the end. We cut off the digits we don't need. For instance, 45.679 rounded to the nearest tenth would be 45.7. It would be wrong to write 45.70, because the zero suggests that the number has been rounded to the nearest hundredth. Rounded to the nearest hundredth, 45.679 becomes 45.68.

An *exact* number such as 1 can be written as 1., 1.0, 1.00, etc. However, a *rounded* number cannot have these zeros added to the right. For example, when 2.95 is rounded to a whole number, we get 3. This rounded 3 is *not the same* as 3.00.

EXAMPLE

Round 32.698 to the nearest hundredth.

Solution   32.698 becomes 32.70. Here we must keep the last zero, since we were asked to round off to the hundredths place.

RULE

ROUNDING

If the digit to the right of the rounded digit is less than 5, leave the digit the same.
If the digit to the right of the rounded digit is 5 or more, increase the rounded digit by one.
Discard all digits to the right of the rounded digit.

Note, you must always start with the given number; do not round an already rounded number. If you did, your errors would be larger than when you round the original number.

EXAMPLE

Round 3.498 to the nearest (a) tenth,   (b) hundredth,   (c) whole number.

> **Solution**
> (a) 3.498 rounded to the nearest tenth is 3.5.
> (b) 3.498 rounded to the nearest hundredth is 3.50.
> (c) 3.498 rounded to the nearest whole number is 3.

EXAMPLE

Round 0.564 to the nearest (a) tenth,   (b) hundredth,   (c) whole number.

> **Solution**
> (a) 0.564 becomes 0.6 (6 > 5).
> (b) 0.564 becomes 0.56 (4 < 5).
> (c) 0.564 becomes 1.

In practical applications we are not told when to round a number. Numbers representing money are usually rounded to the nearest cent.

A general rule for rounding is that we round the answer so that it has the same number of digits as the numbers given in the problem. For example, a whole number divided by a whole number is a whole number. An answer in decimal form is rounded to the fewest decimals in the numbers you have worked with.

EXERCISE 3.1.3

Round to the nearest tenth, hundredth, and whole number.

1. 45.943
2. 0.845
3. 3.899
4. 0.08
5. 0.967
6. 1.529
7. 5.282
8. 10.625
9. 8.624
10. 15.975

3.2 BASIC OPERATIONS WITH DECIMALS

Addition and Subtraction

The procedure for adding and subtracting numbers containing decimal points is the same as for whole numbers. We line up the numbers with the same place values in the same column. The easiest way to do this is to line up the decimal points. (Remember, if there is no decimal point, it is understood to be at the end of the number. 2 = 2., for example.)

EXAMPLE

Add: $2 + 0.3 + 1.15 + 0.009$

Solution

$$\begin{array}{r} 2. \\ 0.3 \\ 1.15 \\ +0.009 \end{array}$$

You are allowed to fill in the empty spaces with zeros if you wish. Many people think it is easier to rewrite the problem like this:

$$\begin{array}{r} 2.000 \\ 0.300 \\ 1.150 \\ +0.009 \\ \hline 3.459 \end{array}$$

In examples in math books, the numbers are supposed to be exact (not rounded). In real life the numbers are rounded before addition.

In subtraction it is advisable to fill in the zeros in cases like $3 - 0.04$, which then becomes

$$\begin{array}{r} 3.00 \\ -0.04 \\ \hline 2.96 \end{array}$$

Otherwise it is easy to get lost when you borrow. You can ignore the zeros at the end of the decimals, but you cannot ignore the zeros that are between the decimal and the first nonzero digit. 0.04 is equal to .04 and also to 0.0400, but not to 0.4! Practice addition and subtraction on paper, and then check the answers by calculator.

RULE

ADDITION AND SUBTRACTION OF DECIMAL NUMBERS

1. Place the numbers with the decimal points below each other.
2. Proceed as with whole numbers.

EXAMPLE

Add: (a) $-2.3 + 4.5$ (b) $-0.43 + (-0.2)$

Solution   Follow the rules for addition of integers.
(a) Subtract absolute values: $4.5 - 2.3 = 2.2$. The sign of the answer is the sign of the larger absolute value.

$$-2.3 + 4.5 = 2.2$$

(b) Add absolute values:

$$\begin{array}{r} 0.43 \\ +0.20 \\ \hline 0.63 \end{array}$$

The sign of the sum is the sign of the numbers added.

$$-0.43 + (-0.2) = -0.63$$

EXAMPLE

Subtract: (a) $13 - 0.4$ (b) $0.59 - 1$

Solution
(a) Place the numbers so that the decimal points line up.

$$\begin{array}{r} 13.0 \\ -\ \ 0.4 \\ \hline 12.6 \end{array}$$

(b) Subtract absolute values.

$$\begin{array}{r} 1.00 \\ -0.59 \\ \hline 0.41 \end{array}$$

The sign of the difference is the sign of the largest absolute value. $0.59 - 1 = -0.41$.

EXERCISE 3.2.1

Solve.

1. $1 + 2.4$
2. $16 + 3.8$
3. $12.34 + 8.85 + 9.3 + 10.2$
4. $0.099 + 0.93 + 1.2$
5. $15 + 2.9 + 0.87 + 4.583$
6. $3.4 - 2.9$
7. $12 - 8.63$
8. $15.09 - 12$
9. $0.99 - 0.099$
10. $16 - 8.5 + 2.36 - 3.75$
11. $1.15 - (-2.3)$
12. $-13.18 - 45.32$
13. $4.00 - 0.004 - 4.02$
14. $-0.179 - 2.85 - (-0.0003)$
15. $2.007 - 32.41 - 91.63 - (-0.016)$
16. $81.369 - 7.45 + (-13.026) - (-11.2)$

Look at these examples:

$$5 \times 10 = 50$$

$$5 \times 100 = 500$$

$$5 \times 1000 = 5000$$

What will the answer be when we multiply 5 by 100,000?

We know that division is the opposite of multiplication. Now look at these examples:

$$50 \div 10 = 5$$

$$500 \div 100 = 5$$

$$5000 \div 1000 = 5$$

Can you make a general rule for multiplication and division by powers of 10 (10, 100, 100, etc.)?

We have mentioned that when we multiply a whole number by 10, we add a 0 to the right of the number. This is the same as moving the number one place to the left ($2 \times 10 = 20$). We can just as easily move the decimal point one step to the right.

$$2. \times 10 = 20.$$

$$0.2 \times 10 = 2.$$

$$0.03 \times 10 = 0.3$$

To multiply by 100, we add two zeros to the right of the whole number or move the decimal point two steps to the right.

$$2. \times 100 = 200.$$

$$0.2 \times 100 = 20.$$

$$0.002 \times 100 = 0.2$$

To divide by 10, we move the decimal point to the left.

$$200. \div 10 = 20.$$

$$20. \div 10 = 2.$$

$$2. \div 10 = 0.2$$

To divide by 100, we move the decimal point two steps to the left:

$$200. \div 100 = 2.$$

$$2. \div 100 = 0.02$$

EXAMPLE

(a) $2500 \div 1000$ (b) $2500 \div 10,000$

Solution   (a) $2500 \div 1000 = 2.5$ (b) $2500 \div 10,000 = 0.25$

Since $10 = 10^1$, $100 = 10^2$, $1000 = 10^3$, and so on, we can state a general rule for multiplying and dividing by powers of 10.

> **RULE**
>
> **MULTIPLICATION AND DIVISION BY POWERS OF 10**
>
> To multiply a number by 10 raised to a whole number, move the decimal point to the right the same number of steps as the exponent.
> To divide a number by 10 raised to a whole number, move the decimal point to the left the same number of steps as the exponent.

EXERCISE 3.2.2

Solve by multiplying or dividing in your head. Check your answer with the calculator.

1. 0.2×100
2. 0.035×10
3. $7583 \div 1000$
4. $572 \div 100$
5. $0.00239 \times 10,000$
6. $0.006 \times 1,000,000$
7. $17.8 \div 100$
8. $950 \div 100,000$

Multiplication

We know already that multiplication and division are related.

$$
\begin{array}{ll}
10 \times 0.1 = 1 & 10 \div 10 = 1 \\
100 \times 0.01 = 1 & 100 \div 100 = 1 \\
1000 \times 0.001 = 1 & 1000 \div 1000 = 1
\end{array}
$$

EXAMPLE

Multiply: (a) 300×0.01 (b) 4562×0.0001

Solution
(a) Instead of multiplying by 0.01, we can divide by 100. Move the decimal point two places to the left.

$$300 \times 0.01 = 3.00 = 3$$

(b) Instead of multiplying by 0.0001, we can divide by 10,000. Move the decimal point four steps to the left.

$$4562 \times 0.0001 = 0.4562$$

What if we want to multiply 4×0.03? We could use addition:

$$0.03 + 0.03 + 0.03 + 0.03 = 0.12$$

But it would be more convenient to change 0.03 to 3×0.01 and get

$4 \times 0.03 = 4 \times 3 \times 0.01 = 12 \times 0.01 = 0.12$ (Move the decimal point two places to the left.)

EXAMPLE

Multiply: (a) 13×0.2 (b) 156×0.004

Solution
(a) $13 \times 0.2 = 13 \times 2 \times 0.1 = 26 \times 0.1 = 2.6$
(b) $156 \times 0.004 = 156 \times 4 \times 0.001 = 624 \times 0.001 = 0.624$

When you multiply two decimal numbers, such as 1.25×0.03, you can first ignore the decimals and multiply 125×3, which is 375. Then count the decimal places in the original problem and move the decimal point that many places to the left. In this case you move the decimal point $2 + 2 = 4$ places to the left:

$$0375. \qquad \text{or} \qquad 0.0375$$

Let's analyze what we are actually doing:

$$1.25 \text{ is the same as } 125 \times 0.01$$

and

$$0.03 \text{ is } 3 \times 0.01$$

Therefore,

$$
\begin{aligned}
1.25 \times 0.03 &= 125 \times 0.01 \times 3 \times 0.01 \\
&= 125 \times 3 \times 0.01 \times 0.01 \\
&= 375 \times 0.01 \times 0.01 \\
&= 375 \times 0.0001 \\
&= 0.0375
\end{aligned}
$$

(Multiplication is commutative, so we can switch the order of the factors.) Multiplication by 0.0001 tells us to move the decimal point four places to the left.

EXAMPLE

Multiply: (a) 3.2×0.04 (b) 15.04×0.003

Solution
(a) $3.2 \times 0.04 = 32 \times 0.1 \times 4 \times 0.01 = 128 \times 0.1 \times 0.01$
Move the decimal point a total of three places to the left:
$$3.2 \times 0.04 = 0.128$$
(b) $15.04 \times 0.003 = 1504 \times 0.01 \times 3 \times 0.001 = 4512 \times 0.01 \times 0.001$
Move the decimal point $2 + 3 = 5$ places to the left:
$$15.04 \times 0.003 = 0.04512$$

You should, of course, continue to use the shortcut of counting the decimals (or use a calculator) when you multiply, but it is often good to understand *why* and not only *how* we determine the position of the decimal point.

RULE

MULTIPLICATION OF DECIMAL NUMBERS

1. Ignore decimals and multiply as with whole numbers.
2. Place decimal point in the product. The number of decimal places in the product is the sum of the numbers of decimal places in the factors.

EXAMPLE

Multiply: (a) 1.3×3.4 (b) 2.45×0.06

Solution
(a) $13 \times 34 = 442$
There are two decimal places, so $1.3 \times 3.4 = 4.42$.
(b) $245 \times 6 = 1470$
In the original problem, there were four decimal places, so

$$2.45 \times 0.06 = 0.1470 = 0.147 \text{ (rounded)}$$

When you multiply using a calculator, it is important to estimate the answer in your head. It is very easy to forget to press down the decimal point key; your answer could be 10 or 100 times too big or too small—or even more!

In order to estimate an answer, we first round off the number we are using. For example, 0.4×3.9 is approximately $0.4 \times 4 = 1.6$. A calculator gives the correct answer of 1.56.

0.8×7.3 can be estimated as 1 times 7. The correct answer is 5.84, but our value is good enough for a quick approximation. If you had missed one or both of the decimals, your answer would have been 58.4 or 584. The estimate of 7 would help you recognize an error in solving the problem.

EXAMPLE

Estimate: 29×0.13

Solution 29×0.13 should be estimated as $30 \times 0.1 = 3$, not $30 \times 0 = 0$. *Do not use zero as a factor when you estimate a product.* Imagine that you need to estimate how many box cars of wheat you need to ship. The answer 3 is very different from 0!

In the following problems, first estimate the answer, then do it by hand and also by calculator. If you still have trouble with any multiplication facts, review Chapter 1. Estimating the answer is extremely important as a check. You don't want to lose hundreds of dollars because the decimal point is in the wrong place.

Become comfortable with developing the "hunches" that can be warning messages when you make mistakes in calculations. You don't need to do long calculations by hand, but you do need to analyze the problems to get an idea of what the answer should look like.

EXERCISE 3.2.3

Estimate the answer, do the work by hand, then use your calculator.

1. 5.3×4.8
2. 0.032×0.25
3. 1.05×0.39
4. $0.0009 \times 5,000,000$
5. 42.9×51.4
6. 4.29×0.15
7. 16.3×0.87

8. 278.9 × 45.5

9. 39.96 × .0062

10. 0.0015 × 0.0053

Division

We described long division with whole numbers in Chapter 1.

EXAMPLE

Solve: 1.16 ÷ 2

Solution   Fill in the "box" first with 1.16. Now we have 2 into 1.16, or $2\overline{)1.16}$. Mark the decimal point above the division box, and then divide:

$$
\begin{array}{r}
0.58 \\
2\overline{)1.16} \\
-1\,0 \\
\hline
16 \\
-16 \\
\hline
0
\end{array}
$$

$2\overline{)1.16} \rightarrow$

4000 ÷ 400 is the same as 400 ÷ 40 or 40 ÷ 4. It is also the same as 4 ÷ 0.4 and 0.4 ÷ 0.04. (As you can see when you check this with your calculator, the answer in all cases is 10.)

Similarly, 4.56 ÷ 0.4 is the same as 45.6 ÷ 4. Since it is easier to divide by a whole number than with a decimal, we always change the outer number (divisor) to a whole number, and at the same time we move the decimal point of the inner number (the dividend) the same number of steps.

EXAMPLE

Divide: 4.56 ÷ 0.4

Solution   We have $0.4\overline{)4.56}$.

To change 0.4 to a whole number, we move the decimal point one step to the right. When we move the decimal point in the divisor, we must move it the same number of places in the dividend. Thus,

$$0.4\overline{)4.56} = 4\overline{)45.6}$$

Now long division! First mark the decimal point for the answer so it doesn't get lost:

$$
4\overline{)45.6} \rightarrow
\begin{array}{r} 1\;. \\ 4\overline{)45.6} \\ -4\downarrow \\ \hline 5 \end{array}
\rightarrow
\begin{array}{r} 11. \\ 4\overline{)45.6} \\ -4\; \\ \hline 5 \\ -4\downarrow \\ \hline 16 \end{array}
\rightarrow
\begin{array}{r} 11.4 \\ 4\overline{)45.6} \\ -4\; \\ \hline 5 \\ -4\downarrow \\ \hline 16 \\ -16 \\ \hline 0 \end{array}
$$

EXAMPLE

Solve: (a) 3.5 ÷ 0.02   (b) 16.48 ÷ 1.6

Solution

(a)

$$3.5 \div 0.02 \rightarrow 0.02.\overline{)3.50.}$$

$$\begin{array}{r} 175. \\ 0.02.\overline{)3.50.} \\ \underline{-2} \\ 15 \\ \underline{-14} \\ 10 \\ \underline{-10} \\ 0 \end{array}$$

(b)

$$16.48 \div 1.6 \rightarrow 1.6\overline{)16.48} \rightarrow 16\overline{)164.8}$$

$$\begin{array}{r} 10.3 \\ 16\overline{)164.8} \\ \underline{-16} \\ 4\,8 \\ \underline{-4\,8} \\ 0 \end{array}$$

RULE

DIVISION OF DECIMAL NUMBERS

1. Move the decimal point in the divisor to obtain a whole number.
2. Move the decimal point the same number of places in the dividend.
3. Proceed as with whole numbers.
4. Mark the decimal point in the quotient directly above the decimal point in the dividend.

Most of the time you will not want to do long division by hand but will prefer to use your calculator instead. As with all calculator work, it is important that you estimate your answer. Look at the numbers before you start. In the example above, $3.5 \div 0.02$, we want to know how many times 0.02 goes into 3.5. 0.02 is a small number, so the answer must be pretty large. It is approximately 200.

In $35.1 \div 0.3$, for example, we suggest that you first estimate the answer by thinking: 35.1 is close to 30, and 30 divided by 0.3 is the same as 300 divided by 3, or 100.

Observe that here we round off 35.1 to 30 for convenience (to divide by 0.3) instead of the usual 40. You could instead round 35.1 to 36 and get $360 \div 3$, which gives 120 as an estimate; both 100 and 120 are fine as estimates.

By hand you would get

$$0.3\overline{)35.1} \rightarrow 3\overline{)351} \rightarrow 3\overline{)351}$$

$$\begin{array}{r} 117 \\ 3\overline{)351} \\ \underline{-3} \\ 5 \\ \underline{-3} \\ 21 \\ \underline{-21} \\ 0 \end{array}$$

EXAMPLE

Divide: 43.45 ÷ 0.18

Solution   43.45 ÷ 0.18 is approximately 40 ÷ 0.2 or 400 ÷ 2, which is 200. The correct answer is 241.39.

EXERCISE 3.2.4

For practice, estimate the answer. Then do the division on paper and check with your calculator. As a matter of routine, do the problem twice when you use a calculator. It is very easy to make mistakes.

1. 2 ÷ 0.5
2. 5 ÷ 0.2
3. 4.54 ÷ 0.04
4. 1.32 ÷ 0.03
5. 0.025 ÷ 2
6. 0.063 ÷ 3
7. 2.1 ÷ 0.02
8. 55 ÷ 0.005
9. 2.8 ÷ 0.014
10. 56 ÷ 0.005
11. 11.52 ÷ 0.9
12. 23.67 ÷ 0.09
13. 0.045 ÷ 0.0009
14. 135.95 ÷ 2.719

Terminating and Nonterminating Decimals

In the examples so far, the division came out exactly. This is often not the case. In elementary school, most of us learned to write the "remainder" when the division did not come out exactly. For example, 46 ÷ 7 would give an answer of 6 R4. In practical life we never use this notation. When the quotient in a division comes out exactly, such as in 1 ÷ 5 = 0.2, we have a *terminating decimal number*. However, in 1 ÷ 6, for example, we could continue the division forever. In mathematics we say that we have a *nonterminating decimal number*. The notation for a decimal that never ends is either . . . , which we already know means "goes on forever," or a bar above the repeating decimal. Here the digit 6 is repeated, so this decimal number is also *repeating*.

$$0.1666\ldots \text{ can be written } 0.1\overline{6}$$

The quotient of any two integers is either an integer, a terminating decimal, or a repeating nonterminating decimal number. There exist numbers that are nonterminating and also nonrepeating but they cannot be found by division of two integers. They will be discussed briefly in the next section.

EXAMPLE

Divide: (a) $1 \div 6$ (b) $46 \div 7$ (c) $1 \div 8$

Solution

(a) $1 \div 6$
$$
\begin{array}{r}
0.166... \\
6\overline{)1.0000} \\
-6 \\
\hline
40 \\
-36 \\
\hline
40 \\
-36 \\
\hline
4
\end{array}
$$
$= 0.1\overline{6}$ Nonterminating (repeating)

(b) $46 \div 7$
$$
\begin{array}{r}
6.57142857... \\
7\overline{)46.00000000} \\
-42 \\
\hline
4\,0 \\
-3\,5 \\
\hline
50 \\
-49 \\
\hline
10
\end{array}
$$
$= 6.\overline{571428}$ Nonterminating (repeating)

(c) $1 \div 8$
$$
\begin{array}{r}
0.125 \\
8\overline{)1.0} \\
-8 \\
\hline
20 \\
-16 \\
\hline
40 \\
-40
\end{array}
$$
Terminating

EXERCISE 3.2.5

Divide by hand and check with calculator.

1. $0.1 \div 1.8$

2. $2.4 \div 0.72$

3. $0.05 \div 0.6$

4. $4.9 \div 0.77$

5. $2 \div 1.1$

6. $0.1 \div 0.07$

7. $3.56 \div 4.8$

8. $0.99 \div 6.5$

Irrational Numbers

Some numbers are both nonterminating and nonrepeating. Examples of such numbers are π (pi) and $\sqrt{2}$. They are called *irrational* numbers.

The value of π cannot be expressed as an exact decimal number. To seven decimal places, $\pi = 3.1415926$, but even that is not exact. However, for our work we can use the approximation $\pi = 3.14$. Mathematicians have calculated π to many decimal places and are still working to determine more.

$\sqrt{2}$ also cannot be expressed as either a terminating decimal or a nonterminating decimal with repeating groups. $\sqrt{2}$ is approximately equal to 1.4. A better approximation is 1.4142, the square of which equals 1.99996.

EXAMPLE

Find a decimal approximation for $\sqrt{15}$.

Solution   We know that $\sqrt{16} = 4$, so $\sqrt{15} < 4$

Guess	$\sqrt{15} \approx 3.9$	$3.9^2 = 15.21$	Too high
Guess	$\sqrt{15} \approx 3.85$	$3.85^2 = 14.83$	Too low
Guess	$\sqrt{15} \approx 3.88$	$3.88^2 = 15.05$	O.K.

By calculator we get $\sqrt{15} \approx 3.8729833$, which we can round to 3.87.

EXERCISE 3.2.6

Approximate to two decimal places by trial and error.

1. $\sqrt{8}$

2. $\sqrt{13}$

3. $\sqrt{18}$

4. $\sqrt{26}$

Applications

Here are some word problems from an 1858 arithmetic book (edited slightly for today's language and spelling).

EXERCISE 3.2.7

1. Bought 14.75 yards of sheeting at 14 cents per yard. What was the cost of the piece? Express answer in dollars.

2. Bought land at $62.50 per acre, and then sold it at $75 per acre, thereby making $846.875. How many acres were bought?

3. How much will 5625 feet of lumber cost at $15.525 per thousand feet? (Round answer to nearest hundredths.)

4. How many bushels of onions at $0.82 per bushel can be bought for $112.24? (Round answer to a whole number.)

5. A merchant deposits $687.25 in a bank, and then later he deposits $943.64. If he draws out $875.29, how much will remain in the bank?

The following problems are more recent.

6. On successive holes a golfer drives a golf ball 205.4, 197.5, 182.75, and 220.25 yards. Find the total number of yards on the four holes.

7. A tank holds 300 gallons. If a pipe empties 0.25 of the tank in 1 hour, how many gallons will be left at the end of 2 hours?

8. If 0.1 inch on a map represents 49 miles, how many miles are represented by 3 inches on the map?

9. If a steel tape expands 0.00016 in. for each inch when heated, how much will a tape 100 feet long expand?

10. If your car averages 37 miles per gallon, approximately how many gallons will you need to drive across the United States from San Francisco to New York City (2934 miles)? Is $400 enough to budget for gas across and back assuming a gas price average of $1.19 a gallon?

3.3 PERCENTS

Conversions from Percent to Decimal

The word "percent" means "per hundred." Ten percent (10%) is 10 per 100, for instance. One hundred percent (100%) is 100 per 100, so it is the "whole thing" or 1. "Per" tells us to divide. To change a percent into a decimal, replace "%" with "divided by 100." This in turn leads us to move the decimal point two steps to the left.

EXAMPLE

Express as a decimal:
(a) 1%   (b) 10%   (c) 100%   (d) 1000%   (e) 5%   (f) 35%
(g) 0.4%   (h) 216%

Solution
(a) 1% = 0.01   (b) 10% = 0.10   (c) 100% = 1   (d) 1000% = 10
(e) 5% = 0.05   (f) 35% = 0.35   (g) 0.4% = 0.004   (h) 216% = 2.16

> **RULE**
>
> To convert a percent into a decimal, divide by 100. Move the decimal point two steps to the left.

EXERCISE 3.3.1

Convert percent to decimal.

1. (a) 25%   (b) 75%
2. (a) 2.5%   (b) 8.25%
3. (a) 200%   (b) 743%
4. (a) 0.3%   (b) 0.5%
5. (a) 0.01%   (b) 0.05%

Conversions from Decimal to Percent

Any number can be written as a percent. 0.5 is equal to 50%. (Working backwards we see that $50\% = 50 \div 100 = 0.5$. Similarly, $0.01 = 1\%$. Again we see that $1\% = 1 \div 100 = 0.01$.)

To change any number into a percent, simply multiply by 100%. For example, $5 = 500\%$. To multiply a decimal number by 100, we move the decimal point two steps to the right:

$$0.02 = 0.02 \times 100\% = 2\%$$
$$0.2 = 0.2 \times 100\% = 20\%$$
$$2 = 2 \times 100\% = 200\%$$

RULE

To express any number as a percent, multiply it by 100%. Move the decimal point two steps to the right.

EXERCISE 3.3.2

Express the number as a percent.

1. (a) 0.25   (b) 0.68

2. (a) 1.3   (b) 1.05

3. (a) 5   (b) 200

4. (a) 2.5   (b) 6.2

5. (a) 0.0075   (b) 0.0091

"Percent of"

Most of the time when we use percent, we deal with the "percent of" something. The tip is 15% of the total bill. The security deposit is 150% of the monthly rent. 40% of the voters turned out for the election. The couple donated 10% of their income to charity.

Many percent problems can be rewritten in the form

A certain *percent of* a *number* equals *what number*?

That is, we know the percent and the first number, and we need to find the other number. In solving such problems, we recognize that the phrase "percent of" tells us to *multiply* by the percent.

EXAMPLE

50% of 200 is what number?

> **Solution**   First guess the answer. You may already know that 50% is half, so 50% of 200 is 100. But what if you don't? This problem tells you to multiply 50% by 200. The easiest way to do this is to convert 50% to a decimal number, then multiply. Thus,
>
> $$50\% \text{ of } 200 = 0.50 \times 200 = 100$$

EXAMPLE

What is 15% of 35?

> **Solution**   Try to guess the answer. How much tip would you leave on a $35 restaurant check? About $5.00? We change 15% to the decimal number 0.15. Now,
>
> $$15\% \text{ of } 35 = 0.15 \times 35 = 5.25$$
>
> So if you liked the service, you'd probably leave a little more than $5.

Sometimes we are asked to solve for the percent. What percent of the voters turned out for the election? If you were charged $3.56 tax on an $89 tapedeck, what percent tax were you charged? If you received a $120 raise on a monthly salary of $1600, what percent raise did you receive?

Here the percent problem can be written in the form

What percent of a *number* equals another *number*?

In this case, we have the two numbers and solve to find the percent. For example, what percent of 200 is 100? We know from the first example above that 50% of 200 = 100. Figure 3.2 helps illustrate this problem.

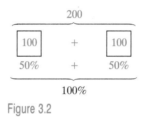

Figure 3.2

We see that when 200 is split into two equal parts, each part is 100, and when 100% is split into two equal parts, each part is 50%.

We can find the percent by treating this problem as a straightforward division problem using decimal numbers. In the case of "what percent of 200 is 100?," we put the problem in the form "100 is what percent of 200?" and divide 100 by 200.

$$100 \div 200 = 0.50$$

Now we convert 0.50 to a percent by multiplying it by 100%:

$$0.50 \times 100\% = 50\%$$

EXAMPLE

What percent of 20 is 5?

> **Solution**   This fits the form: 5 is what percent of 20? This leads us to
>
> $$5 \div 20 = 0.25 = 25\%$$
>
> The figure shows that if the whole (100%) is 20, then 5 represents 25%.

EXAMPLE

What percent of 40 is 8?

> **Solution** $8 \div 40 = 0.20 = 20\%$
> Again, the figure shows that if 100% is 40, then 8 is 20%.

EXAMPLE

What percent of 80 is 20?

> **Solution** $20 \div 80 = 0.25 = 25\%$

EXAMPLE

What percent of 230.60 is 40.355?

> **Solution** $40.355 \div 230.60 = 0.175 = 17.5\%$

EXAMPLE

Let's go back to the tapedeck problem. What percent of $89 is $3.56?

> **Solution** $3.56 \div 89 = 0.04 = 4\%$
> The sales tax was 4%.

EXAMPLE

If you received a $120 raise on a monthly salary of $1600, what percent raise did you receive?

> **Solution**   What percent of $1600 is $120?

$$120 \div 1600 = 0.075 = 7.5\%$$

The raise in salary was 7.5%.

The third form that problems like these can take helps answer still other questions we might have. How much money should be deposited to earn interest of $1000 every year if the interest rate is 5%? If the mortgage company requires a 20% down payment and you have $16,000 available, what is the maximum price you can afford for a home (assuming you can meet the monthly payments)? Such problems take the form illustrated by the following examples.

EXAMPLE

How much money should be deposited to earn interest of $1000 every year if the interest rate is 5%?

> **Solution**   This is really asking "$1000 is 5% of what amount?" Draw a picture.

$$100\% \div 5\% = 20 \quad \text{and} \quad 20 \times \$1000 = \$20,000$$

You need to deposit $20,000 to earn $1000 interest per year.

EXAMPLE

If the mortgage company requires a 20% down payment and you have $16,000 available, what is the maximum price you can afford for a home?

> **Solution**   Even though you'll pay the $16,000 only once, the following diagram can help solve the problem:

$$100\% \div 20\% = 5 \quad \text{and} \quad 5 \times \$16,000 = \$80,000$$

We will need a quick way of solving this type of percent problem. We can do this using division. This time, however, we will divide by the percent, after it's been converted to a decimal number. Thus, in the interest example, "5% of what number" leads us to

$$1000 \div 5\% = 1000 \div 0.05 = 20,000$$

EXAMPLE

50% of what number is 2?

> **Solution** $2 \div 50\% = 2 \div 0.50 = 4$

EXAMPLE

25% of what number is 3?

> **Solution** $3 \div 25\% = 3 \div 0.25 = 12$

EXAMPLE

10% of what number is 6?

> **Solution** $6 \div 10\% = 6 \div 0.10 = 60$

EXAMPLE

18.5% of what number is 61.05?

> **Solution** $61.05 \div 18.5\% = 61.05 \div 0.185 = 330.00$

This type of percent problem takes the form

> A certain *percent* of *what number* equals another *number*?

The following example shows the relationship of these three forms.

EXAMPLE

(a) 39% of 46 is what number?
(b) What percent of 46 is 17.94?
(c) 39% of what number is 17.94?

> **Solution**
> (a) $39\% \times 46 = 17.94$
> (b) $17.94 \div 46 = 0.39 = 39\%$
> (c) $17.94 \div 39\% = 17.94 \div 0.39 = 46$

RULE

SOLVING A PERCENT PROBLEM

Rewrite the problem in one of these forms:

1. What percent of a certain number equals another number? Solve by division.
2. A certain percent of what number equals another number? Solve by division.
3. A certain percent of a certain number equals what number? Solve by multiplication.

EXERCISE 3.3.3

Translate each problem into one of the forms of "some percent of a number is another number" and solve. Round your answers to the nearest hundredth.

1. 8 is what percent of 32?
2. $13 is what percent of $6.50?
3. 1% of what number is 30?
4. 25% of what number is 8?
5. Find 12.5% of 40.

6. Find 7.4% of 195.

7. 12 is what percent of 10?

8. 24% of what number is 120?

9. 50 is 125% of what number?

10. 0.4% of what number is 5?

11. The enrollment at a local community college is 3500. Of these, 30% are liberal arts majors. How many liberal arts majors are there at the college?

12. Lisa spends $175 of her monthly take-home pay for rent. If her monthly salary is $700, what percent does she spend for rent?

13. Suppose 84 people out of 150 interviewed planned to vote for the Democrats in the next election. What percent is that?

14. Suppose 120 students out of 150 passed a chemistry course. What percent is that?

15. Earl lost 10 pounds in 3 months from his original weight of 160 pounds. Find the percent of his decrease in weight.

16. The sales tax on a used car is $294. If the purchase price is $4200, find the sales tax rate.

17. If the sales tax is $1.17 on a purchase of a book priced at $29.25, find the sales tax rate.

18. The price of a cookbook is reduced 12%. If the discount is $1.44, find the original price of the book.

19. Ms. Smith announced that the average test score on a 25-question test was 72%. How many correct answers is this on the average?

20. 5 out of 2 million people with lottery tickets won. What percent is that?

SPECIAL NOTE ON CALCULATORS

Working with percent on calculators can be tricky. Some calculators have special percent keys; others do not. In either case, you can always convert a percent to a decimal number and continue as you would with any other decimal number. (Be sure to convert final answers back to percent, if needed.)

Even the calculators with percent keys have different ways of using them, depending on the internal programming for that model of calculator. If you wanted to solve 100 + 10% − 10%, you might get two different answers. We know that 10% = 0.10, so

$$100 + 10\% - 10\% = 100 + 0.10 - 0.10 = 100$$

However, some calculators read 100 + 10% as 100 + 10% of 100, which equals

$$100 + 0.10 \times 100 = 100 + 10 = 110$$

Now we try to subtract 10% and the calculator is programmed to calculate

$$110 - (10\% \text{ of } 110) = 110 - 11 = 99$$

With the common use of calculators in jobs and school, there has been a decline of skills in mental arithmetic. As we have pointed out several times, it is important to be able to approximate answers quickly in real life. For example, there are known cases of patients in hospitals who have died because the medicine they were given was ten times too strong!

Complete the following example as fast as you can in your head. Check your answers by reworking them with the calculator.

EXAMPLE

Approximate: (a) 253.4×0.05 (b) $568.9 \div 6.7$.

Solution
(a) 0.05 is much smaller than 0.1, so the product must be smaller than 25, and we also know it must be larger than 2.5 ($0.01 \times 253.4 = 2.53$). The calculator answer is 12.67.
(b) $568.9 \div 6.7$ should be around 100 ($600 \div 6$); actually it must be somewhat smaller than 100 because 568.9 is less than 600 and 6.7 is more than 6. The calculator answer rounds to 84.9.

EXERCISE 3.3.4

Choose the answer closest to the correct answer.
1. $106.9 \div 1.5$ (a) 0.07   (b) 0.7   (c) 7   (d) 70
2. $48 \div 0.002$ (a) 240   (b) 2400   (c) 24,000   (d) 240,000
3. 304.2×0.16 (a) 0.06   (b) 0.6   (c) 6   (d) 60
4. $0.03\% \times 50$ (a) 0.015   (b) 0.15   (c) 1.5   (d) 15
5. $0.35 \div 60$ (a) 0.0005   (b) 0.005   (c) 0.05   (d) 0.5
6. $0.88 \div 2.2$ (a) 0.04   (b) 0.4   (c) 4   (d) 40
7. $0.04 \times 500 \div 0.25$ (a) 0.8   (b) 8   (c) 80   (d) 800
8. $0.2\% \times 5,000,000$ (a) 1000   (b) 10,000   (c) 100,000   (d) 1,000,000
9. $5.1 \div 0.003$ (a) 1.7   (b) 17   (c) 170   (d) 1700
10. $0.25 \div 0.5 \div 5$ (a) 0.001   (b) 0.01   (c) 0.1   (d) 1

Sometimes we must use addition or subtraction to find information needed in problems involving percent. This is especially true in problems with sales discounts or mark-ups.

EXAMPLE

A dress tag was changed from $80 to $60. What was the discount rate?

Solution   Rate is the percent, so we are asked to solve the following problem: What percent of 80 is the discount?

The discount in dollars is the difference between the original price and the sale price. In this case, the discount is $80 − $60 = $20. The problem now becomes

What percent of 80 is 20?

So we have

$$20 \div 80 = 0.25 = 25\%$$

The dress is discounted 25%.

EXAMPLE

A cassette tape sold for $7.50. The price was then reduced by 20%. What was the new price of the tape?

Solution   Here we have two possible solution strategies.

First, 20% of 7.50 is 1.50. Since the reduction is 1.50, the sale price of the tape is $7.50 − $1.50 = $6.00.

Second, since there is a 20% reduction, we have to pay 80%.

$$80\% \text{ of } \$7.50 = 0.80 \times \$7.50 = \$6.00$$

Choose the method you find more comfortable.

EXERCISE 3.3.5

1. A house originally sold for $90,000 but was reduced to $84,000. Find the percent discount.

2. Rebecca earns $34,000 per year as an accountant. If 28% of her salary is withheld for taxes, what is her take-home pay?

3. Fran is a nurse with an annual salary of $27,000. If she receives a 6% raise, what is her new salary?

4. Malcolm's monthly salary as a teacher increased from $2400 to $2520. Find the percent increase.

5. During a period of one month, the price of eggs rose from 77¢ a dozen to 88¢ a dozen. Find the percent increase to the nearest whole percent.

6. During a period of one month, the price of eggs decreased from 88¢ a dozen to 77¢ a dozen. Find the percent decrease to the nearest whole percent.

7. The enrollment at Coles Junior College increased from 950 to 1080 students. Find the percent increase to the nearest whole percent.

8. During one year the price of a stock went from $87.5 per share to $37.5 per share. Find the percent decrease to the nearest whole percent.

9. Zack paid $7800 for a new car. After 18 months he finds that its value is only $5200. Find the percent decrease.

10. If 36 students out of a school population of 900 were absent one day, what percent were present?

11. A refrigerator was on sale for $250. This gave the store a profit of $45. What was the percent profit to the store?

12. 72% of those taking the driver's test pass. If 2100 took the test, how many failed?

13. Of 2840 entering freshmen, 90% admitted to having math anxiety. How many students did not dread their required math courses?

14. The selling price of a TV was $675. The sales tax was 6%, and the discount was 8.5%. Should you have the clerk figure out the sales tax before or after the discount? How much would the tax be in the different cases?

SUMMARY

Definitions

Place values: Ones, tenths, hundredths, thousandths, ten-thousandths, ...

Rules

Reading a decimal number: Read the decimal point as "and." Read the decimal part of the number as a whole number followed by the name of the rightmost place.

Ordering decimal numbers: Arrange the numbers in a column with the decimal points below each other. Compare the place values going from left to right.

Rounding a decimal number: If the digit to the right of the rounded digit is less than 5, leave the digit the same. If the digit to the right of the rounded digit is 5 or more, increase the rounded digit by one. Discard all digits to the right of the rounded digit.

Adding or subtracting decimal numbers: Place the numbers with the decimal points below each other. Proceed as with whole numbers.

Multiplying and dividing by powers of 10: To multiply a number by 10 raised to a whole number, move the decimal point to the right the same number of steps as the exponent.

To divide a number by 10 raised to a whole number, move the decimal point to the left the same number of steps as the exponent.

Multiplying decimal numbers: Multiply as with whole numbers. The number of decimals in the product is the sum of the number of decimals of the factors.

To divide decimals, move the decimal point in the divisor to obtain a whole number. Move the decimal point the same number of places in the dividend. Proceed as with whole numbers. Mark the decimal point in the quotient directly above the decimal point in the dividend.

To change a percent into a decimal, divide by 100. Move the decimal point two steps to the left.

To change any number into a percent, multiply by 100%. Move the decimal point two steps to the right.

Rewrite percent problems as

1. What percent of some number equals another number? Solve by division.

2. Some percent of what number equals another number? Solve by division.

3. Some percent of some number equals what number? Solve by multiplication.

VOCABULARY

Decimal point: A point that shows where the whole number ends.

Decimals: Digits to the right of the ones place.

Exact number: A number that is not rounded.

Nonterminating decimal: A decimal number that can't be expressed exactly.

Percent: A number expressed as "per hundred."

Repeating decimal: A decimal number that consists of repeating groups.

Terminating decimal: A decimal number that can be expressed exactly.

CHECK LIST

Check the box for each topic you feel you have mastered. If you are unsure, go back and review.

☐ Reading decimals
☐ Ordering decimals
☐ Rounding decimals
  Basic operations with decimals and percents
    ☐ Addition
    ☐ Subtraction
    ☐ Multiplication
    ☐ Division
☐ Multiplication by powers of 10
☐ Division by powers of 10
☐ Estimating an answer
☐ Using the calculator
☐ Applications
☐ Conversions from percent to decimal
☐ Conversions from decimal to percent
☐ Percent of

REVIEW EXERCISES

1. In which place is the digit 3?
   (a) 3.712
   (b) 41.308
   (c) 93,128.6
   (d) 0.00135

2. Write the number in words.
   (a) 7.002
   (b) 1.785
   (c) 0.0105
   (d) 200.310

3. Arrange the numbers from smallest to largest.
   (a) $0.3, 0.291, 0.4002$
   (b) $2.003, 0.973, 1.58$

4. Perform the indicated operations.
   (a) $0.3 + 0.02 + 10.162 + 0.005$
   (b) $5 - 0.071$
   (c) $1.15 - (-20.98)$
   (d) $0.888 - 0.089$

5. Perform the indicated operation.
   (a) 0.3×1000
   (b) 0.416×100
   (c) $32 \div 1000$
   (d) $6508 \div 10$

6. Estimate, then calculate by hand. Check with calculator.
   (a) 8.2×6.7
   (b) 0.076×3827
   (c) $7 \div 0.003$
   (d) $1.99 \div 0.072$
   (e) 2.55×0.0713

(f) 0.003×3000

(g) $0.375 \div 0.091$

(h) $4952 \div 0.00023$

7. Express as percent.

(a) 0.72

(b) 0.8

(c) 0.123

(d) 1.35

(e) 26

8. Express as a decimal.

(a) 31%

(b) 7%

(c) 45.5%

(d) 0.062%

(e) 257%

9. 10% of 356 is what?

10. What percent of 88 is 22?

11. 17% of what is 51?

12. Jack is paying $27.51 a month on a loan of $826.40. How long will it take to pay off the loan?

13. Pat owes $584.32 on his snowmobile. He wants to pay for it in 12 months. What should the monthly payments be? (Round off to two decimal places.)

14. At the end of June, Paula's bank balance was $285.64. During July she wrote checks for $18.93, $116.59, $71.28, $223.85, $56.00, $15.91, $93.84, and $116.17 and made deposits of $189.64, $201.00, and $191.17. What was her bank balance at the end of July?

15. The enrollment of a Community College increased from 4732 to 5625. What was the percent increase?

16. The population of a town increased 33.7% over a 5-year period. If the population was 18,513 five years ago, what is it now? (Answer as a whole number.)

17. Lee's monthly check is $2635.78. If 21.5% is deducted for income tax, what is his total yearly take-home pay?

READINESS CHECK

Solve the problems to satisfy yourself that you have mastered Chapter 3.

1. Which digit is in the hundredths place in 178.29563?

2. Write 15.078 in words.

3. Insert < or > between 0.9 and 0.0956.

4. Round 4.5968 to the nearest hundredth.

5. How many places and in what direction do you move the decimal point when you multiply by one thousand?

6. Change 0.05% to a decimal.

7. Change 1.23 to a percent.

8. What is 8.25% of 12?

9. 108% of what number is 270?

10. The price of eggs decreased from $1.25 to 95¢ per dozen. What was the decrease in percent?

FRACTIONS

The broken numbers

By approximately 2000 B.C., the Egyptians had developed a system of fractions. A top number of 1 using the symbol for mouth, ⬭, to mean "part," was written over the regular numeric symbols for the bottom numbers, as shown in Figure 4.1.

Figure 4.1

When the top number was not 1, they expressed the fraction as the sum of *different* fractions, each one with 1 as the top number. For example, $\frac{3}{4}$ was not written as $\frac{1}{4} + \frac{1}{4} + \frac{1}{4}$ but as $\frac{1}{2} + \frac{1}{4}$. $\frac{9}{16}$ was written as $\frac{1}{2} + \frac{1}{16}$, and $\frac{5}{7}$ was written as $\frac{1}{7} + \frac{1}{2} + \frac{1}{14}$.

When you get to addition of fractions in this chapter, try working out some more fractions, such as $\frac{4}{9}$ and $\frac{2}{3}$.

4.1 VOCABULARY

As always when we learn a language—and mathematics *is* a language—there are many words we need to become familiar with. You may already know many of the words having to do with fractions, but we mention them here to refresh your memory and perhaps teach you some new ones.

A common fraction consists of three parts: the *numerator* (the top number), a *fraction bar* or a slash, and the *denominator* (the bottom number).

In $\frac{3}{4}$, 3 is the numerator and 4 is the denominator. The denominator says how many parts the whole is divided into and the numerator how many of these parts.

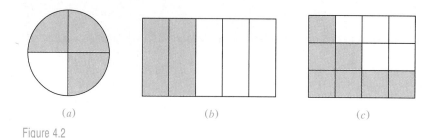

Figure 4.2

A whole number such as 6 can be written in fractional form as $\frac{6}{1}$ or 6/1.

According to James S. Eaton in *Treatise on Arithmetic* (1858), a fraction is nothing more than a division that hasn't been carried out yet. "This is *the key* to a knowledge of fractions."

Fractions are often illustrated with pictures as in Figure 4.2.

EXAMPLE

(a) What fraction tells us how many pieces of pizza have been eaten (the shaded part) in Figure 4.2(a)?
(b) What fraction of Figure 4.2(b) is shaded?
(c) What fraction of Figure 4.2(c) is shaded?

Solution   (a) $\frac{3}{4}$ of the pizza is eaten.   (b) $\frac{2}{5}$ (c) $\frac{7}{12}$

EXAMPLE

If [ ] represents $\frac{1}{8}$, how is the whole represented?

Solution   We must combine 8 of the rectangles representing $\frac{1}{8}$ to get the whole:

$$\frac{1}{8} \quad \frac{1}{8} \quad \frac{1}{8} \quad \frac{1}{8} \quad \frac{1}{8} \quad \frac{1}{8} \quad \frac{1}{8} \quad \frac{1}{8}$$

If the numerator is smaller than the denominator, as in $\frac{6}{13}$, we have a *proper fraction*. If the numerator is larger than the denominator, as in $\frac{13}{6}$, we have an *improper fraction*.

All improper fractions can be changed into *mixed numbers*. A mixed number consists of an integer and a proper fraction, such as $2\frac{1}{6}$.

To change an improper fraction such as $\frac{13}{6}$ into a mixed number, we could first think in terms of pizzas divided into sixths as in Figure 4.3.

Here we have two whole pizzas sliced into sixths and one piece of another pizza that was also sliced into sixths.

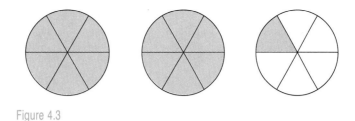

Figure 4.3

Numerically we find out how many times 6 goes into 13 (how many "wholes" can we get?). We get 2, since $2 \times 6 = 12$, with $\frac{1}{6}$ left over.

$\frac{13}{6}$ can also be illustrated with fractions lined up like this:

$$\underbrace{1/6 \ 1/6 \ 1/6 \ 1/6 \ 1/6 \ 1/6}_{\substack{6/6 \\ 1}} \quad \underbrace{1/6 \ 1/6 \ 1/6 \ 1/6 \ 1/6 \ 1/6}_{\substack{6/6 \\ 1}} \ 1/6$$

$$1/6$$

The mixed number is $2\frac{1}{6}$.

EXAMPLE

Change the following improper fractions into mixed numbers: (a) $\frac{5}{2}$ (b) $\frac{17}{11}$

Solution
(a) $5 \div 2 = 2$ with remainder 1, or $\frac{5}{2} = 2\frac{1}{2}$.
(b) $\frac{17}{11}$ is $17 \div 11 = 1$ with remainder 6, or $1\frac{6}{11}$.

> **RULE**
>
> To change an improper fraction into a mixed number, divide the numerator by the denominator. The quotient is the whole number part of the mixed number. The remainder is the new numerator.

To go from a mixed number to an improper fraction, we work the other way. In $2\frac{1}{6}$, the 2 wholes give us $2 \times 6 = 12$ sixths. We add these to the other one-sixth and get 13 sixths, or $\frac{13}{6}$.

EXAMPLE

Change $4\frac{2}{5}$ to an improper fraction.

Solution   The 4 wholes give us $4 \times 5 = 20$ fifths. Add these to the other 2 fifths to get 22 fifths, or $\frac{22}{5}$.

$$4\frac{2}{5} = \frac{22}{5}$$

> **RULE**
>
> To change a mixed number into an improper fraction, multiply the denominator by the whole number. Add the numerator. This is the new numerator. Keep the denominator.

EXERCISE 4.1.1

1. Change the improper fraction into a mixed number.

(a) $\frac{7}{3}$

(b) $\frac{12}{5}$

(c) $\frac{15}{2}$

(d) $\frac{22}{7}$

(e) $\frac{13}{3}$

(f) $\frac{58}{9}$

(g) $\frac{197}{32}$

(h) $\frac{106}{16}$

2. Change the mixed number into an improper fraction.

(a) $2\frac{1}{3}$

(b) $5\frac{1}{2}$

(c) $3\frac{3}{4}$

(d) $5\frac{2}{3}$

(e) $10\frac{1}{11}$

(f) $25\frac{1}{3}$

(g) $7\frac{4}{5}$

(h) $100\frac{2}{7}$

4.2 EQUIVALENT FRACTIONS AND SIMPLIFYING

Equivalent Fractions

When we cut a pie evenly into two pieces, each piece is half the pie, as in Figure 4.4(a). If we had cut the pie into four equal pieces, half the pie would consist of two-fourths of the pie. Similarly, if we cut the pie into six equal pieces, half the pie would be the same as three-sixths of the pie.

In other words, a fraction can be expressed in many different ways. $\frac{1}{2}$, $\frac{2}{4}$, and $\frac{3}{6}$ are· *equivalent fractions*. Although their names are different, they all represent the same number.

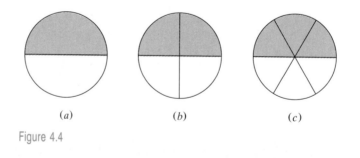

(a)   (b)   (c)

Figure 4.4

To find a fraction equivalent to $\frac{1}{3}$ and with a denominator of 15, think of the pie again. If you had one pie divided into 3 equal parts and you needed 15 equal parts, what would you do? You would cut each piece of pie into five equal parts, as in Figure 4.5. Here it is easy because one of the three parts equals five of the 15 parts. You think "3 into 15 is 5." Now, if the division doesn't come out evenly, you have to trust your calculator and round off the numbers. The principle is still the same!

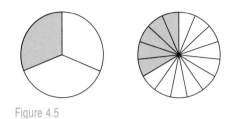

Figure 4.5

In other words,

$$\frac{1}{3} \text{ is equivalent to } \frac{5}{15}$$

What about a fraction equivalent to $\frac{2}{5}$ but with a denominator of 15?

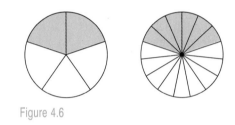

Figure 4.6

Here we divide each of 5 pieces of pie into 3 equal parts. Two of the original pieces give us 6 smaller parts. In other words,

$$\frac{2}{5} = \frac{6}{15}$$

Now,

$$\frac{1}{2} = \frac{1}{2} \times \frac{2}{2} = \frac{2}{4}$$

$$\frac{1}{2} = \frac{1}{2} \times \frac{3}{3} = \frac{3}{6}$$

$$\frac{1}{3} = \frac{1}{3} \times \frac{5}{5} = \frac{5}{15}$$

$$\frac{2}{5} = \frac{2}{5} \times \frac{3}{3} = \frac{6}{15}$$

In each example we have multiplied the numerator and denominator by the same number to get the equivalent fraction.

If we want to change a fraction to an equivalent fraction where the numerator and denominator are smaller, we *divide* the numerator and the denominator by the same number. For example,

$$\frac{3}{6} = \frac{3 \div 3}{6 \div 3} = \frac{1}{2}, \qquad \frac{7}{21} = \frac{7 \div 7}{21 \div 7} = \frac{1}{3}, \qquad \text{and} \qquad \frac{12}{18} = \frac{12 \div 6}{18 \div 6} = \frac{2}{3}$$

Find the equivalent fraction.

1. $\dfrac{4}{5} = \dfrac{}{40}$

2. $\dfrac{3}{7} = \dfrac{}{63}$

3. $\dfrac{2}{5} = \dfrac{18}{}$

4. $\dfrac{1}{3} = \dfrac{}{90}$

5. $\dfrac{3}{7} = \dfrac{}{350}$

6. $\dfrac{12}{24} = \dfrac{2}{}$

7. $\dfrac{39}{52} = \dfrac{3}{}$

8. $\dfrac{100}{300} = \dfrac{}{15}$

9. $\dfrac{9}{45} = \dfrac{}{5}$

10. $\dfrac{60}{120} = \dfrac{}{4}$

Reducing to Lowest Terms

When we find an equivalent fraction with a smaller numerator and denominator, we are *reducing* the fraction. *Reducing a fraction to lowest terms* is a special case of writing equivalent fractions. In this case we want to find equivalent fractions in which the numerator and denominator are as small as possible. For example,

$$\frac{50}{100} = \frac{25}{50} = \frac{5}{10} = \frac{1}{2}$$

$\frac{50}{100}$, $\frac{25}{50}$, $\frac{5}{10}$, and $\frac{1}{2}$ are equivalent fractions, but $\frac{1}{2}$ is written in lowest terms. There is no number (except 1) that you can divide into both the numerator and denominator.

EXAMPLE

$\dfrac{36}{42} = \dfrac{18}{21} = \dfrac{6}{7}$. Which is in lowest terms?

Solution   Again, $\frac{36}{42}$, $\frac{18}{21}$, and $\frac{6}{7}$ are *equivalent* fractions, but $\frac{6}{7}$ is in *lowest terms*.

Divisibility by 2, 3, and 5

Sometimes it is not immediately obvious what number, if any, divides into both the numerator and the denominator. We call any number that divides into two or more numbers a *common factor* of those numbers. To reduce a fraction to lowest terms, we must divide both the numerator and denominator by all the factors common to both. To find common factors, it might be helpful to use a list of divisibility rules. Here are a few to help you.

Numbers divisible by 2:   All even numbers $(20, 52, 18, 96, 134, \dots)$

Numbers divisible by 3:   The sum of the digits is divisible by 3 $(27: 2 + 7 = 9$, which is divisible by 3; $14946: 1 + 4 + 9 + 4 + 6 = 24$; $2 + 4 = 6$, which is divisible by 3)

Numbers divisible by 5:   Number ends in 0 or 5 $(15, 40, 85, 830, 1925, \dots)$

EXAMPLE

Reduce $\frac{51}{93}$ to lowest terms.

> **Solution**   At first, neither number looks factorable because neither one has turned up in our multiplication tables. Neither the numerator nor the denominator is divisible by 2 because they are not even numbers.
>
>   What about 3?

$$51: 5 + 1 = 6 \qquad \text{divisible by 3}$$
$$93: 9 + 3 = 12 \qquad \text{divisible by 3}$$
$$51 \div 3 = 17, \qquad 93 \div 3 = 31$$

So $\frac{51}{93} = \frac{17}{31}$.

>   17 and 31 are both prime numbers. They cannot be factored further. Remember that a prime number has no factors other than 1 and itself.

EXAMPLE

Reduce $\frac{324}{828}$ to lowest terms.

> **Solution**   Both the numerator and the denominator are even. Divide them by 2.

$$\frac{324}{828} = \frac{162}{414}$$

Both numerator and denominator are even, so divide by 2 again:

$$\frac{162}{414} = \frac{81}{207}$$

Check for divisibility by 3: $8 + 1 = 9$ and $2 + 0 + 7 = 9$.

$$\frac{81}{207} = \frac{27}{69}$$

Again, check 3:

$$\frac{27}{69} = \frac{9}{23}$$

23 is a prime number and therefore has no factor in common with 9.

It is not necessary for one of the numbers to be a prime number. For example, $\frac{4}{9}$, $\frac{6}{25}$, and $\frac{39}{44}$ are all in lowest terms. In each case, the numerator and denominator have no common factor.

Prime Factorization

Prime factorization is a particular factorization where each factor is a prime number. Look at the number 12. 12 can be factored as 1×12, 12×1, 2×6, 6×2, 4×3, 3×4. 2×6 can be rewritten as $2 \times 3 \times 2$, and 3×4 can be rewritten as $3 \times 2 \times 2$, which has exactly the same factors. Since multiplication is commutative, the order in which the factors are written doesn't matter. It is customary (but not necessary) to write the factors in order of size from left to right. The *prime* factors of 12 are 2, 2, and 3.

   Another way to determine the prime factors is to use a *factor tree*. Two factor trees for 12 are shown in Figure 4.7; we have circled each prime factor.

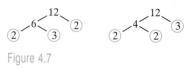

Figure 4.7

Again we see that $12 = 2 \times 3 \times 2$ or $12 = 2 \times 2 \times 3$. Since multiplication is commutative, these mean the same thing. The fact that a number can be factored into prime factors in only one way is called the *fundamental theorem of arithmetic*.

FUNDAMENTAL THEOREM OF ARITHMETIC

Every composite number can be factored uniquely into prime factors, if the factors are arranged in order of size.

EXAMPLE

Find the prime factors of 30 and 45.

Solution $30 = 2(3)(5)$ and $45 = 3(3)(5)$.

EXAMPLE

Find the prime factors of 324 and 828.

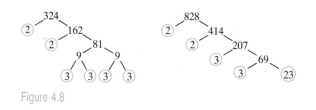

Figure 4.8

Solution   See Figure 4.8.

$$324 = 2(2)(3)(3)(3)(3) \quad \text{and} \quad 828 = 2(2)(3)(3)(23)$$

The Greatest Common Factor

The largest factor that divides evenly into both of two or more numbers is called their *greatest common factor* (abbreviated GCF).

EXAMPLE

Find the greatest common factor of 30 and 42.

Solution $30 = 2 \times 3 \times 5 \quad \text{and} \quad 42 = 2 \times 3 \times 7$

The numbers have the factors 2 and 3 in common.
The greatest common factor (GCF) of 30 and 42 is $2 \times 3 = 6$.

EXAMPLE

Find the GCF of 60 and 72.

Solution $60 = 2 \times 2 \times 3 \times 5$ and $72 = 2 \times 2 \times 2 \times 3 \times 3$

The GCF is $2 \times 2 \times 3 = 12$. Note that in calculating the greatest common factor you use each prime factor as many times as it appears in *both* numbers.

EXAMPLE

Find the greatest common factor of 324 and 828.

Solution $324 = 2(2)(3)(3)(3)(3)$ and $828 = 2(2)(3)(3)(23)$.

The two numbers have the factors $2, 2, 3, 3$ in common.
$2(2)(3)(3) = 36$, so 36 is the greatest common factor.

DEFINITION

The **greatest common factor (GCF)** of two numbers is the largest number that can divide evenly into these two numbers.

EXAMPLE

Find the greatest common factor of 16 and 24.

Solution

$$16 = 2 \times 2 \times 2 \times 2$$
$$24 = 2 \times 2 \times 2 \times 3$$

The largest factor common to 16 and 24 is $2 \times 2 \times 2$ or 8, so 8 is the GCF.

EXAMPLE

Find the greatest common factor of 6, 15, and 18.

Solution

$$6 = 2 \times 3$$
$$15 = 3 \times 5$$
$$18 = 2 \times 3 \times 3$$

The only factor common to the three given numbers is 3, so 3 is the GCF.

It is often convenient to find the greatest common factor of the numerator and the denominator before we reduce a fraction to lowest terms. For example, since the greatest common factor of 324 and 828 is 36, we divide both 324 and 828 by 36 when we reduce to lowest terms:

$$\frac{324}{828} = \frac{324 \div 36}{828 \div 36} = \frac{9}{23}$$

EXAMPLE

Reduce $\frac{51}{93}$ to lowest terms.

> **Solution** $51 = 3(17)$ and $93 = 3(31)$. The greatest common factor is 3.

$$\frac{51}{93} = \frac{51 \div 3}{93 \div 3} = \frac{17}{31}$$

EXERCISE 4.2.2

1. Which of the following numbers are prime numbers?
  2   13   15   37   49   91   111   189

2. Factor into prime factors.
  (a) 24
  (b) 38
  (c) 56
  (d) 96
  (e) 144
  (f) 260

3. Find the greatest common factor.
  (a) 12, 15
  (b) 24, 42
  (c) 27, 45
  (d) 56, 98
  (e) 85, 170
  (f) 144, 256

4. Reduce to lowest terms.
  (a) $\frac{9}{15}$
  (b) $\frac{24}{144}$
  (c) $\frac{6}{54}$
  (d) $\frac{16}{18}$
  (e) $\frac{27}{72}$
  (f) $\frac{90}{150}$

Applications

If 8 pounds of apples costs \$5, how much does 2 pounds of apples cost? One way to solve this is to use equivalent fractions. The apples cost \$5 for 8 lb. We need an equivalent fraction for the price for 2 pounds:

$$\frac{\$5}{8 \text{ lb}} = \frac{?}{2 \text{ lb}}$$

Since $8 \div 2 = 4$, you can say

$$\frac{5 \div 4}{8 \div 4} = \frac{1.25}{2}.$$

Therefore 2 pounds of apples costs \$1.25.

EXAMPLE

A motorist uses 18 gallons of gasoline in driving 270 miles. At that rate, how many miles can he go on 12 gallons of gasoline?

Solution   There are 270 miles per 18 gallons. We need an equivalent fraction for the number of miles per 12 gallons:

$$\frac{270}{18} = \frac{?}{12} \qquad 18 \div 12 = 1.5 \qquad \text{and} \qquad 270 \div 1.5 = 180$$

The motorist gets 180 miles on 12 gallons of gasoline. You could just as well write the fraction as $\frac{18}{270} = \frac{12}{?}$ The answer is the same.

EXAMPLE

12 out of every 1000 cars in a certain city were stolen. If 1625 cars were stolen, what was the total number of cars in the city to begin with?

Solution   12 per 1000 or $\frac{12}{1000}$ were stolen. We need an equivalent fraction for 1625 stolen cars per total or

$$\frac{12}{1000} = \frac{1625}{?}$$

Use a calculator to find the numbers.

$$12 \div 1625 = 0.0073846$$
$$1000 \div 0.0073846 = 135{,}417, \text{ which rounds to } 135{,}000$$

EXERCISE 4.2.3

1. Greg saves $18 in 8 weeks. How long will it take him to save $81 at the same rate?

2. Mr. Smith pays $110 taxes on a property assessed at $3800. What will the taxes be on a property assessed at $6350 if the same rate is used?

3. A man travels 152 miles in 4 hours. At that rate, how long will it take him to travel 247 miles?

4. A car uses 25 gallons of gasoline for 350 miles. How much gasoline does it use for 462 miles?

4.3 MULTIPLICATION AND DIVISION OF FRACTIONS

Multiplication

Here is a square with length and width equal to 1 unit:

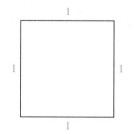

The area, which is length × width, is 1 × 1 or 1 square unit.

Here on the left we have shaded $\frac{1}{2}$ of the square, and on the right we have shaded $\frac{1}{3}$ of the square.

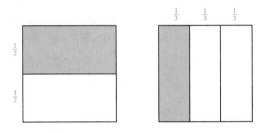

To find $\frac{1}{3} \times \frac{1}{2}$ put one square on top of the other. Remember that the area of a rectangle equals length times width.

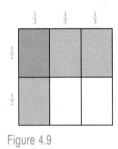

Figure 4.9

As Figure 4.9 shows, $\frac{1}{2} \times \frac{1}{3}$ is one of the six rectangles, or $\frac{1}{6}$.

Now try $\frac{1}{3} \times \frac{1}{4}$:

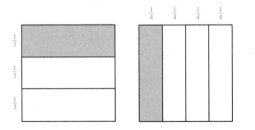

Putting the two squares together, we get Figure 4.10.

$\frac{1}{3} \times \frac{1}{4}$ is one of the 12 equal rectangles, or $\frac{1}{12}$ of the square.

$\frac{2}{3} \times \frac{1}{5} = ?$ $\frac{2}{3} \times \frac{1}{5}$ is 2 of the 15 rectangles, or $\frac{2}{15}$.(Fig. 4.11)

$\frac{2}{3} \times \frac{4}{5} = ?$ $\frac{2}{3} \times \frac{4}{5}$ is 8 of the 15 rectangles, or $\frac{8}{15}$.(Fig. 4.12)

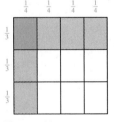

Figure 4.10

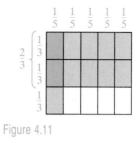

Figure 4.11

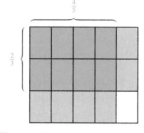

Figure 4.12

We have now seen these examples:

$$\frac{1}{3} \times \frac{1}{2} = \frac{1}{6}, \qquad \frac{1}{3} \times \frac{1}{4} = \frac{1}{12}$$

$$\frac{2}{3} \times \frac{1}{5} = \frac{2}{15}, \qquad \frac{2}{3} \times \frac{4}{5} = \frac{8}{15}$$

Do you see any easier way of arriving at the answers?

$$\frac{1}{3} \times \frac{1}{2} = \frac{1 \times 1}{3 \times 2} = \frac{1}{6}, \qquad \frac{1}{3} \times \frac{1}{4} = \frac{1 \times 1}{3 \times 4} = \frac{1}{12}$$

$$\frac{2}{3} \times \frac{1}{5} = \frac{2 \times 1}{3 \times 5} = \frac{2}{15}, \qquad \frac{2}{3} \times \frac{4}{5} = \frac{2 \times 4}{3 \times 5} = \frac{8}{15}$$

RULE

To multiply two fractions, multiply the numerators and multiply the denominators.

EXAMPLE

Multiply: (a) $\frac{7}{11} \times \frac{5}{8}$ (b) $\frac{3}{7} \times 4$

Solution

(a) $\dfrac{7}{11} \times \dfrac{5}{8} = \dfrac{7 \times 5}{11 \times 8} = \dfrac{35}{88}$

(b) Rewrite 4 as $\frac{4}{1}$. The example becomes

$$\frac{3}{7} \times \frac{4}{1} = \frac{3 \times 4}{7 \times 1} = \frac{12}{7} = 1\frac{5}{7}$$

Multiplication by fractions is often expressed using the word "of." In practical life, we can see this. For example, you might have one-third of a bag of 6 cookies, which is 2 cookies. To find one-third of 6 cookies, you multiply $\frac{1}{3} \times 6$. (But it's unlikely you would say you ate $\frac{1}{3}$ times 6 cookies.)

$$\frac{1}{3} \times 6 = \frac{1}{3} \times \frac{6}{1} = \frac{6}{3} = 2$$

EXAMPLE

Find: (a) $\frac{2}{3}$ of 15   (b) $\frac{3}{4}$ of $\frac{7}{9}$

Solution
(a) $\frac{2}{3} \times \frac{15}{1} = \frac{30}{3} = 10$
(b) $\frac{3}{4}$ of $\frac{7}{9} = \frac{3}{4} \times \frac{7}{9} = \frac{21}{36} = \frac{7}{12}$

When multiplying mixed numbers such as $2\frac{3}{5} \times 5\frac{1}{3}$, we must first change the mixed numbers to improper fractions.

$$2\frac{3}{5} = \frac{13}{5} \qquad \text{and} \qquad 5\frac{1}{3} = \frac{16}{3}$$

EXAMPLE

Multiply: (a) $2\frac{3}{5} \times 5\frac{1}{3}$ (b) $\frac{3}{5} \times \frac{15}{18} \times \frac{4}{12}$ (c) $\left(-\frac{2}{5}\right) \times \frac{3}{7}$

Solution
(a) $2\frac{3}{5} \times 5\frac{1}{3} = \frac{13}{5} \times \frac{16}{3} = \frac{13 \times 16}{5 \times 3} = \frac{208}{15}$, or $13\frac{13}{15}$
(b) $\frac{3}{5} \times \frac{15}{18} \times \frac{4}{12} = \frac{3 \times 15 \times 4}{5 \times 18 \times 12} = \frac{180}{1080} = \frac{1}{6}$

It is often easier to reduce before we multiply. For example,

$$\frac{2}{3} \times \frac{3}{4} = \frac{2 \div 2}{3 \div 3} \times \frac{3 \div 3}{4 \div 2} = \frac{1 \times 1}{1 \times 2} = \frac{1}{2}$$

Thus,

$$\frac{3}{5} \times \frac{15}{18} \times \frac{4}{12} = \frac{3}{5} \times \frac{15 \div 3}{18 \div 3} \times \frac{4 \div 4}{12 \div 4} = \frac{\cancel{3}^{1}}{\cancel{5}_{1}} \times \frac{\cancel{5}^{1}}{6} \times \frac{1}{\cancel{3}_{1}} = \frac{1}{6}$$

(c) $-\frac{2}{5} \times \frac{3}{7} = -\frac{6}{35}$

EXAMPLE

Solve: (a) $\left(\frac{2}{5}\right)^{3}$ (b) $\left(-\frac{3}{4}\right)^{2}$

Solution   Exponential notation implies repeated multiplication.
(a) $\left(\frac{2}{5}\right)^{3} = \frac{2}{5} \times \frac{2}{5} \times \frac{2}{5} = \frac{2 \times 2 \times 2}{5 \times 5 \times 5} = \frac{8}{125}$
(b) $\left(-\frac{3}{4}\right)\left(-\frac{3}{4}\right) = \frac{9}{16}$

EXERCISE 4.3.1

Multiply. Reduce your answers to lowest terms.

1. $3 \times \frac{2}{5}$

2. $\frac{2}{3} \times \frac{1}{5}$

3. $\frac{3}{7} \times \frac{2}{11}$

4. $\frac{4}{9} \times \frac{1}{3}$

5. $\frac{2}{5}$ of $\frac{3}{4}$

6. $\frac{9}{10} \times \frac{7}{16}$

7. $\frac{1}{2} \times 2$

8. $\frac{1}{5}$ of 15

9. $\left(\frac{2}{3}\right)^2$

10. $\left(-\frac{1}{2}\right)^3$

11. $-\left(1\frac{1}{2}\right)^3$

12. $\left(1\frac{3}{5}\right)\left(2\frac{1}{2}\right)$

13. $\left(2\frac{3}{4}\right)^2$

14. $\left(\frac{2}{-3}\right)\left(\frac{-5}{8}\right)$

15. $-\left(-\frac{1}{3}\right)\left(\frac{5}{8}\right)\left(\frac{8}{15}\right)$

16. $\left(1\frac{1}{2}\right)^2\left(\frac{7}{9}\right)\left(-\frac{4}{7}\right)$

Division

$$6\left(\tfrac{1}{2}\right) = 3 \qquad 6 \div 2 \text{ is also } 3.$$

$$8\left(\tfrac{1}{4}\right) = 2 \qquad 8 \div 4 = 2$$

Number pairs such as 2 and $\frac{1}{2}$ or 4 and $\frac{1}{4}$ are called *reciprocals*, or sometimes *inverses*. To find the reciprocal of a fraction, invert the fraction (turn it over) so that the denominator becomes the numerator and the numerator becomes the denominator. Thus, the reciprocal (or inverse) of $\frac{4}{5}$ is $\frac{5}{4}$. Remember that a whole number can be expressed as a fraction with a denominator of 1. Thus, the reciprocal of 7 is $\frac{1}{7}$. If you multiply a pair of reciprocal numbers, you get 1 as the product.

For example,

$$3\left(\tfrac{1}{3}\right) = 1, \qquad \left(\tfrac{1}{5}\right)(5) = 1, \qquad \tfrac{4}{5} \times \tfrac{5}{4} = 1, \qquad \tfrac{-2}{3} \times \tfrac{3}{-2} = 1$$

You know from experience that \$6 contains 24 quarters. We can find the number of quarters mathematically in two ways:

$$6 \div \tfrac{1}{4} = 24 \qquad \text{and} \qquad 6 \times 4 = 24$$

EXAMPLE

Solve: (a) $3 \div \frac{3}{5}$ (b) $\frac{2}{3} \div 2$

Solution

(a) How many times does $\frac{3}{5}$ go into 3?

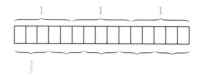

We can fit 5 of the $\frac{3}{5}$s into 3 wholes, so $3 \div \frac{3}{5} = 5$.
The answer is 5 times.

(b) $\frac{2}{3} \div 2$:

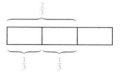

If we divide $\frac{2}{3}$ into two equal parts, each part equals $\frac{1}{3}$.

$$\frac{2}{3} \div 2 = \frac{1}{3}$$

In each example in the left-hand column below, we *divide* one number by another. In the right-hand column we *multiply* the first number at the left by the reciprocal of the second. Compare the results.

$$56 \div 8 = 7 \qquad 56 \times \tfrac{1}{8} = 7$$

$$50 \div 10 = 5 \qquad 50 \times \tfrac{1}{10} = 5$$

$$3 \div \tfrac{3}{5} = 5 \qquad 3 \times \tfrac{5}{3} = 5$$

$$\tfrac{2}{3} \div 2 = \tfrac{1}{3} \qquad \tfrac{2}{3} \times \tfrac{1}{2} = \tfrac{1}{3}$$

RULE

To divide a number by a fraction, multiply it by the reciprocal of the fraction.

If a, b, c, and d are any numbers, and b, c, and $d \neq 0$, then

$$\frac{a}{b} \div \frac{c}{d} = \frac{a}{b} \times \frac{d}{c}$$

EXAMPLE

Solve: (a) $5 \div 5$ (b) $5 \times \frac{1}{5}$ (c) $24 \div 4$ (d) $24 \times \frac{1}{4}$
(e) $32 \div 16$ (f) $32 \times \frac{1}{16}$

Solution
(a) $5 \div 5 = 1$ (b) $5 \times \frac{1}{5} = 1$ (c) $24 \div 4 = 6$
(d) $24 \times \frac{1}{4} = 6$ (e) $32 \div 16 = 2$ (f) $32 \times \frac{1}{16} = 2$

We change mixed numbers to improper fractions before we invert and multiply.

EXAMPLE

Solve: $2\frac{1}{2} \div 1\frac{1}{4}$

Solution $2\frac{1}{2} \div 1\frac{1}{4} = \frac{5}{2} \div \frac{5}{4} = \frac{5}{2} \times \frac{4}{5} = \frac{20}{10} = 2$

EXERCISE 4.3.2

Divide.

1. $\frac{4}{5} \div \frac{8}{15}$
2. $\frac{1}{2} \div \frac{3}{4}$
3. $\frac{1}{2} \div \frac{1}{4}$
4. $\frac{2}{3} \div \frac{5}{7}$
5. $\frac{3}{4} \div 2$
6. $\frac{1}{2} \div 3$
7. $2 \div \frac{2}{7}$
8. $3 \div \frac{2}{3}$
9. $\frac{4}{5} \div \frac{2}{3}$
10. $\frac{5}{8} \div \frac{25}{32}$
11. $1\frac{4}{5} \div \frac{3}{10}$
12. $3\frac{1}{2} \div \frac{1}{4}$
13. $4 \div 1\frac{1}{7}$
14. $9 \div 22\frac{1}{2}$
15. $5\frac{1}{6} \div 6\frac{8}{9}$
16. $3\frac{1}{2} \div 1\frac{2}{3}$

4.4 ADDITION AND SUBTRACTION OF FRACTIONS

In this section, you will add fractions, first with the same denominators and then with different denominators. Later, you will add and subtract mixed numbers as well as subtract fractions.

Addition

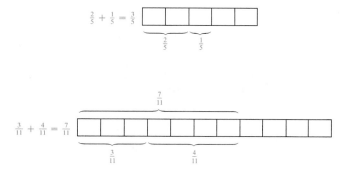

In both of these examples we are adding the same "things," fifths in the first example and elevenths in the second. In other words, the *denominators* are the same. The rules for adding fractions with the same denominator are straightforward. We add the numerators of the fractions and write the sum over the common denominator. Thus

$$\frac{2}{5} + \frac{1}{5} = \frac{2+1}{5} = \frac{3}{5}$$

RULE

If a, b, and c stand for any number and $c \neq 0$, then

$$\frac{a}{c} + \frac{b}{c} = \frac{a+b}{c}$$

What happens if the denominators are different? For example, add

$$\frac{1}{3} + \frac{1}{4}$$

We need to change these two fractions to equivalent fractions with the same denominator (a common denominator). We know that $\frac{1}{3} = \frac{2}{6} = \frac{3}{9} = \frac{4}{12} = \frac{5}{15}$, etc. Also, $\frac{1}{4} = \frac{2}{8} = \frac{3}{12} = \frac{4}{16}$. Instead of adding $\frac{1}{3} + \frac{1}{4}$, we can add the equivalent fractions $\frac{4}{12} + \frac{3}{12}$. Thus,

$$\frac{1}{3} + \frac{1}{4} = \frac{4}{12} + \frac{3}{12} = \frac{7}{12}$$

In order for us to add $\frac{1}{5} + \frac{1}{6}$, we have to search for equivalent fractions with the same denominator. Both 5 and 6 divide evenly into 30. We say that 30 is a multiple of 5 as well as of 6.

Multiples of 5 are $5, 10, 15, 20, 25, 30, 35, 40, 45, 50, 55, 60, 65, \ldots$.

Multiples of 6 are $6, 12, 18, 24, 30, 36, 42, 48, 54, 60, \ldots$.

$30, 60, 90, \ldots$ are common multiples of 5 and 6, but 30 is the smallest. The smallest common multiple is usually referred to as the *least common multiple* and written LCM or lcm.

EXAMPLE

Find the least common multiple of 12 and 16.

Solution
Multiples of 12: $12, 24, 36, 48, 60, 72, 84, 96, \ldots$.
Multiples of 16: $16, 32, 48, 64, 80, 96, \ldots$.
The least common multiple is 48.

DEFINITION

The **least common multiple (LCM)** of two numbers is the smallest number that is a multiple of both numbers (the smallest number that the two numbers can be divided into evenly).

A different way of finding the least common multiple is to use prime factorization of the numbers:

$$10 = 2 \times 5 \quad \text{and} \quad 15 = 3 \times 5$$

The least common multiple of 10 and 15 must contain the factors 2, 3, and 5. $2 \times 3 \times 5 = 30$. So the LCM of 10 and 15 is 30.

EXAMPLE

Find the least common multiple of 15 and 20.

Solution $15 = 3 \times 5 \quad \text{and} \quad 20 = 2 \times 2 \times 5$
Here we need the factors 2, 3, and 5. However, we need *two* factors of 2 to get the multiple of 20. The least common multiple is

$$2 \times 2 \times 3 \times 5 = 60$$

60 is the smallest number that can be divided evenly by both 15 and 20.

The least common multiple for denominators is called the *least common denominator* (LCD).

EXAMPLE

Add: $\dfrac{3}{10} + \dfrac{4}{15}$

Solution $10 = 2 \times 5 \quad \text{and} \quad 15 = 3 \times 5$
The least common denominator is $2 \times 3 \times 5 = 30$.

$$\frac{3}{10} = \frac{9}{30} \quad \text{and} \quad \frac{4}{15} = \frac{8}{30}$$

$$\frac{3}{10} + \frac{4}{15} = \frac{9}{30} + \frac{8}{30} = \frac{17}{30}$$

EXAMPLE

Add: $\dfrac{5}{12} + \dfrac{7}{16}$

Solution $12 = 2 \times 2 \times 3 \quad \text{and} \quad 16 = 2 \times 2 \times 2 \times 2$
The LCD is $2 \times 2 \times 2 \times 2 \times 3 = 48$.

$$\frac{5}{12} = \frac{5 \times 4}{12 \times 4} = \frac{20}{48} \quad \text{and} \quad \frac{7}{16} = \frac{7 \times 3}{16 \times 3} = \frac{21}{48}$$

$$\text{so} \quad \frac{5}{12} + \frac{7}{16} = \frac{20}{48} + \frac{21}{48} = \frac{41}{48}$$

RULE

To add fractions with different denominators, rewrite the fractions as equivalent fractions with a common denominator. Then follow the rule on page 110.

Although we don't need to find the *least* common denominator to add fractions, our work will generally be easier if we do. It is more important to remember that fractions *must* have a *common* denominator before you can add the numerators.

Adding Mixed Numbers

To add mixed numbers, first add the whole numbers together. Then add the fractions together. If the sum of the fractions is an improper fraction, then change it to a mixed number.

EXAMPLE

Add: $2\frac{3}{5} + 1\frac{4}{5}$

Solution

$$\begin{array}{r} 2\frac{3}{5} \\ +1\frac{4}{5} \\ \hline 3\frac{7}{5} \end{array} = 3 + 1\frac{2}{5} = 4\frac{2}{5}$$

If the denominators of the mixed numbers are different, first find equivalent numbers with a common denominator. Then proceed as above.

EXAMPLE

Add: $4\frac{1}{3} + 2\frac{3}{5}$

Solution   Rewrite these with a common denominator. 15 is the least common multiple of 3 and 5.

$$\begin{array}{r} 4\frac{1}{3} = \quad 4\frac{5}{15} \\ +2\frac{3}{5} = +2\frac{9}{15} \\ \hline 6\frac{14}{15} \end{array}$$

Subtracting

The rules for subtracting fractions are essentially the same as for adding fractions. If the denominators are already the same, subtract the numerators and write the difference over the common denominator. Thus, $\frac{3}{5} - \frac{1}{5} = \frac{2}{5}$.

EXAMPLE

Subtract: $\dfrac{7}{11} - \dfrac{4}{11}$

Solution

$$\frac{7}{11} - \frac{4}{11} = \frac{7-4}{11} = \frac{3}{11}$$

If the denominators are not the same, first rewrite the fractions as equivalent fractions with a common denominator. Then follow the above steps.

EXAMPLE

Subtract: $\dfrac{5}{6} - \dfrac{1}{4}$

Solution   The least common denominator for $\frac{5}{6}$ and $\frac{1}{4}$ is 12.

$$\dfrac{5}{6} = \dfrac{10}{12}$$
$$-\dfrac{1}{4} = -\dfrac{3}{12}$$
$$\overline{\phantom{-\dfrac{1}{4} = -}\dfrac{7}{12}}$$

Subtracting Mixed Numbers

Here again, the rules for subtracting mixed numbers are similar to those for addition. However, say you want to subtract $1\frac{4}{5}$ from $3\frac{1}{5}$, that is, $3\frac{1}{5} - 1\frac{4}{5}$. You cannot subtract $\frac{4}{5}$ from $\frac{1}{5}$. If the fractional part of the number you are subtracting is larger than the fractional part of the number you are subtracting from, there are two ways you can proceed. One is to rewrite the first mixed number by borrowing from the whole number.

For example,

$$3\tfrac{1}{5} = 2 + 1\tfrac{1}{5} = 2\tfrac{6}{5}$$

Then

$$3\tfrac{1}{5} = 2\tfrac{6}{5}$$
$$-1\tfrac{4}{5} = -1\tfrac{4}{5}$$
$$\overline{\phantom{-1\tfrac{4}{5} = -}1\tfrac{2}{5}}$$

The second method is to change both numbers to improper fractions with a common denominator.

$$3\tfrac{1}{5} - 1\tfrac{4}{5} = \tfrac{16}{5} - \tfrac{9}{5} = \tfrac{7}{5} = 1\tfrac{2}{5}$$

EXAMPLE

Subtract: (a) $5\frac{3}{11} - 3\frac{1}{11}$ (b) $4\frac{1}{3} - 2\frac{3}{5}$ (c) $3 - \frac{4}{5}$ (d) $1\frac{1}{2} - 3\frac{2}{5}$

Solution

(a) First method:
$$5\tfrac{3}{11}$$
$$-3\tfrac{1}{11}$$
$$\overline{\phantom{-3\tfrac{1}{11}}2\tfrac{2}{11}}$$

Second method:
$$5\tfrac{3}{11} = \tfrac{58}{11}$$
$$-3\tfrac{1}{11} = -\tfrac{34}{11}$$
$$\overline{\phantom{-3\tfrac{1}{11} = -}\tfrac{24}{11}} = 2\tfrac{2}{11}$$

(b) Rewrite both numbers with a common denominator.
First method (borrowing):
$$4\tfrac{1}{3} = 4\tfrac{5}{15} = 3\tfrac{20}{15}$$
$$-2\tfrac{3}{5} = -2\tfrac{9}{15} = -2\tfrac{9}{15}$$
$$\overline{\phantom{-2\tfrac{3}{5} = -2\tfrac{9}{15} = -}1\tfrac{11}{15}}$$

Second method (changing to improper fractions):

$$
\begin{array}{rcl}
4\frac{5}{15} &=& \frac{65}{15} \\
-2\frac{9}{15} &=& -\frac{39}{15} \\
\hline
&& \frac{26}{15} = 1\frac{11}{15}
\end{array}
$$

(c) First method:

$$
\begin{array}{rcl}
3 &=& 2\frac{5}{5} \\
-\frac{4}{5} &=& -\frac{4}{5} \\
\hline
&& 2\frac{1}{5}
\end{array}
$$

Second method:

$$
\begin{array}{rcl}
3 &=& \frac{15}{5} \\
-\frac{4}{5} &=& -\frac{4}{5} \\
\hline
&& \frac{11}{5} = 2\frac{1}{5}
\end{array}
$$

(d) First method:

$$
\begin{array}{rcccl}
1\frac{1}{2} &=& 1\frac{5}{10} &=& 1\frac{5}{10} \\
-3\frac{2}{5} &=& -3\frac{4}{10} &=& -2\frac{14}{10} \\
\hline
&&&& -1\frac{9}{10}
\end{array}
$$

Here we subtracted the smaller absolute value from the larger absolute value, that is $2\frac{14}{10} - 1\frac{5}{10} = 1\frac{9}{10}$
Second method:

$$
\begin{array}{rcccl}
1\frac{1}{2} &=& 1\frac{5}{10} &=& \frac{15}{10} \\
-3\frac{2}{5} &=& -3\frac{4}{10} &=& -\frac{34}{10} \\
\hline
&&&& -\frac{19}{10} = -1\frac{9}{10}
\end{array}
$$

Again, we subtract the absolute values, $\frac{34}{10} - \frac{15}{10} = \frac{19}{10}$

EXERCISE 4.4.1

1. Add. Reduce your answers to lowest terms.
   (a) $\frac{1}{9} + \frac{2}{9}$
   (b) $\frac{1}{5} + \frac{4}{5}$
   (c) $\frac{4}{11} + \frac{2}{11}$
   (d) $\frac{5}{16} + \frac{7}{16}$
   (e) $3\frac{3}{7} + 2\frac{1}{7}$
   (f) $1\frac{7}{9} + 5\frac{4}{9}$

2. Subtract. Simplify your answer.
   (a) $\frac{6}{13} - \frac{4}{13}$
   (b) $\frac{5}{12} - \frac{1}{12}$
   (c) $\frac{8}{9} - \frac{5}{9}$
   (d) $5\frac{2}{5} - 3\frac{4}{5}$
   (e) $2\frac{1}{4} - 1\frac{3}{4}$
   (f) $-\frac{1}{2} - \frac{5}{2}$

3. Find a common denominator; add or subtract.

(a) $\frac{2}{10} + \frac{1}{100}$

(b) $\frac{1}{100} - \frac{1}{1000}$

(c) $\frac{3}{10} - \frac{4}{1000}$

(d) $\frac{5}{100} + \frac{6}{10000}$

4. Find the least common multiple.

(a) $6, 8$

(b) $15, 20$

(c) $12, 18, 36$

(d) $24, 40, 48$

5. Find a common denominator; add or subtract.

(a) $\frac{2}{3} + \frac{5}{6}$

(b) $\frac{5}{6} + \frac{5}{8}$

(c) $\frac{7}{3} - \frac{3}{7}$

(d) $\frac{9}{10} - \frac{11}{35}$

(e) $\frac{1}{5} + \frac{2}{10}$

(f) $\frac{2}{5} + \frac{4}{15}$

6. Subtract.

(a) $2 - \frac{1}{3}$

(b) $3 - 1\frac{2}{5}$

(c) $4 - 2\frac{3}{4}$

(d) $5 - 4\frac{4}{7}$

(e) $6 - 3\frac{1}{2}$

(f) $8 - 7\frac{1}{5}$

7. Solve.

(a) $10\frac{5}{6} - 8\frac{7}{15}$

(b) $5\frac{4}{5} + 4\frac{1}{8}$

(c) $6\frac{3}{5} + 2\frac{13}{15}$

(d) $7\frac{2}{11} - 4\frac{1}{8}$

(e) $2\frac{1}{3} + 3\frac{1}{8}$

(f) $\frac{4}{5} - 3$

4.5 APPLICATIONS

Solving application problems with fractions is much the same as solving problems with whole numbers. One of your first steps is to figure out what operation to use.

For example, if 1 pound of veal costs $\$4\frac{1}{2}$, what is the cost of $3\frac{1}{4}$ pounds of veal?

One pound costs $4\frac{1}{2}$ dollars. We multiply the amount of veal by $4\frac{1}{2}$ to get the total cost of veal. If we buy $3\frac{1}{4}$ pounds, we multiply this number by $4\frac{1}{2}$.

$$3\frac{1}{4} \times 4\frac{1}{2}$$

Change to improper fractions:

$$\frac{13}{4} \times \frac{9}{2} = \frac{117}{8} = 14\frac{5}{8}$$

$3\frac{1}{4}$ pounds of veal costs $14\frac{5}{8}$ dollars, or $14.63.

EXAMPLE

A merchant who owned $\frac{3}{8}$ of a ship sold $\frac{4}{6}$ of his share for $3000. What is the value of the ship?

Solution   First we figure out what fraction of the ship the merchant sold.

$$\frac{4}{6} \text{ of } \frac{3}{8} = \frac{4}{6} \times \frac{3}{8} = \frac{12}{48} = \frac{1}{4}$$

$\frac{1}{4}$ of the ship was sold for $3000. If we multiply $\frac{1}{4}$ by 4 we will have 1 whole ship. If we multiply $3000 by 4, we have the value of the whole ship. $3000 \times 4 = $12,000$.

Many of these exercises are taken from books published more than 100 years ago.

EXERCISE 4.5.1

1. If a ton of hay costs $16\frac{3}{4}$, how much will 22 tons cost?

2. Required: the cost of 60 yards of muslin at $35\frac{3}{8}$ cents a yard.

3. At $5\frac{1}{3}$ a bushel, how much clover seed can be bought for $\frac{8}{9}$?

4. If $8\frac{2}{3}$ quarts of strawberries can be bought for $1\frac{39}{50}$, what is the price of a quart?

5. If 1 rod of fence requires $74\frac{1}{4}$ ft of boards, how many rods will require $1811\frac{7}{10}$ ft?

6. Bell metal is composed of $\frac{4}{5}$ copper and $\frac{1}{5}$ tin. How much of each of these metals is there in a church bell that weighs $\frac{87}{100}$ ton?

7. A can of frozen lemonade contains 6 fluid ounces. The directions tell you to mix 4 cans of water with the contents. How many fluid ounces will you have? There are 32 fluid ounces in a liquid quart. What part of a quart does one can of lemonade concentrate make?

8. On his hardware bill John saw an item of $12 for $1\frac{1}{2}$ quarts of varnish. What was the cost per quart?

9. Judy spent $\frac{1}{4}$ of the day sleeping, $\frac{1}{6}$ of the day in school, $\frac{1}{3}$ of the day working, and $\frac{1}{6}$ of the day commuting. What part of the day was left over for Judy to do as she wished?

10. Jack and Betsey are redoing their living room. It is $6\frac{2}{3}$ yards long and $5\frac{1}{2}$ yards wide.
    (a) How much longer is it than wide?
    (b) What is its area (length times width)?
    (c) What is the perimeter (sum of the four sides)?

11. It took Sandy and Jean seven days to drive from Augusta, Maine, to Seattle, Washington. They kept track of the driving time: Monday $8\frac{1}{3}$ hours, Tuesday $9\frac{3}{4}$, Wednesday $8\frac{2}{3}$, Thursday $10\frac{1}{4}$, Friday $7\frac{3}{4}$, Saturday $11\frac{1}{4}$, and Sunday $5\frac{1}{2}$. How many hours of driving time did it take in all?

12. Casey and Fiori ate half of the cookies, Aaron and Omm ate $\frac{1}{2}$ of what was left, and Alex had the rest. If there were 36 cookies to begin with, how many did Alex have?

4.6 RELATIONSHIPS BETWEEN DECIMALS, FRACTIONS, AND PERCENTS

Decimals to Fractions

Decimals can be changed into fractions. By reading the decimal number properly, we get the fraction form immediately.

For instance, 0.02 is "two hundredths" or 2/100; 0.135 is "one hundred thirty-five thousandths" or 135/1000. Of course, both of these can be reduced to lowest terms; $\frac{2}{100} = \frac{1}{50}$, and $\frac{135}{1000} = \frac{27}{200}$.

EXAMPLE

Convert to a fraction and simplify: (a) 0.5   (b) 0.25   (c) 0.0003

Solution
(a) $0.5 = \frac{5}{10} = \frac{1}{2}$
(b) $0.25 = \frac{25}{100} = \frac{1}{4}$
(c) $0.0003 = \frac{3}{10000}$

With mixed numbers we keep the whole part of the number, since this doesn't change. Only the places to the right of the decimal point are changed to fractions. Thus, 2.5 becomes "2 and $\frac{5}{10}$" or $2\frac{5}{10}$. This in turn reduces to $2\frac{1}{2}$.

EXAMPLE

Change to mixed numbers: (a) 3.08   (b) 21.196

Solution
(a) $3.08 = 3\frac{8}{100} = 3\frac{2}{25}$
(b) $21.196 = 21\frac{196}{1000} = 21\frac{49}{250}$

EXERCISE 4.6.1

Translate decimals into fractions. Express in lowest terms.

1. 0.37

2. 1.25

3. 12.3

4. 17.892

5. 245.245

6. 1.045

7. 0.02

8. 0.005

9. 0.000075

10. 0.00030

Fractions to Decimals

The fraction bar has the same effect as a division sign.

EXAMPLE

Rewrite the following in decimal form: (a) $\frac{3}{4}$ (b) $\frac{5}{8}$

Solution

(a) $\frac{3}{4}$ can be read as "3 divided by 4" and written

$$3 \div 4 \quad \text{or} \quad 4\overline{)3}$$

Carrying out the division, we get

$$
\begin{array}{r}
0.75 \\
4\overline{)3.00} \\
2\,8 \\
\hline
20 \\
20 \\
\hline
0
\end{array}
$$

So $\frac{3}{4} = 0.75$.

(b) $\frac{5}{8}$ can be read "5 divided by 8," $5 \div 8$, or $8\overline{)5}$. Doing the division, we get

$$
\begin{array}{r}
0.625 \\
8\overline{)5.000} \\
4\,8 \\
\hline
20 \\
16 \\
\hline
40 \\
40 \\
\hline
0
\end{array}
$$

So $\frac{5}{8} = 0.625$.

EXAMPLE

Rewrite in decimal form: (a) $1\frac{4}{5}$ (b) $13\frac{5}{8}$

Solution

(a) To change a mixed number such as $1\frac{4}{5}$ into a decimal, we only need to change the fraction part because the whole part stays the same. Since $\frac{4}{5} = 0.8$, $1\frac{4}{5} = 1.8$.

(b) 13 doesn't change, and since $\frac{5}{8} = 0.625$, $13\frac{5}{8} = 13.625$.

$\frac{1}{3}$ is the same as $1 \div 3$. This division does not come out evenly; it equals $0.3333\ldots$, which we recognize as a nonterminating repeating decimal. Show this by writing $0.33\ldots$ or $0.\overline{3}$. Again, the points and the bar above the 3 mean that 3 is repeated forever.

EXAMPLE

Change (a) $\frac{4}{11}$ and (b) $\frac{5}{33}$ into decimals.

> **Solution**
> (a) $\frac{4}{11} = 0.363636\ldots = 0.\overline{36}$ 36 forms a repeating group.
> (b) $\frac{5}{33} = 0.1515\ldots = 0.\overline{15}$ The repeating group is 15.

RULE

To change a terminal decimal number to a fraction:

1. Read the decimal number with the last place spelled out.
2. Write it in fractional form.
3. Reduce, if possible.

To change a fraction to a decimal, divide the numerator by the denominator. Round, if needed.

EXERCISE 4.6.2

Translate each fraction into a decimal. If the division does not come out evenly, try to determine which numbers form a repeating group.

1. $\frac{1}{5}$

2. $\frac{2}{25}$

3. $\frac{2}{9}$

4. $\frac{1}{7}$

5. $\frac{3}{11}$

6. $2\frac{1}{3}$

7. $3\frac{2}{3}$

8. $\frac{2}{13}$

Percents into Fractions

To change a number expressed as a percent into a fraction, divide the number by 100. Remember, % means per hundred.

EXAMPLE

Write as a fraction: (a) 5%   (b) $5\frac{1}{2}\%$ (c) 45.5%

> **Solution**
> (a) $5\% = \frac{5}{100} = \frac{1}{20}$
> (b) $5\frac{1}{2}\% = \frac{11}{2} \div 100 = \frac{11}{200}$
> (c) $45.5\% = \frac{45.5}{100} = \frac{455}{1000} = \frac{91}{200}$

EXERCISE 4.6.3

Change percents to fractions.

1.   (a) 8%
     (b) 30%
     (c) 33%
     (d) 105%

2.   (a) $7\frac{1}{2}\%$
     (b) $8\frac{1}{3}\%$
     (c) $2\frac{2}{6}\%$
     (d) $8\frac{3}{4}\%$

Fractions into Percents

To change a fraction into a percent, multiply by 100. Remember that 100% equals 1, and we can multiply any number by 1 without changing the value of the number.

For example,

$$\frac{3}{5} = \frac{300}{5}\% = 60\%$$

EXAMPLE

Write as percent: (a) $2\frac{1}{2}$ (b) $\frac{1}{3}$

Solution
(a) $2\frac{1}{2} = \frac{5}{2} = \frac{500}{2}\% = 250\%$
(b) $\frac{1}{3} = \frac{100}{3}\% = 33\frac{1}{3}\%$

RULE

To change a number into percent, multiply the number by 100%. To change a percent into a fraction, divide by 100. To change a percent into a decimal, move the decimal point two places to the left.

EXERCISE 4.6.4

Convert from fractions to percents:

1.   (a) $\frac{5}{6}$
     (b) $\frac{11}{15}$
     (c) $\frac{6}{11}$
     (d) $1\frac{1}{3}$

2.   (a) $\frac{5}{111}$
     (b) $7\frac{1}{3}$
     (c) $15\frac{2}{3}$
     (d) $36\frac{4}{11}$

Common fractions can be arranged in order by the following methods:

1. Change the fractions to decimals and proceed as before for ordering decimals or

2. Change the fractions to equivalent fractions with the same denominator, then compare the numerators.

EXAMPLE

Order $\frac{4}{5}$ and $\frac{7}{9}$.

Solution
Method 1. $\frac{4}{5} = 0.8$ and $\frac{7}{9} = 0.\overline{7}$. 0.8 is larger than $0.77\ldots$. The first fraction, $\frac{4}{5}$, is larger than $\frac{7}{9}$ (or $\frac{4}{5} > \frac{7}{9}$).
Method 2. Both 5 and 9 go evenly into 45. Change to equivalent fractions with 45 as denominator:

$$\frac{4}{5} = \frac{36}{45} \quad \text{and} \quad \frac{7}{9} = \frac{35}{45}$$

$$\frac{4}{5} > \frac{7}{9}$$

EXERCISE 4.7.1

Arrange each group of fractions in order from the smallest to the largest.

1. $\frac{2}{3}, \frac{3}{5}, \frac{4}{7}$
2. $\frac{8}{15}, \frac{4}{5}, \frac{3}{10}$
3. $\frac{5}{12}, \frac{7}{16}, \frac{9}{18}$
4. $\frac{5}{19}, \frac{3}{14}, \frac{4}{17}$

SUMMARY

Definitions

The greatest common factor (GCF) of two numbers is the largest number that can divide evenly into both numbers.

The least common multiple (LCM) of two numbers is the smallest number that is a multiple of both numbers (the smallest number that both numbers can be divided into evenly).

Rules

To change an improper fraction into a mixed number, divide numerator by denominator. The quotient is the whole number part of the mixed number. The remainder is the new numerator of the fraction part.

To change a mixed number into an improper fraction, multiply the denominator by the whole number. Add the numerator. This is the new numerator. Keep the denominator.

To multiply two fractions, multiply the numerators and multiply the denominators.

To divide one fraction by another fraction, multiply by the reciprocal of the divisor.

To add fractions with the same denominators, add the numerators and keep the denominators.

To add fractions with different denominators, rewrite the fractions as equivalent fractions with common denominators. Then follow the rule for addition of fractions with the same denominators.

To change a number in decimal form to a fraction, read the decimal number with the last place spelled out and write it in fractional form. Reduce, if possible.

To change a fraction to a decimal, divide the numerator by the denominator.

To change a number into percent, multiply the number by 100%.

To change a percent into a fraction, divide by 100.

To change a percent into a decimal, move the decimal point two spaces to the left.

Fundamental theorem. A number can be factored into only one set of prime factors, if the order of the factors is disregarded.

VOCABULARY

Common factors: Numbers that are factors of two or more numbers.

Denominator: The bottom part of a fraction.

Divisibility rules: Divisible by 2: Even numbers.
Divisible by 3: The sum of the digits is divisible by 3.
Divisible by 5: The number ends in 0 or 5.

Equivalent fractions: Fractions that have the same value.

Factor tree: A way of showing how a number is factored into prime factors.

Fraction bar: The dividing line between the numerator and the denominator.

Greatest common factor (GCF): The largest factor that is a factor of two or more numbers.

Improper fraction: A fraction in which the numerator is larger than the denominator.

Inverse numbers: Number pairs that give the identity element in operations, for example, $2 + (-2) = 0$ and $2 \times \frac{1}{2} = 1$.

Lowest terms: The numerator and denominator have no factors in common.

Mixed number: A number composed of an integer and a common fraction.

Multiple: The product of a certain number and other numbers; for example, $5, 10, 15, \ldots$ are multiples of 5.

Numerator: The top number in a fraction.

Proper fraction: A fraction in which the numerator is smaller than the denominator.

Reciprocal: The product of a number and its reciprocal equals 1. $\frac{1}{2}$ and 2, $\frac{3}{4}$ and $\frac{4}{3}$ are reciprocals. Reciprocals are also called multiplicative inverses.

Reducing a fraction: Dividing the numerator and denominator by one or more common factors.

CHECK LIST

Check the box for each topic you feel you have mastered. If you are unsure, go back and review.

☐ Changing improper fractions into mixed numbers
☐ Changing mixed numbers into improper fractions

☐ Equivalent fractions
☐ Reducing to lowest terms
☐ Prime factorization
☐ Greatest common factor
☐ Least common multiple
Basic operations with fractions
☐ Addition
☐ Subtraction
☐ Multiplication
☐ Division
☐ Conversions between fractions and decimals
☐ Conversions between fractions and percents
☐ Ordering fractions
☐ Applications

REVIEW EXERCISES

1. Change into mixed numbers.
   (a) $\frac{9}{4}$
   (b) $\frac{12}{5}$
   (c) $\frac{-58}{7}$
   (d) $\frac{15}{-6}$

2. Change to improper fractions.
   (a) $1\frac{1}{3}$
   (b) $5\frac{3}{4}$
   (c) $10\frac{4}{9}$
   (d) $-7\frac{2}{5}$

3. Change to equivalent fractions.
   (a) $\frac{3}{5} = \frac{}{40}$
   (b) $\frac{5}{7} = \frac{}{56}$
   (c) $\frac{10}{24} = \frac{}{12}$
   (d) $\frac{8}{48} = \frac{2}{}$

4. Which are prime numbers?

   2   3   15   19

   57   83   101   153

5. Factor into prime factors.
   (a) 36
   (b) 39
   (c) 56
   (d) 120

6. Find the greatest common factor.
   (a) 12, 28
   (b) 18, 45
   (c) 36, 48
   (d) 63, 70
   (e) 45, 56

7. Reduce to lowest terms.
   (a) $\frac{14}{21}$
   (b) $\frac{32}{144}$
   (c) $\frac{39}{52}$
   (d) $\frac{45}{405}$

8. Multiply or divide as indicated. Express all answers in lowest terms.
   (a) $\frac{3}{5} \cdot \frac{2}{9}$
   (b) $5 \cdot \frac{7}{10}$
   (c) $\frac{2}{5} \cdot \frac{10}{16}$
   (d) $\frac{-9}{10} \cdot \frac{5}{8}$
   (e) $\frac{9}{10} \div \frac{6}{15}$
   (f) $\frac{2}{3} \div \frac{5}{7}$
   (g) $\frac{-9}{16} \div \frac{3}{4}$
   (h) $\frac{-15}{18} \div \frac{-5}{12}$
   (i) $1\frac{1}{5} \cdot 1\frac{1}{3}$
   (j) $(3\frac{1}{2})(-2\frac{1}{7})$
   (k) $(3\frac{1}{4}) \div (-1\frac{1}{8})$
   (l) $(-1\frac{1}{2}) \div (-4\frac{5}{8})$

9. Find the least common multiple.
   (a) 12, 15
   (b) 18, 42
   (c) 24, 36, 48
   (d) 18, 63, 81

10. Find a common denominator and solve.
   (a) $\frac{2}{3} + \frac{5}{9}$
   (b) $\frac{1}{2} + \frac{3}{5}$
   (c) $\frac{3}{8} - \frac{1}{12}$
   (d) $2 - \frac{1}{3}$

11. Find the answer. Reduce to lowest terms.
    (a) $3\frac{1}{2} \times 1\frac{2}{5}$
    (b) $1\frac{3}{4} \div 1\frac{1}{3}$
    (c) $7\frac{4}{5} + 2\frac{1}{2}$
    (d) $8\frac{1}{6} - 3\frac{1}{4}$

12. Express as a fraction and a percent. Reduce the fractions to lowest terms.
    (a) 0.25
    (b) 0.005
    (c) 1.23
    (d) 2

13. Express as a decimal and a percent.
    (a) $\frac{1}{2}$
    (b) $\frac{3}{4}$
    (c) $\frac{5}{8}$
    (d) 3

14. Express as a fraction and a decimal. Reduce fractions to lowest terms.
    (a) 16%
    (b) 100%
    (c) 3%
    (d) $6\frac{1}{2}\%$

15. A $5\frac{3}{4}$-lb fish loses $1\frac{1}{2}$ lb in cooking. How much will the cooked fish weigh?

16. One side of a tape has four songs. The times for the songs are as follows: $3\frac{1}{3}$, $4\frac{7}{8}$, $2\frac{5}{6}$, and $4\frac{1}{4}$ minutes. How long is the tape?

17. A utility stock listed at $18\frac{7}{8}$ on Wednesday rose $2\frac{1}{2}$ points on Thursday and dropped $1\frac{3}{8}$ on Friday. What was it worth when the market closed on Friday?

18. If the height marked on an overpass is 17 ft 9 in. and your truck is $14\frac{2}{3}$ feet high, what will your clearance be when you drive under the overpass?

19. Four electrical resistors have resistances of $\frac{1}{8}$, $\frac{1}{4}$, $\frac{2}{5}$, and $\frac{1}{3}$ ohm. What is the total resistance of the four resistors?

20. A rectangular garden has dimensions of $15\frac{1}{2}$ ft wide and $25\frac{2}{3}$ ft long.
    (a) What is the area of the garden?
    (b) David wants to put a fence around it to keep the raccoons out. How much fencing does he need?
    (c) If David buys 100 ft of fencing, how much will he have left over?

READINESS CHECK

Solve the problems to satisfy yourself that you have mastered Chapter 4.
1. Change into a mixed number: $\frac{15}{8}$.
2. Change into an improper fraction: $2\frac{4}{7}$.
3. Change to an equivalent fraction: $\frac{4}{5} = \frac{}{45}$.
4. Reduce to lowest terms: $\frac{62}{93}$.
5. Add: $5\frac{1}{12} + 4\frac{3}{8}$.
6. Subtract: $5 - 3\frac{4}{5}$.
7. Multiply: $3\frac{3}{4} \times 2\frac{2}{5}$.
8. Divide: $2\frac{2}{3} \div 6$.
9. Convert 0.125 into a common fraction.
10. Insert $<$ or $>$ between $\frac{7}{13}$ and $\frac{5}{11}$.

APPLICATIONS

Using mathematics

Word problems have been around for centuries, as you can see in this application from *Grownde of Artes*, published in 1541.

> Then what say you to this equation? If I sold unto you an horse having 4 shoes, and in every shoe 6 nayles, with this condition, that you shall pay for the first nayle one ob: for the second nayle two ob: for the third nayle foure ob: and so forth, doubling untill the end of all the nayles, now I ask you, how much would the price of the horse come unto?

Say that the unit "ob" is worth 1 cent. Then we can figure out how much money these 24 nails would cost.

First nail: 1¢
Second nail: 2¢
Third nail: 4¢ = 2^2¢
Fourth nail: 8¢ = 2^3¢
Fifth nail: 16¢ = 2^4¢
...
Twenty-fourth nail: 2^{23}¢

The total cost would be

$$1 + 2 + 2^2 + 2^3 + \cdots \ + 2^{23} \text{ cents.}$$

This sum can be calculated using a formula (to be studied in Chapter 7) that gives us 16,777,215¢. This is approximately \$200,000!

In this chapter we will use the techniques we learned in Chapters 1 to 4 in practical applications. Go back to the appropriate chapter for rules and examples when you need to.

SOLVING A WORD PROBLEM

1. Read the problem through.
2. Try to estimate what the answer will be.
3. Decide whether to add, subtract, multiply, divide, or use a combination of several operations.
4. Get an answer.
5. Check whether the answer makes sense.

Do use your calculator!

5.1 PERCENT ADDED OR SUBTRACTED

Say that you know that the price of an item with 8% tax included is $43.20. What was the price of the item before taxes? Can you guess the answer? Since $4 \times 8 = 32$, a good guess would be that the price was $40. But how do you solve it mathematically? Some people might suggest that you take 8% of $43.20 and subtract it from $43.20. That is wrong, however, because the tax was calculated on the original price, not on the price that already includes the tax!

When we analyze the problem, we see that the original cost is 100% of itself. If we add 8% tax, we have 108%, which is the cost plus the tax. That fits the pattern "108% of what number is 43.20?" that we saw in Chapter 3. Thus

$$43.20 \div 108\% = 43.20 \div 1.08 = 40.00$$

and $40.00 is the original item price.
Check: 8% of $40.00 = $0.08 \times $40.00 = 3.20
$40 + $3.20 = 43.20

EXAMPLE

A dress costs $40.00 after a reduction of 20%. What was the original price?

Solution $40.00 represents 80% ($100\% - 20\%$) of the original price. We can rewrite the problem as "80% of what number is 40?"
Then we have

$$40 \div 80\% = 50$$

The original price of the dress is $50.00.

Is it true that 20% of $50 gives a reduction of $10? Check it.

EXAMPLE

What number increased by 6% of itself is 2544?

Solution $100\% + 6\% = 106\%$. 106% of what number is 2544?

$$2544 / 106\% = 2544 / 1.06 = 2400$$

Check: 6% of 2400 = 144, and 2400 + 144 = 2544.

EXAMPLE

If the price of an item after discount is $22.50 and the discount is 10%, what was the original price?

> **Solution**   Here we have subtracted, so the original price will be higher.
> $100\% - 10\% = 90\%$ of the original price is $22.50, so the question is "90% of what number is 22.50?"

$$22.50 \div 90\% = 22.50 \div 0.90 = 25.00$$

The original price was $25.00.
Check: 10% of $25.00 = 2.50, and $25.00 - $2.50 = $22.50

EXAMPLE

In Sweden, foreign tourists get a rebate of the 24% sales tax when they leave the country. For a purchase of $500.65 (sales tax included), what percent of the total purchase price (item + tax) is given back?

> **Solution**   You will probably want to solve this problem in several steps.
> *Step 1.* How much was the purchase itself without the tax?

$$124\% \text{ of what number is } 500.65?$$

$$500.65 \div 1.24 = 403.75$$

Step 2. How much was the tax?

$$500.65 - 403.75 = 96.90$$

Step 3. What % of 500.65 is the tax?

$$96.90 \div 500.65 = 0.1935483 = 19.35\%, \text{ rounded to the nearest hundredth}$$

Check: Is it true that 24% of $403.65 is the same as 19.35% of $500.65?

$$24\% \times \$403.75 = \$96.90$$

$$19.35\% \times \$500.65 = \$96.88, \text{ due to rounding of } 19.35.$$

Tourists leaving Sweden are often surprised that they get back less than 20% for the tax when they paid 24%. Some people assume that it is some kind of handling charge, but as you can see it is because the rebate is a percentage of the *purchase + tax*, while the 24% tax is a percentage of the *purchase only*.

RULE

Solving percent problems when the percent is already included.

To find the original number:

1. If a% is already added to the number, divide the number by $(100 + a)$%.
2. If a% is already subtracted from the number, divide the number by $(100 - a)$%.

a can be any number.

EXERCISE 5.1.1

1. A speculator bought stocks and later sold them for $4815, making a profit of 7%. How much did the stocks cost him?

2. Complete the following table:

	Item + Tax	Tax	Item
(a)	$56.16	8%	
(b)	$43.30	8.25%	
(c)	$79.50	6%	
(d)	$82.40	3%	
(e)	$107.10	19%	

3. Complete the chart below. The price is given with a certain percent deducted. Find the original price.

	Price − Discount	Percent Discount	Original Price
(a)	$13.50	10%	
(b)	$20.40	15%	
(c)	$28.00	20%	
(d)	$96.00	25%	
(e)	$150.08	$33\frac{1}{3}\%$	

4. When a foreign tourist leaves Sweden, what percent of his purchases does he get back if he paid (a) $399.50, (b) $253.27 (item plus 24% tax)?

5.2 THE HOTEL BUSINESS

To solve many problems, you must know how to calculate *averages*. Suppose Anna earns $42 per day, Brett earns $35, Chris earns $51, and Donna earns $47. To find their average earnings per day, we first add 42, 35, 51, and 47. The sum is 175. We then divide the sum by 4, since we added four numbers. 175 ÷ 4 = 43.75. Their average earnings per day are $43.75.

Suppose five people contributed a total of $350 to a charity. To find out what the average contribution was, we divide the total, 350, by the number of contributions, 5. 350 ÷ 5 = 70. Thus, the average contribution was $70.

In the problems of Exercise 5.2.1, your answers will have to be rounded to two decimals, since the problems deal with money. Remember the rules for rounding:

35.983 becomes 35.98

2.678 becomes 2.68

16.895 becomes 16.90.

The following problems are taken from the records of a large hotel.

Table 5.1

Server	Number of People Served	Sales	Average Check
Barb	35	$354.60	$ _____
Nancy	33	$335.30	$ _____
Cis	33	$328.20	$ _____
Halina	30	$319.60	$ _____
Lee	30	$305.25	$ _____
Luis	34	$342.85	$ _____
Hugo	32	$324.20	$ _____
Total	_____	$ _____	$ _____

EXERCISE 5.2.1

1. (a) Find the cost of this inventory of the hotel's wine cellar:

Bottles on Hand	Item	Price per Bottle	Total Cost
15	Mouton Cadet	4.98	_____
2.5	Chateau Ricon	4.85	_____
7	Mise Joseph	5.25	_____
9	Volnay	16.91	_____
12	Valpolicella	3.35	_____
36	Mt Vender	13.00	_____
Total			_____

(b) What was the average cost of a bottle of wine?

2. Table 5.1 shows one day's sales records for breakfasts served at the hotel.
(a) What is the size of the average breakfast check for each server?
(b) How much does the average guest spend for breakfast at the hotel?

3. Table 5.2 shows how many hours each server worked waiting on tables and how much money each one took in on sales.
(a) Find the average sales per hour for each person.
(b) Find the average of the averages from part (a).

Table 5.2

Name	Hours Worked	Total Sales	Average Sales per Hour
Bill	8	$673.95	$ _____
Hung	8	$469.15	$ _____
Bob	8	$850.00	$ _____
Mary	8.5	$784.45	$ _____
Maureen	6	$702.40	$ _____
Lisa	8.5	$861.70	$ _____
Alfred	7.5	$513.52	$ _____
Total	_____	$ _____	

Average sales/ hr $ _____

Table 5.3

Name	Hourly Wage	Week 1		Week 2		Week 3		Total Pay for 3 weeks
		Hours	Pay	Hours	Pay	Hours	Pay	
Dominick	6.05	40	____	44	____	40	____	____
Lucas	6.67	48	____	40	____	40	____	____
Ortiz	9.18	8	____	48	____	40	____	____
Rodriguez	6.05	48	____	48	____	40	____	____
Spearman	6.50	48	____	40	____	40	____	____
Jiminez	6.05	51.8	____	47.2	____	40	____	____
Mitchell	6.33	55.7	____	48	____	40	____	____
Brown	10.57	32	____	40	____	32	____	____
Azzarano	7.08	58.8	____	40	____	40	____	____
Borras	5.93	16	____	20	____	16	____	____
Krykes	9.88	48	____	40	____	40	____	____
Manos	6.18	56	____	40	____	40	____	____
Tran	7.07	48	____	40	____	40	____	____
White	6.30	48	____	40	____	40	____	____
Bloomer	6.05	55.6	____	40	____	40	____	____
McKenna	9.18	48	____	48	____	40	____	____
Totals			____		____		____	____

Total payroll for 3 weeks = $_____

(c) Find the average sales per person per hour by first finding the total hours worked and the total sales.

4. Table 5.3 is a 3-week payroll. Use your calculator to help answer the questions that follow.
    (a) Calculate how much each person earns each week for 3 weeks.
    (b) Add these weekly salaries to find the total amount for each person for 3 weeks.
    (c) Find the total payroll for each week by adding each "Pay" column.
    (d) Add the totals for the 3 weeks from (c). You should get the same answer as you get when you add the 3-week total wages for each employee.

Table 5.4 is a liquor inventory from the hotel. The table states the inventory according to the accounting books. The bars had been issued a certain amount of liquor, and the hotel stewards had sold some wine. All remaining bottles were counted for the closing inventory and listed as "Physical."

When the book value is more than the physical value, the hotel is short. "Short" is noted with a type of bracket ($\langle$ $\rangle$) that signifies a negative number.

EXAMPLE

If the opening inventory was $300.00 for liquor, $50.00 worth was issued to the bar, and $120.00 worth was sold by the wine stewards, what was the closing book inventory?

Solution Issues to the bar and steward sales amount to a total of $50.00 + $120.00 = $170.00. The closing inventory should be

$$\$300.00 - \$170.00 = \$130.00$$

Table 5.4

	Liquor	Wine	Beer & Misc.	Total
Opening Inventory:				
	$38,315.24	$51,866.55	$16,666.87	_____
Issues:				
Bars	$18,898.68	$20,226.04	$9,557.18	_____
Steward sales		$327.64		_____
Closing Inventory:				
Book	_____	_____	_____	_____
Physical	$19,669.49	$32,663.94	$6,944.77	_____
Over ⟨Short⟩	_____	_____	_____	_____

If the physical inventory (the value determined by counting the bottles) is $128 and the book inventory $130, the difference between them is $128.00 − $130.00 = −$2.00. The hotel is $2.00 short, which is written ⟨$2.00⟩.

EXERCISE 5.2.2

Refer to Table 5.4.

1. Find the total value of alcoholic beverages at the opening inventory.
2. (a) How much was issued to the bars?
    (b) How much was sold through the steward department?
    (c) How much should be available after these issues? (This is the book closing total inventory.)
3. Referring to Table 5.4 and your answers to Problem 1, find the difference between the actual (physical) value at the closing inventory and the calculated value.

5.3 OTHER BUSINESS PROBLEMS

Stock Market

The following examples are taken from areas many people are exposed to in the business world.

EXAMPLE

If the last price of a stock reported was 73 and the change was $-\frac{1}{2}$, what was the price the day before?

Solution   Since the stock went *down* to 73, it was worth *more* earlier. The higher price was $73\frac{1}{2}$.

EXERCISE 5.3.1

The following table is an excerpt from a stock market table in a newspaper. "Last" means the closing price. "Previous" is the price the day before. Complete the table.

	Previous	Net Change	Last
1.	_____	$-\frac{7}{8}$	$84\frac{3}{4}$
2.	$72\frac{1}{4}$	_____	$71\frac{3}{4}$
3.	$96\frac{5}{8}$	_____	$97\frac{1}{2}$
4.	_____	$-\frac{1}{2}$	76
5.	$71\frac{3}{8}$	$-\frac{3}{4}$	_____
6.	$93\frac{1}{8}$	$+\frac{1}{8}$	_____

Depreciation

When a piece of equipment is used, its value decreases every year. This is called *depreciation*. In the "straight-line" method of calculating depreciation, the following formula is used:

$$\frac{\text{Cost} - \text{salvage value}}{\text{Estimated useful life}} = \text{depreciation}$$

The *salvage value* or *scrap value* is an estimate of what the machine's value will be at the end of its useful life. A machine is "written off" after this time.

Book value is the value of the equipment after a certain time. For example, if a $400 office fax machine depreciates $50 per year, the book value after 2 years is $400 − 2 × $50 = $300.

EXAMPLE

A truck was purchased for $15,000 and has a salvage value of $600 and an estimated useful life of 3 years.
(a) Find the depreciation per month.
(b) What is the book value after 1 year?

Solution
(a) For monthly depreciation we need to express the estimated useful life in months. The equation becomes

$$\text{Depreciation} = \frac{15,000 - 600}{36 \text{ months}} = \frac{14,400}{36} = \$400 \text{ per month}$$

(b) Book value after 12 months is the original cost minus 12 months of depreciation.

$$15,000 - 12 \times 400 = 15,000 - 4800 = 10,200.$$

The book value after 1 year is $10,200.

EXERCISE 5.3.2

1. A typewriter costs $650. It is to be written off in 5 years with a scrap value of $50. What is the depreciation per year?

2. An adding machine costing $420.00 had an estimated life of 6 years. At the end of that time its scrap value equaled 12% of its original cost.
(a) How much was the depreciation for the third year?
(b) What was the book value at the end of the second year?

5.4 PERSONAL FINANCE

Bank Accounts

Most adults have bank accounts, need to calculate postage rates, and pay taxes.

EXAMPLE

Karen has a "checking plus" bank account that lets her overdraw on her account (write checks for more money than there is in the account). Overdrawn money is noted with a minus sign. Karen starts out with −$15 in her account. She takes out $100 from the cash machine twice. What's her account balance?

> **Solution** −$15 − $100 − $100 = −$215 is one way of writing out these transactions. You could also write
>
> $$-\$15 - 2(\$100) \quad \text{or} \quad -\$15 + 2(-\$100) = -\$215$$

EXERCISE 5.4.1

Find the new balance.

	Starting Balance	Deposits	Checks Written		New Balance
1.	$254	$346	$25,	$406	
2.	−$54	$45	$23,	$45	
			$105,	$14	

Postage

It costs 29¢ for the first ounce to mail a first class letter. Each additional ounce or part of an ounce is 23¢.

EXAMPLE

How much does it cost to mail a 3.5-ounce letter?

> **Solution** The first ounce costs 29¢. The next 2.5 ounces cost as much as 3 ounces, or 3 × 23¢ = 69¢. The total cost is 29¢ + 69¢ = 98¢, or $0.98.

Table 5.5 shows the rates for priority mail. Use Table 5.6 to find the zones where the priority mail will go. For example, to mail a package from a town with Zip code 11201 to a place with Zip code 10436, you pay the local rate. But to mail from Zip code 11201 to Zip code 20562, you pay according to zone 3.

Table 5.5

Weight, up to but not exceeding — pounds(s)	Zones					
	Local 1, 2, and 3	4	5	6	7	8
1	2.90	2.90	2.90	2.90	2.90	2.90
2	2.90	2.90	2.90	2.90	2.90	2.90
3	4.10	4.10	4.10	4.10	4.10	4.10
4	4.65	4.65	4.65	4.65	4.65	4.65
5	5.45	5.45	5.45	5.45	5.45	5.45
6	5.55	5.75	6.10	6.85	7.65	8.60
7	5.70	6.10	6.70	7.55	8.50	9.65
8	5.90	6.50	7.30	8.30	9.40	10.70
9	6.10	7.00	7.95	9.05	10.25	11.75
10	6.35	7.55	8.55	9.80	11.15	12.80
11	6.75	8.05	9.20	10.55	12.05	13.80

PRIORITY MAIL POSTAGE RATES

Table 5.6

ZONE CHART FOR MAIL ORIGINATING FROM ZIP CODES
BEGINNING WITH 070–076, 079, 090–104, 110–113, OR 116

Zip Code Prefixes of Destination	Zone
150–153	4
200–209	3
508–516	6

EXAMPLE

What will it cost to mail an 8.9-pound priority mail package from a post office that has a Zip code beginning with 100 to a location with a Zip code beginning with (a) 101   (b) 510?

Solution
(a) Since the Zip code begins with 100, the 101 Zip code is local. 8.9 lb is more than 8 lb but less than 9 lb, so find the cost by going down the weight column in Table 5.5 to 9 and then across to the "local" column. The cost is $6.10.
(b) From Table 5.6 we find that the 510 Zip code is in zone 6. The package costs $9.05.

EXERCISE 5.4.2

1. How much does it cost to mail a 9.3-ounce letter first class?

2. Airmail to Europe costs 50¢ for the first half-ounce, 45¢ for the second half-ounce, and 39¢ for each additional half-ounce. How much does it cost to mail a 5.25-ounce letter from New York to Paris by air mail?

3. The foreign surface rate is 50¢ for the first half-ounce, 20¢ for the second half-ounce, and 25¢ for each additional ounce. How much does it cost to send the letter in problem 2 to Paris by surface mail?

4. Fill out the following table. Assume that the packages are being mailed from a post office with a Zip code beginning with 100 to locations in the indicated Zip code area.

	Zip Code Prefix	Zone	Weight (lb)	Cost ($)
(a)	152	——	5.3	——
(b)	208	——	1.4	——
(c)	510	——	10.8	——
(d)			Total cost	——

Consumer Finance

The following information comes from a Sunday newspaper advertisement.

London broil $1.70 per pound
Cheese $1.79 half a pound
Ginger ale $1.19 each (+5¢ deposit each)
Eggplant $.69 per lb
Lite ground round $3.99 per pound
Tomatoes $.99 per basket
Belgian endive $1.89 per pound
Milk $1.19 per half gallon
Orange juice $1.19 per can
Butter $2.39 per pound
Cookies $1.00 per pound

EXERCISE 5.4.3

1. Use the advertisement to determine how much money you would have left from a $50 bill after you purchased the following items.

$3\frac{3}{4}$ pound of London broil
1.75 lb cheese
2 bottles ginger ale
1.4 lb eggplant
1.82 lb ground round
1 basket tomatoes
1.7 lb Belgian endive
$\frac{1}{2}$ gallon milk
6 cans orange juice
$\frac{1}{2}$ lb butter
2 lb cookies

Best Buys

The following examples and exercises are based on information in newspaper advertisements.

EXAMPLE

One brand of mushrooms sells for 90¢ for a 12-ounce can, and another is 79¢ for a 10-ounce can. Which is the best buy?

Solution   90¢ for 12 ounces is equivalent to $\frac{90}{12} = 7.5$ or 7.5¢ per ounce. 79¢ for 10 ounces is equivalent to $\frac{79}{10}$ or 7.9¢ per ounce. The best buy is the 12-ounce can for 90¢.

Now that many delicatessens have started to use digital scales, we need to know how to convert between fractions and decimals. For example, if you ask for $\frac{1}{2}$ lb of ham, what should the scale read? $\frac{1}{2} = 0.5$, so the scale should read about 0.5 lb.

EXAMPLE

Is 3 ounces of salami the same weight as 0.25 pound?

Solution   16 ounces (oz) = 1 pound (lb), so 3 oz = $\frac{3}{16}$ lb = 0.1875 lb = 0.19 lb. The answer is no. 0.25 lb is more than 3 oz.

EXERCISE 5.4.4

1. A certain oil treatment product comes in two can sizes. The regular 12-fl oz can costs 65¢, and the large economy size, 20 fl oz, sells for $1. Is the economy size the better buy?

2. Which of the following is the best buy? (There are 16 oz in 1 lb and 128 fluid ounces in 1 gallon.) The symbol @ means "at."
   (a) Swiss cheese sliced @ 99¢ for 5 oz or a piece of Swiss cheese @ $1.59 for $\frac{1}{2}$ pound.
   (b) Orange juice @ $1.59 per half gallon or $1.19 for 12 fl oz.
   (c) Mushrooms @ 89¢ for 12 oz or $1.24/lb.

3. In this problem, we give you the digital weight and you decide if that is what you asked for. If the weight is not what you want, is it too high ($+$) or too low ($-$)? Try to do this in your head!

	You Want	Scale Reading	OK? (yes if OK, $+$ if too high, $-$ if too low)
(a)	2 oz	0.125	yes $\left(\frac{2}{16} = 0.125\right)$
(b)	4 oz	0.250	
(c)	5 oz	0.30	
(d)	6 oz	0.4	
(e)	7 oz	0.5	
(f)	8 oz	0.5	
(g)	9 oz	0.75	
(h)	10 oz	0.60	

Table 5.7

DISTRIBUTION OF THE $768.88 TAX	
Landfill	$ 14.61
Roads	$157.62
Education	$442.87
General administration	$ 63.82
Fire protection	$ 12.30
County tax	$ 31.52
Miscellaneous	$ 46.14

Property Tax

A tax bill in the country had the following information:

$$\text{Rate: } 14.00 \text{ mils } (1 \text{ mil} = 1/1000)$$

$$\text{Valuation: } \$54,920 \qquad \text{Amount of Tax: } \$768.88$$

From this information we see that 14.00 mils times $54,920 is $0.014 \times 54,920 = 768.88$, which is the tax.

EXAMPLE

If the tax is reduced by 24%, what is the new tax for landfill?

Solution   Landfill:

$$14.61 - 24\% \text{ of } 14.61 = 14.61 - 0.24 \times 14.61$$

$$= 14.61 - 3.51 = 11.10$$

The new tax for landfill is $11.10.
        You could also say $100\% - 24\% = 76\%$, and 76% of $14.61 = $11.10.

EXERCISE 5.4.5

1.   Find the amount if each of the other tax items in Table 5.7 is reduced by 24%.

Income Tax

Table 5.8 is an excerpt from the 1991 Federal Tax Rate Schedules. The amounts in the first column are taxable income—the total earnings minus allowed deductions.
        To determine what a single taxpayer with taxable income of $30,000 would pay, we look in the appropriate schedule (Schedule X). $30,000 is over $20,350 but not over $49,300, so the tax is $3052.50 + 28% of the amount over $20,350.
        Our taxpayer has $30,000 - $20,350 = $9,650 over $20,350. Thus the tax is

$$\$3,052.50 + 28\% \times \$9,650 = \$5754.50$$

Married persons filing a joint return for $30,000 taxable income would pay 15% of $30,000, or $4500.

Table 5.8

SCHEDULE X SINGLE TAXPAYER			
Over	But not over	Tax	of the amount over
$0	$20,350	15%	$0
$20,350	$49,300	$3,052.50 + 28%	$20,350
$49,300	...	$11,158.50 + 31%	$49,300

SCHEDULE Y-1 MARRIED FILING JOINT RETURNS			
Over	But not over	Tax	of the amount over
$0	$34,000	15%	$0
$34,000	$82,150	$5,100.00 + 28%	$34,000
$82,150	...	$18,582.00 + 31%	$82,150

SCHEDULE Y-2 MARRIED FILING SEPARATE RETURNS			
Over	But not over	Tax	of the amount over
$0	$17,000	15%	$0
$17,000	$41,075	$2,550.00 + 28%	$17,000
$41,075	...	$9,291.00 + 31%	$41,075

SCHEDULE Z HEAD OF HOUSEHOLD			
Over	But not over	Tax	of the amount over
$0	$27,300	15%	$0
$27,300	$70,450	$4,095.00 + 28%	$27,300
$70,450	...	$16,138.00 + 31%	$70,450

If a married person files a separate return, a taxable income of $30,000 would require a tax of

$$\$2550.00 + 28\%(\$30,000 - \$17,000) = \$2550.00 + \$3640.00$$

$$= \$6190.00$$

An unmarried head of household (Schedule Z) would pay

$$\$4095.00 + 28\% \times (\$30,000 - \$27,300) = \$4095.00 + \$756.00 = \$4851.00$$

EXERCISE 5.4.6

Complete the tax table:

	Income	Single	Married		Head of Household
			Joint	Separate	
1.	$22,500				
2.	$36,000				
3.	$52,000				

5.5 MISCELLANEOUS PROBLEMS

Money Problems

EXERCISE 5.5.1

Certain gold coins come in various sizes: 1 ounce, $\frac{1}{2}$ ounce, $\frac{1}{4}$ ounce, and $\frac{1}{10}$ ounce. If gold is $308 per ounce,

1. How much would it cost to buy five $\frac{1}{4}$-ounce coins?

2. How many $\frac{1}{4}$-ounce coins can you buy for $500?

3. Could you buy any $\frac{1}{10}$-ounce coins for what is left of the $500?

Time Problems

Table 5.9 is a train schedule for a run between New Haven, Connecticut, and Springfield, Massachusetts. "p" means P.M.

Table 5.9

New Haven Dp	12 54 p
North Haven	1 08 p
Wallingford	1 16 p
Meriden	1 25 p
Berlin	1 36 p
Hartford	1 49 p
Windsor	1 58 p
Windsor Locks	2 04 p
Enfield	2 14 p
Springfield	2 29 p

EXAMPLE

How long does the train take from Hartford to Enfield?

Solution   The train leaves Hartford at 1:49 P.M. and arrives at Enfield at 2:14 P.M. This is a subtraction problem and could be solved by using fractions:

$$2\tfrac{14}{60} - 1\tfrac{49}{60} = 1\tfrac{74}{60} - 1\tfrac{49}{60} = \tfrac{25}{60} \qquad \tfrac{25}{60} \text{ of an hour is 25 minutes}$$

You could also figure out how many minutes are left until 2 o'clock (11 minutes) when the train leaves Hartford. Then it takes 14 minutes more to get to Enfield. So the whole trip takes 25 minutes.

Be careful with using time in decimal notation. Remember that 1 minute is $\frac{1}{60}$ of an hour.

EXAMPLE

Convert 0.1 hr to minutes.

Solution $0.1 \text{ hr} = \tfrac{1}{10} \text{ hr} = \tfrac{6}{60} \text{ hr} = 6 \text{ min.}$

EXERCISE 5.5.2

1. From Table 5.9, find the time it takes to go by train from New Haven to (a) Meriden, (b) Hartford, (c) Springfield.

2. On a particular day in January, the sun rises at 6:56 A.M. and sets at 5:11 P.M. In July at the same place, the sun rises at 5:03 A.M. and sets at 7:05 P.M. How long is the day in (a) January and (b) July?

3. Translate the following times in decimal form into hours and minutes.
   (a) 0.4 hr
   (b) 4.8 hr
   (c) 3.5 hr
   (d) 5.25 hr
   (e) 7.75 hr

   The 24-hour clock is used in scientific work, in Europe, and in the U.S. Armed Forces. The day begins at midnight as 0000 and ends as 2400. (For example, 1705 is the same as 1705 − 1200 = 5:05 P.M.)

Standard Time		
Eastern Standard Time	1200	12 noon
Oslo	1800	6 P.M.
Karachi	2200	10 P.M.
London	1700	5 P.M.

EXAMPLE

If it is 5 A.M. in London, what time is it in (a) New York   (b) Karachi?

   Solution   When it is 12:00 noon in New York it is 5 P.M. in London, which is therefore 5 hours ahead.
   (a) 5 A.M. minus 5 hours is 0. It is midnight in New York.
   (b) 2200 − 1700 = 500. Karachi is 5 hours ahead of London. 5 A.M. plus 5 hours is 10 A.M. It is 10 in the morning in Karachi.

EXERCISE 5.5.3

1. If it is 5 P.M. in Oslo, what time is it in (a) New York   (b) Karachi?
   Think of the time zones as being on a circular number line going around the world. −12 and 12 meet at the international date line, which passes through Fiji in the Pacific Ocean and between Alaska and Asia.

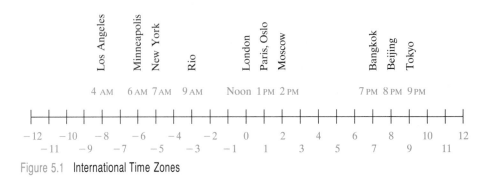

Figure 5.1   International Time Zones

EXAMPLE

Ingrid leaves Copenhagen at 1625 local time. She is due to arrive in New York City at 1855 New York time. How long is the flight?

> **Solution** Copenhagen is in the same time zone as Oslo and Paris, 6 hours ahead of New York. The time difference between 1855 and 1625 is 2 hours and 30 minutes. Add the 6-hour time difference. The trip should take 8 hours and 30 minutes.

EXERCISE 5.5.4

1. All times are given as local time.
    (a) Brita is leaving New York City at 4:50 P.M. for Los Angeles. She will arrive at 9:20 P.M. local time. How long is the flight?
    (b) Ed will leave from Los Angeles for New York City at 11:50 A.M. If his flight takes as long as Brita's, at what time will he arrive in New York?

2. Several cousins are flying from San Francisco to Minneapolis taking different airlines. Some leave at 3:45 P.M. and arrive at 9:08 P.M., while others leave at 1320 and arrive at 2128. Which party has the shortest flying time?

> Daylight savings time starts the first Sunday in April and ends the last Sunday in October in the United States. Europe is generally on daylight savings time (DST) from the last Sunday in March to the last Sunday in September.

EXAMPLE

It is 7 A.M. in Boston. What time is it in Paris on each of the following dates?
(a) January 1
(b) the week between the last Sunday in March and the first Sunday in April
(c) July 1
(d) October 1

> **Solution** See Figure 5.2. (a) and (c) Both countries are on the same system. According to Figure 5.1, Paris is 6 hours ahead of New York. So 7 A.M. in New York means 1 P.M. in Paris.
> (b) The United States is on standard time and France is on daylight savings time, so it is 2 P.M. in Paris.
> (d) The United States is on daylight savings time and France is on standard time, so it's 12 noon in Paris.

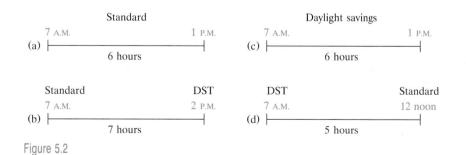

Figure 5.2

EXERCISE 5.5.5

1. Gisela leaves Minneapolis at 8:20 A.M. on a direct flight to Paris. The flight takes 8 hours. At what time does she expect to arrive in Paris, if she flies (a) in May? (b) on October 15?

2. At what time should Alex in Rio de Janeiro call his cousin in Moscow if he wants to reach him at 10 o'clock in the morning?

SUMMARY

Formula

Straight-line method for depreciation:

$$\text{Depreciation} = \frac{\text{cost} - \text{salvage value}}{\text{estimated useful life}}$$

Strategy for solving word problems:

1. Read the problem through.
2. Try to estimate what the answer will be.
3. Decide whether to add, subtract, multiply, or divide or a combination of several operations.
4. Get an answer.
5. Check whether the answer makes sense.

Rule

Percent added or subtracted: Let a stand for any number: To find the original number, if $a\%$ is already added to the number, divide the number by $(100 + a)\%$, if $a\%$ is subtracted from the number, divide the number by $(100 - a)\%$

VOCABULARY

Average: The sum of a group of numbers divided by the number of numbers.

Book value: Cost minus depreciation.

Depreciation: How much the value of an item is lowered after a certain time.

Mil: 1/1000

Salvage value: The value of an item at the end of its useful life.

Scrap value: Same as salvage value.

Time zone: The earth is divided into 24 time zones, each differing from the next by 1 hour.

CHECK LIST

Check the box for each topic you feel you have mastered. If you are unsure, go back and review.

☐ Percent added or subtracted
☐ Averages
☐ Hotel problems

☐ Stock market problems
☐ Depreciation problems
☐ Bank account problems
☐ Postage problems
☐ Consumer finance problems
☐ Best buys problems
☐ Tax problems
☐ Time problems

REVIEW EXERCISES

1. In a certain year your income is $47,000 and you are filing as head of household. Your federal tax is $4095.00 + 28% of the amount over $27,300, and your state tax is 5% of your federal tax.
   (a) What is your federal tax?
   (b) What is your state tax?
   (c) How much must you pay for federal and state taxes together?

2. The price of a windsurfer with a 20% end-of-season markdown was $840. What was the price before the markdown?

3. The following list gives the circulation figures (number of copies distributed) for seven magazines.

Modern Maturity	22,103,887
Reader's Digest	16,264,547
National Geographic	10,189,703
Ebony	1,810,668
Newsweek	3,211,958
Business Week	894,053
PC	786,041

   What is their average circulation?

4. The following table gives the average teacher's pay and average number of pupils per teacher for 1989 for six states.

States	Average Pay	Pupils per Teacher
Maine	$26,881	14.1
New Hampshire	28,939	16.2
Vermont	28,798	13.8
Massachusetts	34,225	14.0
Rhode Island	36,057	14.5
Connecticut	40,496	13.1

   (a) What was the average teacher's pay for these states?
   (b) What was the average number of students per teacher?

5. A summer clearance sale had the following prices.

	Regular Price	Sale Price
Grills	$199.99	$119.99
Tennis racquets	89.99	67.49
Beach towels	8.99	4.99
Cordless telephones	79.99	57.99
Air conditioners	319.99	229.99

   (a) Find the percent markdown for each item.
   (b) Find the average percent markdown for the five articles.

6.

	Foreign Currency per U.S. dollar	
	Last Week	Last Year
Japanese yen	124.38	136.41
German mark	1.4583	1.7470
Canadian dollar	1.1889	1.1470
British pound	1.9510	1.6943

   Given the exchange rates shown in the table, find
   (a) the percentage gain or loss for each listed currency and
   (b) the average gain or loss.
   (The table tells you that, for example, last week $1 U.S. was worth 124.38 Japanese yen and last year it was worth 136.71 yen.)

7. Given the following information, find the new balance.

	Starting Balance	Deposits	Checks
(a)	$385.00	$125, $230	$45, $92, $225, $308
(b)	$ − 81	$165, $82, $35	$70, $100, $15, $16

8. It costs 29¢ for the first ounce and 23¢ for each additional ounce or part of an ounce to mail a domestic letter. How much does it cost to mail a 3.6-ounce letter?

9.

Eastern Standard time	12 noon
Oslo	6 P.M.
Karachi	10 P.M.
London	5 P.M.

(a) Philadelphia
(b) Karachi
(c) Paris
(d) Cambridge (England)

If it is 11 P.M. in Oslo, what time is it in each of the following cities?

10. Which of the following is the best buy—3 pounds of peanut butter at $3.69 or $\frac{3}{4}$ pound for 89 cents?

READINESS CHECK

Solve the problems to satisfy yourself that you have mastered Chapter 5.

1. Ed paid $23.00 for a taxi ride, with 15% tip included. How much did the trip itself cost?
2. The office manager got a $325.00 adding machine. It was to be depreciated in 5 years with a scrap value of $25.00. What was the annual depreciation?
3. Joshua's bank account was $350.38 at the end of last month. This month he deposited $15.45, $478.10 twice, and $35.95. He also wrote checks for $120.60, $45.98, $13.95, and $285.00. What is his bank balance now?
4. How much does it cost to mail a 4.2-oz letter first class if it costs 29¢ for the first ounce and 23¢ for each additional ounce?
5. The price of a grill is $120 after a markdown of 20%. What was the price before markdown?
6. The train leaves Irvington at 7:22 A.M. and arrives at Grand Central Station at 8:17 A.M. How long is the trip?
7. If the price of gold is $340 per ounce, how many $\frac{1}{4}$-oz coins can you buy for $600?
8. One U.S. dollar was worth 4630 Brazilian cruzeiros in 1992 and 334 cruzeiros a year earlier. Find the percent increase in the value of the cruzeiro.
9. At Weight Watchers, three people each lost 2 pounds, two people each gained half a pound, and five people lost $2\frac{1}{2}$ pounds each. What was the average weight loss?
10. Elsie, who had a yearly income of $36,000, married Mark, who had the same income. How much more or less does the couple have to pay in federal tax than before if they file a joint return?

Single: $3,052.50 + 28% of the amount over $20,350
Married: $5,100.00 + 28% "   "   "   " $34,000

MEASUREMENTS

Which wrench do I use?

The decimal or metric system of measurement based on multiples of 10 was first proposed in France in 1640, but it took the upheaval of the French Revolution to achieve its acceptance in 1791. The United States inherited the measures we commonly use today from the British, who had their own system. The Trade Act of 1988 called for the federal government to adopt metric specifications by December 31, 1992. The metric system has evolved into the International System (abbreviated SI for its French name Système International d'Unités), which is used throughout the world for scientific measurements.

Thomas Jefferson introduced the dollar as a money unit and at the same time argued that other units should also be changed. Jefferson noted:

> The most *easy ratio* of multiplication and division is that by 10. Everyone knows the facility of Decimal Arithmetic. Everyone remembers, that when learning money arithmetic he used to be puzzled with adding the farthings, taking out the fours and carrying them on; adding the pence, taking out the twelves and carrying them on; adding the shilling, taking out the twenties and carrying them on. But when he came to pounds where he had only the tens to carry forward, it was free and easy from error. The bulk of mankind are school boys through life. And even mathematical heads feel the relief of an easier substituted for a more difficult process. ... If we adopt the Dollar for our unit we should strike 4 coins, one of gold, two of silver, and one of copper.

> 1. A Golden piece equal in value to 10 Dollars
> 2. The unit, or Dollar itself of silver
> 3. The 10th of a Dollar of silver also
> 4. The hundredth of a Dollar, of copper

Jefferson's proposal for uniform coinage was adopted on July 6, 1785. But his attempts to change weights and measures to the decimal system failed.

Many attempts since then have also failed. Although 90% of the people of the world today use the metric system, Americans have stubbornly stuck with the customary system, a newer name for the Old English system. If you work on a car manufactured in any western country but the United States or Great Britain, you will need a second set of wrenches!

This chapter deals with measurements in both the metric and the customary systems. Scientific notation is also included because of its importance in scientific measurements.

6.1 POWERS OF 10

Recall that exponential notation is a short form of multiplication: $10^3 = 10 \times 10 \times 10 = 1000$, for instance.

In Chapter 1 we discussed place values and showed some of the place values in whole numbers:

$$1,000,000 \qquad 100,000 \qquad 10,000 \qquad 1000 \qquad 100 \qquad 10 \qquad 1$$

Instead of writing the numbers out we could have used exponential notation:

$$10^6 \qquad 10^5 \qquad 10^4 \qquad 10^3 \qquad 10^2 \qquad 10^1 \qquad 10^?$$

How do we write 1 in exponential notation with 10 as a base?

Look at the pattern for the exponents in the place values:

$$6 \rightarrow 5 \rightarrow 4 \rightarrow 3 \rightarrow 2 \rightarrow 1 \rightarrow ?$$

If this pattern is followed, the next exponent should be 0.

$$\text{Is } 10^0 \text{ equal to 1?}$$

When we move from left to right among the place values, we divide by 10 each time:

$$10^6 \div 10 = 10^5 \qquad \text{and} \qquad 10^5 \div 10 = 10^4$$

The exponent decreases by 1 every time we divide by 10. Remember that $10 = 10^1$, so $10^5 \div 10 = 10^5 \div 10^1 = 10^{5-1} = 10^4$.

Therefore,

$$10^1 \div 10 = 10^{1-1} = 10^0$$

It is not only 10^0 that equals 1. For example, $5^2 \div 5^2 = 25 \div 25 = 1$. But $5^2 \div 5^2$ is also $5^{2-2} = 5^0$. Therefore, $5^0 = 1$.

DEFINITION

Any nonzero number to the 0th power is equal to 1.

0^0 is not defined.

In Chapter 3 we discussed place values and found the following values to the right of the units place:

$$0.1 \qquad 0.01 \qquad 0.001$$
$$\text{tenth} \quad \text{hundredth} \quad \text{thousandth}$$

If we continue the pattern of exponential notation, we have

$$10^{-1} \qquad 10^{-2} \qquad 10^{-3}$$

Thus,

$$10^{-1} = 0.1, \qquad 10^{-2} = 0.01, \qquad 10^{-3} = 0.001$$

We also note that

$$0.1 = 1 \div 10, \qquad 0.01 = 1 \div 100, \qquad 0.001 = 1 \div 1000$$

It follows that

$$10^{-1} = \frac{1}{10^1} = \frac{1}{10}$$

$$10^{-2} = \frac{1}{10^2} = \frac{1}{100}$$

$$10^{-3} = \frac{1}{10^3} = \frac{1}{1000}$$

EXAMPLE

(a) Write 10^{-7} in fractional and decimal form.
(b) Write 0.0001 in fractional and exponential form.

Solution

(a)
$$10^{-7} = \frac{1}{10^7} = \frac{1}{10,000,000} = 0.0000001$$

(b)
$$0.0001 = \frac{1}{10,000} = \frac{1}{10^4} = 10^{-4}$$

EXERCISE 6.1.1

1. Write as a fraction and as a decimal.
    (a) 10^{-1}
    (b) 10^{-2}
    (c) 10^{-3}
    (d) 10^{-4}

2. Write in exponential notation and as a decimal.
    (a) $1/100,000$
    (b) $1/1,000,000$

3. Write in exponential notation and as a fraction.
    (a) 0.0000000001
    (b) 0.00000000001

Multiplication and Division by Powers of 10

When a number in decimal notation is multiplied by 10, the decimal point is moved one step to the right. When multiplying by 100, the point is moved two steps to the right, and so on.

Instead of writing 10, 100, or 1000 we can use exponential notation.

EXAMPLE

Multiply: (a) 0.053×10^3 (b) 25×10^{-4}

Solution
(a) Move the decimal point three places to the right. $0.053 \times 10^3 = 53$.

(b) $10^{-4} = \dfrac{1}{10^4} = 1 \div 10^4$.

Move the decimal point 4 places to the left. $25 \times 10^{-4} = 0.0025$.

EXAMPLE

Divide: (a) $549 \div 10^3$ (b) $0.3 \div 10^{-2}$

Solution
(a) Move the decimal point three places to the left. $549 \div 10^3 = 0.549$.

(b) $\dfrac{1}{10^{-2}} = 10^2$ so $0.3 \div 10^{-2} = 0.3 \times 10^2 = 30$.

Move the decimal point two places to the right.

RULE

To multiply by	Move the decimal point
10^1	1 place to the right
10^2	2 places to the right
10^3	3 places to the right
10^{-1}	1 place to the left
10^{-2}	2 places to the left

To divide by	Move the decimal point
10^1	1 place to the left
10^2	2 places to the left
10^3	3 places to the left
10^{-1}	1 place to the right
10^{-2}	2 places to the right

EXERCISE 6.1.2

Multiply.

1. 0.23×10

2. 1.456×10^3

3. 24.95×10^4

4. 0.00024×10^5

5. 230×10^{-2}

6. 0.03×10^{-1}

Divide.

7. $450 \div 10^2$

8. $12.2 \div 10^3$

9. $0.059 \div 10$

10. $0.083 \div 10^{-2}$

11. $2.56 \div 10^{-1}$

12. $1247 \div 10^{-3}$

6.2 SCIENTIFIC NOTATION

Scientists often use very large or very small numbers. Avogadro's number, which tells how many molecules there are in certain weights of a substance (18 grams of water, for example) is 602,300,000,000,000,000,000,000. The mass of one electron is 0.0000000000000000000000000009108 g. Instead of writing these long numbers, we can use scientific notation.

A number in scientific notation always has one nonzero whole-number digit. 1.9×10^2 is the number 190 written in scientific notation. If we had written 0.19×10^3, that would also mean 190 (0.19×1000), but 0.19 is not between 1 and 10 so 0.19 is *not* scientific notation. 2.45×10^{25} is in scientific notation, but 24.5×10^6 is not. In scientific notation, Avogadro's number is 6.023×10^{23} and the mass of an electron is 9.108×10^{-28} g.

DEFINITION

Scientific Notation

A number in scientific notation is written as a product of a number between 1 and 10 and a power of 10.

EXAMPLE

Write in standard notation. (Write without a power of 10 as a factor.)
(a) 2×10^2 (b) 3.4×10^{-3}

Solution
(a) $2 \times 10^2 = 200$ (Move the decimal point 2 places to the right.)
(b) $3.4 \times 10^{-3} = 0.0034$ (Move the decimal point 3 places to the left.)

EXAMPLE

Write in scientific notation: (a) 125.6   (b) 0.0039

Solution
(a) We want the decimal after 1, so we divide by 100. To compensate for the division, we multiply by 100 (or 10^2).

We get

$$125.6 = 1.256 \times 10^2$$

Shortcut: Count how many places you move the decimal point. That is the exponent. The exponent is positive if you move to the left. The exponent is negative if you move to the right.
(b) $0.0039 = 3.9 \times 10^{-3}$

EXAMPLE

Write 32.5×10^3 in scientific notation.

Solution $32.5 = 3.25 \times 10$

$$32.5 \times 10^3 = 3.25 \times 10 \times 1000 = 3.25 \times 10,000 = 3.25 \times 10^4$$

$32.5 \times 10^3 = 3.25 \times 10^4$ in scientific notation.

EXAMPLE

The diameter of the planet Mercury is 3100 miles, and its average distance to the sun is 36,000,000 miles. Write these distances in scientific notation.

Solution   3100 miles $= 3.1 \times 10^3$ miles and 36,000,000 miles $= 3.6 \times 10^7$ miles.

EXERCISE 6.2.1

1. Express in standard notation.
   (a) 1.34×10^5
   (b) 4.78×10^3
   (c) 8.2×10^2
   (d) 7.352×10^6
   (e) 4.6×10^{-3}
   (f) 1.8×10^{-5}
   (g) 2.95×10^{-9}
   (h) 9.7×10^{-11}

2. Express in scientific notation.
   (a) 4680
   (b) 839
   (c) 34
   (d) 0.00071
   (e) 0.000000196
   (f) 0.0000000023
   (g) 0.00000000000000000168
   (h) 279140000

3. Write in scientific notation.
   (a) 0.35×10^{-2}
   (b) 12.6×10^3
   (c) 0.03×10^4
   (d) 35.8×10^{-3}
   (e) 198.2×10^{-5}
   (f) 57.5×10^{-4}
   (g) 0.006×10^8
   (h) 198.53×10^6

4. A proton (part of an atom) weighs 1.672×10^{-24} g. How many grams is that in standard notation?

5. A carbon atom weighs 5.3×10^{-23} g. Translate that into decimal notation.

6. The average distance between the earth and the sun is 150,000,000 kilometers (km). Write that in scientific notation.

7. In the Mosquito Range west of Colorado Springs, we walked through approximately 1400 million years of geological history. What would that be in scientific notation?

8. We brought back from Australia fossilized leaves 250,000,000 to 300,000,000 years old. What was the range of their ages in scientific notation?

6.3 THE METRIC SYSTEM

Metric measurements are all around us; many medicines, foods, and beverages are measured in metric units. Some distances are also expressed in metric units. In our neighbors to the north and south, Canada and Mexico, metric units are the rule (although they often translate them for American tourists).

Many people are afraid of the metric system. Why learn something new about measurements when we have grown up with inches, miles, and pounds? The reason people are afraid of the metric system often has to do with the assumption that they would always have to translate any metric measure into a kind of measure they are already used to using.

It is hard to compare the metric and customary systems. Even conversions from one customary unit to another are difficult. There is no consistency among the relationships: 3 feet in a yard, 12 inches in a foot, 1760 yards in a mile. Conversions between the systems can also be hard because there are so many numbers to remember. However, the metric system of measurement is easy to learn. Look at this illustration of the metric system to see how easy it is:

1 liter of water weighs 1 kilogram (1 kg). It fills a box 10 centimeters (cm) long, 10 cm wide, and 10 cm high (volume 1000 cubic centimeters).

So 3 liters of water weighs 3 kg and fills a 3000-cm^3 container.

15 liters of water weighs 15 kg and fills a 15,000-cm^3 container.

62.4 liters of water weighs 62.4 kg and fills a 62,400-cm^3 container.

Compare this metric relationship above with its counterpart in the customary system.

1 quart of water weighs 2.08 pounds and fills a 57.75 cubic inch container.

62.4 qt of water weighs 62.4×2.08, or 129.792 lb, and fills a 62.4×57.75 or 3603.6-cu in. container.

Ever since Thomas Jefferson made the decision, we have used metric money. However, except for scientific measurements, we still use the customary system for almost everything else.

The metric system is a decimal system. To change from one measure to another you multiply or divide by powers of 10.

1 meter = 10 decimeters = 100 centimeters = 1000 millimeters

To get familiar with the metric system, take a metric ruler or tape measure and look at it:

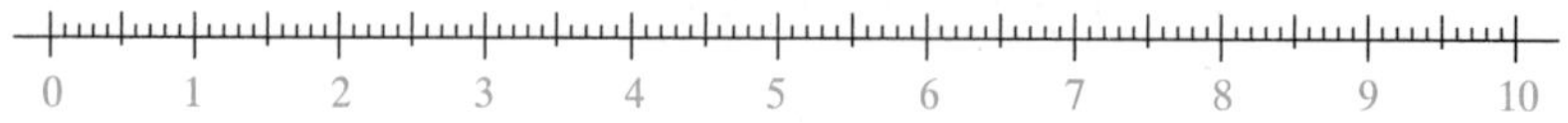

Figure 6.1

The distance from 0 to 10 equals 1 decimeter, from 0 to 1, 1 to 2, and so on, equals 1 centimeter, and the smallest distances, tenths of a centimeter, are millimeters. The basic unit of length is the meter. Other length units are named by attaching a prefix to "meter" that indicates the power of 10. The prefixes from thousand to thousandth are listed in the box below. The same prefixes are used for other types of metric measures also.

For example,

dm = decimeter,   cm = centimeter,   mm = millimeter

Find the following measures on your body:

1 mm = a wrinkle?

1 cm = the width of a nail?

5 cm = the length of your thumb?

1 dm = the width of your hand?

1 m = distance from the tip of your middle finger when you stretch
out your arm to the tip of your nose?

How long is a mosquito?

What is the distance from the floor to the door knob?

How tall are you?

When we measure something, we compare its size (length, weight, volume) to a certain standard measure. The standard for length is the meter. The official meter prototypes are metal rods in Paris and Washington (National Institute of Standards and Technology). The basic unit of weight (or mass) is the kilogram, and the basic unit of volume is the liter. The standard kilogram (a metal cylinder) is kept in Paris and a copy is in Washington. A liter used to be defined as the volume of one kilogram of water at 4° Centigrade.

SOME METRIC PREFIXES

kilo (k) = thousand	10^3
hecto (h) = hundred	10^2
deka (da) = ten	10
deci (d) = tenth	10^{-1}
centi (c) = hundredth	10^{-2}
milli (m) = thousandth	10^{-3}

Table 6.1 shows the relationships between the different length measures and the meter.

Table 6.1

1 kilometer (km) = 1000 m
1 hectometer (hm) = 100 m
1 decameter (dam) = 10 m
1 meter (m) = 1 m
1 decimeter (dm) = 0.1 m
1 centimeter = 0.01 m
1 millimeter = 0.001 m

To convert between units, we multiply or divide by powers of 10. For example, 5 km = 50 hm = 500 dam = 5000 m = 50,000 dm = 500,000 cm = 5,000,000 mm. When we convert from a larger unit to a smaller unit, we multiply. When we convert from a smaller unit to a larger unit, we divide.

EXAMPLE

Convert to meters:
(a) 3.2 km   (b) 10 dm   (c) 15 cm   (d) 300 mm

Solution
(a) To multiply by 1000, we move the decimal point three places to the right. 3.2 km = 3200 m.
(b) Multiply by 0.1 (or divide by 10). That is, move the decimal point one place to the left. 10 dm = 1 m.
(c) Divide by 100; move the decimal point two places to the left. 15 cm = 0.15 m.
(d) Divide by 1000; move the decimal point three places to the left. 300 mm = 0.3 m.

To convert between any length units within the metric system, it is convenient to list the units in order of size:

km   hm   dam   m   dm   cm   mm

For example, convert 2.05 dm to mm. Place the ones digit (2) under the metric unit name (dm). Then write one digit in each place.

km	hm	dam	m	dm	cm	mm
				2.	0	5

From the illustration we see that we have to move the decimal point two places to the right to get millimeters as the new measuring unit. 2.05 dm = 205 mm.

EXAMPLE

Convert: (a) 13.2 dm to cm   (b) 0.5 m to mm   (c) 530 cm to m
(d) 30,600 dm to km

Solution

(a) km	hm	dam	m	dm	cm	mm
			1	3.	2	

Move the decimal one place to the right. 13.2 dm = 132 cm.

(b) km	hm	dam	m	dm	cm	mm
			0.	5		

Move the decimal point three places to the right. 0.5 m = 500 mm.

(c) km   hm   dam   m   dm   cm   mm
                              5   3   0

Move the decimal point two places to the left. 530 cm = 5.30 m.

(d) km   hm   dam   m   dm   cm   mm
     3   0   6   0   0

Move the decimal point four places to the left. 30,600 dm = 3.0600 km.

The units for weight (mass) and volume have the same prefixes as those for length, as shown in Table 6.2.

Table 6.2

1 kilogram (kg) = 1000 g	1 kiloliter (kL) = 1000 L
1 hectogram (hg) = 100 g	1 hectoliter (hL) = 100 L
1 decagram (dag) = 10 g	1 decaliter (daL) = 10 L
1 gram (g) = 1 g	1 liter (L) = 1 L
1 decigram (dg) = 0.1 g	1 deciliter (dL) = 0.1 L
1 centigram (cg) = 0.01 g	1 centiliter (cL) = 0.01 L
1 milligram (mg) = 0.001 g	1 milliliter (mL) = 0.001 L

Some of the units, in particular the decagram and decaliter, are rarely (if ever) used, but we list them anyway in order to make conversions clear.

EXAMPLE

Convert: (a) 3.5 hL to L   (b) 0.02 L to mL   (c) 500 g to kg
          (d) 40 mg to g

Solution

(a) kL   hL   daL   L   dL   cL   mL
            3.   5   0

3.5 hL = 350 L. We must attach the zero to get the place values in the correct position.

(b) kL   hL   daL   L   dL   cL   mL
                             0.   0   2
0.02 L = 20 mL.

(c) kg   hg   dag   g   dg   cg   mg
            5   0   0.

The decimal point is moved from gram to kilogram. 500 g = 0.500 kg or 0.5 kg. We usually drop the zeros at the end of the decimals.

(d) kg   hg   dag   g   dg   cg   mg
                                            4   0
40 mg = 0.04 g.

RULE

Changing Units in the Metric System

To convert to a smaller unit: *multiply*
To convert to a larger unit: *divide*

EXERCISE 6.3.1

1. Convert each metric measure to the smaller unit.
   (a) 5 m =   _____ cm
   (b) 3 g =   _____ cg
   (c) 10 L =   _____ cL
   (d) 25 m =   _____ cm
   (e) 0.4 g =   _____ cg
   (f) 0.05 L =   _____ cL
   (g) 0.00030 m =   _____ mm
   (h) 0.00053 g =   _____ mg

2. Convert each metric measure to the larger unit.
   (a) 200 mm =   _____ m
   (b) 300 mg =   _____ g
   (c) 4500 mL =   _____ L
   (d) 120 hg =   _____ kg
   (e) 13150 dm =   _____ km
   (f) 2000 g =   _____ kg
   (g) 3800 mL =   _____ L
   (h) 80 dag =   _____ kg

Area

Thus far we have discussed units of measure that are a multiple of 10 apart (i.e., 1 cm = 10 mm). A length needs one measurement; we can measure length with a tape measure. An area, however, needs two measurements; we have to measure two lengths and multiply them to get the area. For example, in a rectangle we have length times width; its area is expressed in square units. In fact, all areas are expressed in square units.

If we take a square that is 1 cm in length, its area is length times length or $1 \text{ cm} \times 1 \text{ cm} = 1 \text{ cm}^2$. To convert the area to mm^2, we have to multiply each factor by 10.

$$1 \text{ cm} \times 1 \text{ cm} = 10 \text{ mm} \times 10 \text{ mm} = 100 \text{ mm}^2$$

Thus $1 \text{ cm}^2 = 100 \text{ mm}^2$. From this we can see that the units of area are 100 apart. Now we have

1 square kilometer = 100 square hectometers

1 square hectometer = 100 square decameters

1 square decameter = 100 square meters

Therefore

1 square kilometer = 1,000,000 square meters

We also have

1 square meter = 100 square decimeters

1 square meter = 10,000 square centimeters

1 square meter = 1,000,000 square millimeters

EXAMPLE

Convert 3000 cm^2 to m^2.

> **Solution**   Since there are 100 square units in every adjacent larger square unit, we need *two* places for every square unit. For example, 1 m^2 = 100 dm^2 or

$$\left| \begin{array}{c} m^2 \\ 1 \end{array} \right| \begin{array}{cc} dm^2 \\ 0 \quad 0 \end{array} \left| \begin{array}{c} cm^2 \\ \quad \end{array} \right|$$

So for 3000 cm^2 we can use

$$\left| \begin{array}{c} m^2 \\ \quad \end{array} \right| \begin{array}{cc} dm^2 \\ 3 \quad 0 \end{array} \left| \begin{array}{cc} cm^2 \\ 0 \quad 0 \end{array} \right|$$

The decimal point has to be moved four places to the left. 3000 cm^2 = 0.3 m^2.

EXERCISE 6.3.2

1. Convert.
   (a) 500 cm^2 = _____ dm^2
   (b) 0.3 cm^2 = _____ mm^2
   (c) 163.78 dm^2 = _____ m^2
   (d) 0.289 dm^2 = _____ mm^2

2. Convert.
   (a) 0.314 m^2 = _____ dm^2
   (b) 40,000,000 m^2 = _____ km^2
   (c) 103,000 mm^2 = _____ dm^2
   (d) 0.0049 m^2 = _____ mm^2

Volume

When we multiply length times length times length, we get length^3 or cubic length. Here we need three measures to get volume or capacity. A cube with sides 1 cm long has a volume of 1 cm $\times$ 1 cm $\times$ 1 cm = 1 cm^3. Since 1 cm = 10 mm, we can conclude that 1 cm^3 equals 1000 mm^3. When we convert volumes in the metric system, we move the decimal point *three* places for each consecutive unit.

EXAMPLE

Convert: (a) 1 m^3 to cm^3 (b) 0.0045 dm^3 to cm^3

> **Solution**
> (a) To move from meters to centimeters, we move the decimal point two places to the right, to move from cubic meters to cubic centimeters, we move the decimal point two times three places, or six places to the right.
>
> $$1 \text{ m}^3 = 1,000,000 \text{ cm}^3 \ (10^6 \text{ cm}^3)$$
>
> (b) 0.0045 dm^3 = 4.5 cm^3. Move three places to the right.

EXERCISE 6.3.3

1. Convert.
    (a) 800 cm^3 =   _____ dm^3
    (b) 20 dm^3 =   _____ cm^3
    (c) 0.006 m^3 =   _____ dm^3
    (d) 1675 dm^3 =   _____ m^3

2. (a) 5000 mm^3 =   _____ cm^3
    (b) 0.5 cm^3 =   _____ mm^3
    (c) 0.005 dm^3 =   _____ mm^3
    (d) 3050 mm^3 =   _____ dm^3

Because the steps between these units are so large, it is often convenient to use another measure for volume. If you were to make a cube with a side of 1 dm, fill it with a substance such as sand, and pour the contents of the cube into a liter measure, you would find that 1 dm^3 = 1 liter. To measure liters, we need only one measure and all other measuring units are 10 apart. We know that 1 L = 1 dm^3. But 1 dm^3 also equals 1000 cm^3. We also know

$$1 \text{ L} = 10 \text{ dL} = 100 \text{ cL} = 1000 \text{ mL}$$

In other words,

$$1 \text{ L} = 1 \text{ dm}^3$$
$$1000 \text{ mL} = 1000 \text{ cm}^3$$
$$1 \text{ mL} = 1 \text{ cm}^3$$

In column form we have

m^3				dm^3				cm^3
kL	hL	daL	L		dL	cL		mL

Another name for cm^3 is cc (<u>c</u>ubic <u>c</u>entimeter). It might be helpful to know that 1 cc and 1 mL are the same.

EXAMPLE

Convert 5 L to cm^3.

   Solution   5 L = 5 dm^3 = 5000 cm^3 or   5 L = 5000 mL = 5000 cc

EXAMPLE

1 dL = ? cm^3

   Solution   1 dL = 100 mL = 100 cm^3

EXERCISE 6.3.4

1. Convert.
    (a) 4 L =   _____ dm^3
    (b) 50 mL =   _____ cm^3
    (c) 600 cL =   _____ cm^3
    (d) 5000 cm^3 =   _____ L

2. (a) 0.5 dL =   _____ cm^3
    (b) 0.03 dm^3 =   _____ dL
    (c) 0.004 m^3 =   _____ L
    (d) 0.005 L =   _____ cm^3

Applications

The following exercises are applications of the metric system. The micrometer and Ångström units which are used for very small length measures are explained as problems.

EXERCISE 6.3.5

1. A gasoline tank in a jeep has the form of a box with a base area of 10 dm^2 and a height of 4 dm. How many liters of gasoline does the tank hold?

2. A bathtub has a bottom area of 130 dm^2. How far up does the water go if you put 455 liters into the tub?

3. An ice hockey rink has a rectangular shape with a length of 60.8 meters and width of 29.5 meters. Find (a) the perimeter (the distance all around) of the rink (round to the nearest whole meter) and (b) the area (round to the nearest m^2).

4. The driveway to a house has the shape of a square with a side of 8 m adjacent to a rectangle with width 4.5 m and length 14 m. How much does it cost to pave the driveway if the price is $22.50 per square meter?

5. The dimension of a page in a certain book is 16.6 cm × 23.8 cm. There are 192 pages in the book.
   (a) What is the area of one page? (Round to a whole number.)
   (b) If all the pages are placed next to each other, what area do they cover? Give the answer in dm^2 rounded to a whole number.

6. Henrik is getting a wall-to-wall carpet in his room that measures 3.2 m by 4 m. The carpet costs $7 per square meter. What is the cost of the carpet?

7. The Ångström unit is used for very small length measures such as wavelengths of radiation. It has its name from its inventor, the Swedish scientist Anders Ångström. 1 Å = 1 × 10^{-10} m. How many centimeters are there in 50 Å?

8. The micrometer (abbreviated μm) is also used for small length measures, such as those measured with a microscope. There are 10^6 μm in 1 meter, so 1 μm = 10^{-6} m.
   (a) How many μm are there in 1 mm?
   (b) How many Å are there in 1 μm?

6.4 THE CUSTOMARY SYSTEM OF MEASUREMENT

Conversions among common units in the customary system of measurements are given in the box.

CONVERTING WITHIN THE CUSTOMARY SYSTEM		
Length	Weight	Volume
1 mile = 5280 feet	1 pound = 16 ounces	1 gallon = 4 quarts
1 yard = 3 feet		1 quart = 2 pints
1 foot = 12 inches		1 pint = 2 cups
		16 fluid ounces = 1 pint
		1 cup = 8 fluid ounces

EXAMPLE

What part of a quart is 1 cup?

> **Solution** Since there are 2 cups in 1 pint and 2 pints in 1 quart, there are $2 \times 2 = 4$ cups in 1 quart. 1 cup is 1/4 of a quart, or 1 cup $= \frac{1}{4}$ quart.

EXAMPLE

There is one-half fluid ounce in one tablespoon (1 Tbsp $= \frac{1}{2}$ fl oz). How many tablespoons are there in a cup?

> **Solution**   1 cup $= 8$ fl oz, so 1/8 cup $= 1$ fl oz and 1/16 cup $= 1/2$ fl oz. In turn, since 1 Tbsp $= 1/2$ fl oz, there are 16 tablespoons in a cup.

EXERCISE 6.4.1

1. The ceilings in your apartment are 9 feet high. The floor of a room you want to paint measures 12 ft by 16 ft. One gallon of paint is enough for about 450 square feet. How many gallons of paint will you need to paint all four walls and the ceiling?

2. A room is 24 ft wide and 30 ft long. (a) What is the area of the floor? (b) There must be an electric wall outlet every 6 feet. How many outlets must there be? (c) How many square yards of rug are needed for the room?

3. Table 6.3 is taken from a government publication.
   (a) No numbers were given where the asterisks are. Find these numbers.
   (b) How many teaspoonfuls are there in $1\frac{1}{2}$ tablespoons?
   (c) Your recipe calls for $1\frac{1}{3}$ cups and you have only tablespoons and teaspoons available. How do you measure the $1\frac{1}{3}$ cups?
   (d) 1 tablespoon of vegetable oil has 120 calories. How many calories are there in 1 cup of the oil?

4. In a recipe for crab sandwiches, a restaurant uses $37\frac{1}{2}$ lb of crabmeat, $12\frac{1}{2}$ lb of grated cheese, and 3 qt of mayonnaise. (a) How many oz of grated cheese are there in each sandwich, which contains 3 oz crabmeat? (b) How many fl oz of mayonnaise?

5. A case of lettuce contains about $12\frac{1}{2}$ lb of lettuce. How many 1/2-oz servings can we get from a case?

6. A recipe calls for $1\frac{1}{2}$ lb of peaches and 2/3 cup of apricot jam to make a tart for six people. If you are going to have a party for 50 people, (a) how many pounds of peaches do you need? (b) How many cups of apricot jam?

Table 6.3

EQUIVALENTS OF THE COMMON CAPACITY UNITS USED IN THE KITCHEN

Unit	Teaspoons	Tablespoons	Fluid Ounces	Cups	Pints	Quarts
1 teaspoon	1	1/3	1/6	*	*	*
1 tablespoon	3	1	1/2	1/16	1/32	*
1 fluid ounce	6	2	1	1/8	1/16	1/32
1 cup	48	16	8	1	1/2	1/4
1 pint	*	*	16	2	1	1/2

6.5 CONVERSIONS BETWEEN THE CUSTOMARY AND METRIC SYSTEMS

The English used to say

> A meter measures three foot three
> It's longer than a yard, you see.

To compare measures and values in the customary and metric systems, we need a few approximations:

1 meter is a little more than 1 yard. (1 m = 39.37 in.; 1 yd = 36 in.)

1 kilogram is a little more than 2 pounds. (1 kg = 2.2 lb)

1 liter is a little more than 1 quart. (1 L ≈ 34 fl oz; 1 qt = 32 fl oz)

Some handy conversions are given in the box.

CONVERTING BETWEEN THE CUSTOMARY AND THE METRIC SYSTEMS

Length	Weight	Volume
1 in. = 2.54 cm	1 lb = 454 g	1 qt = 0.946 L
1 km = 0.6 mi	1 kg = 2.2 lb	1 L = 1.06 qt
1 mi = 1.6 km	1 oz = 28 g	

EXAMPLE

(a) How many centimeters are there in 4 in.?
(b) How many inches are there in 25 cm?
(c) How many kilometers are there in 5 miles?

Solution
(a) 1 in. = 2.54 cm, so 4 in. = 4 × 2.54 cm = 10.16 cm
(b) 1 in. = 2.54 cm, so 1 cm = 1/2.54 in. = 0.3937 in. and
     25 cm = 25 × 0.3937 in. = 9.8425 in. = 9.8 in. (rounded)
(c) 1 mi = 1.6 km, so 5 mi = 5 × 1.6 km = 8 km

EXERCISE 6.5.1

1. Jane is 5 feet 5 inches tall. What is her height in centimeters?

2. Greg is 192 cm tall. How tall is he in the customary system?

3. Convert 3 tons to kg. There are 2000 lb in 1 ton.

4. Convert 50 cm to feet.

5. Convert 300 g to pounds.

6. Convert 1 cup to deciliters.

7. I need to drill a hole 4 mm in diameter; the drill bits in my hardware store come only in 1/16 in., 1/8 in., 3/16 in., and 1/4 in. Which one should I buy? (*Hint:* Change inches first into cm. Use the drill bit that is slightly smaller.)

8. Sandy has a 3/8-in. socket wrench. What would this be in millimeters?

9. On the highway from Normandy to Paris, the speed limit is 120 km/hr. What is the limit in miles per hour?

10. Pat weighs 93.2 kg. How much does he weigh in pounds?

11. Jacky wears blue jeans with a 28-in. waist and 34-in. length. What are these measures in centimeters?

12. In the Olympics there is a 5000-m ice skating relay. How far is this in feet? (1 mi = 5280 ft)

Best Buys

We all like a bargain. Newspaper ads often try to convince us that everything is a bargain on a particular day. With the help of a calculator and some knowledge of mathematics, we can readily check the prices.

EXAMPLE

Which is the best buy: four 8.5-fl oz cans of a fruit drink for $1.99 or a 2-liter bottle for the same price?

Solution Make a chart:

4×8.5 fl oz	1.99
2 L	1.99

To compare these prices, we have a choice: we can compare the price per volume, or we can compare the volume per dollar.

It is more common to use the price per unit measure. That is done in stores that use unit pricing. Let's find the price per ounce.

Suppose we buy 4 cans?

$$4 \times 8.5 \text{ fl oz} = 34 \text{ fl oz}$$

$$\$1.99/34 = \$0.0585/\text{oz} = 5.85 ¢/\text{fl oz}$$

Suppose we buy the 2-L bottle?

$$2 \text{ L} = 2 \times 1.06 \text{ qt} = 2 \times 1.06 \times 32 = 67.84 \text{ fl oz}$$

67.84 fl oz costs $1.99, so

$$\$1.99/67.84 = \$0.0293/\text{fl oz} = 2.93 ¢/\text{fl oz}$$

The best buy is the 2-liter bottle.

EXERCISE 6.5.2

Which is the best buy?

1. Six 12-fl oz cans of Coke for $1.89 or a 2-L bottle of Coke for $1.09?

2. Gasoline at 34¢/L or at $1.36/gallon?

3. A 120-g can of sardines for 99¢ or a $4\frac{3}{8}$-oz can for 85¢?

4. 150 g of cookies for $1.25 or 5 oz for $1.15?

Table 6.4

	$ Value per Unit of Foreign Currency	Units of Currency per Dollar
France (Franc)	0.1104	9.0525
Sweden (Krona)	0.1185	8.4325

FOREIGN EXCHANGE
WEDNESDAY, SEPTEMBER 5, 1984

International Best Buys

The following prices were collected from newspaper ads and visits to super-markets in New York City, Paris (France), and Stockholm (Sweden). Here we need to translate between the metric system and the customary system as well as between currencies used in different countries. The exchange rates between the dollar and foreign currencies fluctuate, so we'll use a foreign exchange chart from around the time the study was made. We have translated the items into English. The Europeans write decimals with a comma; we've replaced this with a decimal point to make the numbers more convenient to read.

The value of the dollar has changed a lot during the last ten years, but we decided to use the information we gathered in 1984. If you are interested in today's rates, you can find them in your newspaper. Some are given in the business reports on television.

EXAMPLE

Convert
(a) 5 francs to dollars
(b) 5 kronor to dollars
(c) $5 to francs and kronor

Solution
(a) 5 francs equal $5 \times \$0.1104 = \$0.55$
         or $5/\$9.0525 = \$0.55$
(b) 5 kronor equal $5 \times \$0.1185 = \$0.59$
         or $5/\$8.4325 = \$0.59$
(c) 5 dollars equal 5×9.0525 francs $= 45.26$ francs
         5×8.4325 kronor $= 42.16$ kronor

EXERCISE 6.5.3

Find the best buy for each item. Translate all information to dollars per pound.

			United States (dollar)	France (franc)	Sweden (krona)
1.	(a)	Brie cheese	$3.98/lb	43.40/kg	88.55/kg
	(b)	Mozzarella	$1.49/12 oz	66.80/kg	
	(c)	Scallops	$6.99/lb	126.00/kg	
	(d)	Sole		74.00/kg	78.00/kg
	(e)	Lamb chops		55.00/kg	49.80/kg
	(f)	Steak		51.50/kg	106.15/kg

		United States (dollar)	France (franc)	Sweden (krona)
(g)	Beef kidneys		24.60/kg	19.80/kg
(h)	Chicken	59¢/lb	15.90/kg	15.95/kg
(i)	Chicken wings	89¢/lb	18.00/kg	
(j)	Salami	$2.99/lb	37.80/kg	
(k)	Ham	$2.99/1/2 lb	71.00/kg	
(l)	Coffee		11.95/250 g	27.47/500 g
(m)	Apples	69¢/lb	7.07/kg	9.90/kg
(n)	Pears		12.60/kg	9.80/kg
(o)	Oranges		6.30/kg	4.90/kg
(p)	Lemons		7.20/kg	5.50/kg

2. In the spring of 1988 more prices were obtained in France. The currency exchange was as follows:

	$ Value per Unit of Foreign Currency	Units of Currency Per Dollar
France (franc)	0.1753	5.7045
Sweden (krona)	0.1685	5.9350

Express all prices in dollars per pound.

		United States (dollar)	France (franc)
(a)	Boursin	$3.99/5 oz	7.50/96 g
(b)	Top round	$2.79/lb	69.80/kg
(c)	Veal scallopini	$8.99/lb	72.80/kg
(d)	Pork roast	$1.99/lb	29.80/kg
(e)	Calves liver	$3.99/lb	99.80/kg

SUMMARY

Definitions

Scientific notation: A number in scientific notation is written as a product of a number between 1 and 10 and a power of 10.

Prefixes:
  kilo   (k) = thousand
  hecto (h) = hundred
  deka (da) = ten
  deci   (d) = tenth
  centi (c) = hundredth
  milli (m) = thousandth

The customary system of measurement:

Length	Weight	Volume
1 mile = 5280 feet	1 pound = 16 ounces	1 gallon = 4 quarts
1 yard = 3 feet		1 quart = 2 pints
1 foot = 12 inches		1 pint = 2 cups
		16 fluid ounces = 1 pint
		1 cup = 8 fluid ounces

Conversions:

Length	Weight	Volume
1 in. = 2.54 cm	1 lb = 454 g	1 qt = 0.946 L
1 km = 0.6 mi	1 kg = 2.2 lb	1 L = 1.06 qt
1 mi = 1.6 km	1 oz = 28 g	

Rules

Any nonzero number to the zeroth power is equal to 1. 0^0 is not defined.

Multiplication and division by powers of 10:

To multiply by	Move decimal point
10^1	1 place to the right
10^2	2 places to the right
10^3	3 places to the right
10^{-1}	1 place to the left
10^{-2}	2 places to the left

To divide by	Move decimal point
10^1	1 place to the left
10^2	2 places to the left
10^3	3 places to the left
10^{-1}	1 place to the right
10^{-2}	2 places to the right

Changing units in the metric system: From a larger unit to a smaller: *multiply* by a power of 10. From a smaller unit to a larger: *divide* by a power of 10.

VOCABULARY

Basic units: Metric units that have defined standards, such as the meter, liter, and kilogram.

Customary system: Formerly the English system with feet, inches, pounds, etc.

Metric system: A system of measurement built on powers of 10.

Powers of 10: Exponential notation with 10 as the base.

Scientific notation: A number written as a product of a number between 1 and 10 and a power of 10.

Standard: A tangible object that represents one unit of measurement. An example is the one kilogram metal cylinder kept in Paris.

CHECK LIST

Check the box for each topic you feel you have mastered. If you are unsure, go back and review.

- ☐ Operations with powers of 10
- ☐ Zero as an exponent
- ☐ Negative exponents
- ☐ Scientific notation

The metric system:

- ☐ Length
- ☐ Weight

☐ Area
☐ Volume
☐ Applications
☐ The customary system
☐ Conversions between the customary and metric systems
☐ Conversions within the customary and metric systems
☐ Best buys

REVIEW EXERCISES

1. Convert between powers of 10, fractions, and decimals.

	Powers of 10	Fractions	Decimals
(a)	10^0		
(b)		1/ 10	
(c)	10^{-3}		
(d)			0.00001
(e)		1/ 100	
(f)			100

2. Solve.
   (a) 0.023×10^2
   (b) $17 \div 10^{-1}$
   (c) 1.384×10^4
   (d) $342 \div 10^{-3}$
   (e) 2900×10^4
   (f) 0.00038×10^6
   (g) $0.0106 \div 10^5$
   (h) $1247 \div 10^0$

3. Express in standard notation.
   (a) 4.83×10^4
   (b) 8.2×10^2
   (c) 2.108×10^{-6}
   (d) 3.256×10^{-3}

4. Express in scientific notation.
   (a) 4800
   (b) 316
   (c) 0.000000016
   (d) 38104000

5. Convert: (a) 0.0067 m to millimeters
             (b) 4892 cm^3 to liters

6. Convert: (a) 1/2 pint to fluid ounces
             (b) 45,000 feet to miles

7. Convert: (a) 55 miles to kilometers
             (b) 600 g to pounds

8. 12 g of carbon, 23 g of sodium, and 32 g of sulfur each contain 6.02×10^{23} atoms (Avogadro's number). Find the number of atoms in (a) 36 g carbon, (b) 46 g sodium, (c) 160 g sulfur. Write all answers in scientific notation.

9. A hydrogen atom is made up of one proton (1.672 $\times 10^{-24}$ g) and one electron (9.108×10^{-28} g). What is the total weight of the hydrogen atom? (Round the weight of the electron before you add.)

10. How many 12-fluid ounce glasses can you fill from a 1-liter bottle of soda?

11. The addition to the Jones family's house measured 27′ 6″ by 20′ 4″. What was the area of the addition? (In construction, ′ is often used for feet and ″ for inches.)

READINESS CHECK

Solve the problems to satisfy yourself that you have mastered Chapter 6.
   1. Write 10^{-4} as a decimal.
   2. Write 5260 in scientific notation.
   3. Convert 40 cm to meters.
   4. Convert 5380 g to kilograms.
   5. Convert 0.4 dm^3 to deciliters.
   6. Convert 1/3 ft to inches.
   7. Convert 47 oz to pounds.
   8. Convert 165 cm to feet and inches.
   9. Which is cheaper: 2 liters of ginger ale for 99¢ or twelve 12-fl oz cans for $1.99?
   10. A Pyrex baking dish has the dimensions $8.5 \times 4.5 \times 2.5$ in. If your cake batter measures 1 qt, is this a suitable dish? (*Hint:* Change to the metric system!)

ALGEBRA

CHAPTER 7

BASIC OPERATIONS

Letters for numbers

Now that you have become somewhat more comfortable with numbers, we are going to replace them with letters. The letters will stand for all sorts of numbers and make it possible for you to handle situations you never have been able to deal with before.

Algebra comes from the Arabic *al-jabr*, meaning "the reduction." Diophantus of Alexandria, who lived in the 3rd century, is generally credited with being the "father" of algebra as we know it today. He used signs for the unknown quantity: one like an inverted *h*, the other like an ordinary *s*. Diophantus was probably the first to make a distinction between positive and negative numbers. He used names that meant a "forthcoming" (positive) and a "wanting" (negative). He also stated that a "wanting" multiplied by a "wanting" makes a "forthcoming" and that a "wanting" multiplied by a "forthcoming" makes a "wanting." He used no sign equivalent to the plus sign; instead he indicated addition by writing the terms side by side. He wrote all the positive terms of an expression first and then wrote all the negative terms following the sign ↑ for subtraction.

7.1 VOCABULARY

We use all the operations from arithmetic in algebra, but there are some differences. a and b stand for different numbers, so $a + b$ cannot be simplified. However, $a + a$ can be added: $a + a = 2a$.

In arithmetic, we write 23 and mean 2 tens and 3 ones. In algebra, ab implies multiplication: $ab = a \cdot b$.

The number before a letter, such as the 2 in $2a$, is called the *numerical coefficient* or simply the *coefficient*.

Numbers are called *constants* because 2 is always 2, 15 is always 15, and so on. Letters are called *variables* because a letter can stand for different numbers.

The expression $2 + 3$ consists of two *constant terms*, 2 and 3.

The expression $5x + 7$ consists of two terms, $5x$, which is a variable with a numerical coefficient, and 7, which is a constant.

7.2 ADDITION AND SUBTRACTION

Adding is a matter of counting:

Adding apples:

or money:

$$2¢ + 3¢ = 5¢$$

or anything else:

$$2 \text{ things} + 3 \text{ things} = 5 \text{ things}$$

The only restriction is that the things we add must be *alike*.

Recall that when we added three-fourths and two-thirds in Chapter 4, we could not add them until we made them both twelfths.

$$\tfrac{3}{4} + \tfrac{2}{3} \text{ can be rewritten } \tfrac{9}{12} + \tfrac{8}{12}, \text{ which then becomes } \tfrac{17}{12}$$

Similarly, we can't add $\$3000 + 3$ automobiles or 17 kangaroos + 15 leopards. When we put them together we still have 17 kangaroos and 15 leopards!

So it is with algebraic letters. $2a$ and $3a$ are similar to 2 apples and 3 apples. They describe similar "things." We call these terms *like terms*.

$3a^2b$ and $5a^2b$ are like terms, but $2a$ and $2a^2$ or $3a^2b$ and $3ab$ are *not* like terms. (Remember: 10 and 10^2 are definitely not alike!)

$$2a + 3a = 5a$$

$$3a^2b + 5a^2b = 8a^2b$$

$$3x + x = 4x \qquad (x \text{ is really } 1x)$$

But

$$3x + x^2 = 3x + x^2 \quad (\text{Remember the kangaroos and leopards!})$$

When we add or subtract like terms, we say we are *combining like terms*, and we follow the rules of signs. (See Chapter 2.)

In $-3 + 5 - 4 + 2 - 1$, we combine all numbers with a minus sign before them: $-3 - 4 - 1 = -8$; we combine all numbers with a plus sign before them: $+5 + 2 = 7$; and finally, we subtract: $7 - 8 = -1$.

DEFINITION

Like terms are terms whose variables and exponents match.

$$3a^n b^m \text{ and } 7a^n b^m \text{ are like terms.}$$

$$3a^n b^m \text{ and } 7a^m b^n \text{ are not like terms.}$$

EXAMPLE

Combine like terms:
(a) $3a + 4b - a + 2b + 2a - 3b$
(b) $2ab + 3bc - 4ab - 2bc$
(c) $3x^2 y + 2xy^2$

   Solution
   (a) First combine the a's: $3a - a + 2a = 5a - a = 4a$
      Then combine the b's: $4b + 2b - 3b = 6b - 3b = 3b$
      Thus,

$$3a + 4b - a + 2b + 2a - 3b = 4a + 3b$$

   (b) $2\underline{ab} + 3\underline{\underline{bc}} - 4\underline{ab} - 2\underline{\underline{bc}}$
      $2ab - 4ab = -2ab$ and $+3bc - 2bc = +bc$
      $2ab + 3bc - 4ab - 2bc = -2ab + bc$
   (c) $3x^2 y + 2xy^2$ cannot be combined. The variables do not match; they are not like terms.

EXERCISE 7.2.1

Simplify by combining like terms.

   1. $2y + 5y$

   2. $4b + b$

   3. $8x - 3x$

   4. $6a - 10a$

   5. $4x - 5x$

   6. $3a - 8a + 6a$

   7. $3c - 9c + c$

   8. $-2b + 4b - 6b$

   9. $-2x + 6x - 8x + 4x$

10. $7y - 8y + 11y + 2y - 7y$

11. $4c + 7d + 6c + 9d$

12. $6ab - 5bc + 7ab - bc$

13. $-4y + 5x - 7y - 2x$

14. $7m - 3m^2 + 2m^2$

15. $3a^2b - 2ab^2 - a^2b$

16. $5z^2 - 3z^2 - 8z + 11z$

17. $6a^2y - 3a^2y + 7ay^2 - 7ay^2$

18. $x - 2x^3 - x^2 - 6x + 3x^2 + x^3$

19. $3a + 2b - 6a - 2b + c$

20. $3x^2y + 3x^2 + 5xy^2 - xy^2 + 2x^2y - 3x^2$

7.3 EXPONENTIAL NOTATION

In Chapter 2 we introduced exponential notation for numbers:

$$2^3 = 2 \cdot 2 \cdot 2 \qquad 3^2 = 3 \cdot 3$$

Letters can also be written in exponential notation.

$$a^3 = a \cdot a \cdot a \qquad b^2 = b \cdot b \qquad c^5 = c \cdot c \cdot c \cdot c \cdot c \qquad d^1 = d$$

As with the numbers, here the letter is the *base* and the small number is the *exponent*. *Exponential notation* is simply shorthand for repeated multiplication. Why write $f \cdot f \cdot f \cdot f \cdot f \cdot f \cdot f \cdot f \cdot f \cdot f \cdot f \cdot f \cdot f$ when f^{13} means the same thing?

Be careful with negative factors:

$$(-2)(-2)(-2)(-2) = (-2)^4 = +16$$

while

$$-2^4 = -(2^4) = -1 \cdot 2 \cdot 2 \cdot 2 \cdot 2 = -16$$

Remember, in a multiplication, an *even* number of minus signs give *plus*, and an *odd* number of minus signs give *minus*.

EXAMPLE

Write without parenthesis: (a) $(-n)^4$ (b) $(-x)^3$ (c) $-(-x)^3$

Solution
(a) $(-n)^4 = (-n)(-n)(-n)(-n) = +n^4 = n^4$
(b) $(-x)^3 = (-x)(-x)(-x) = -x^3$
(c) $-(-x)^3 = -(-x)(-x)(-x) = -(-x^3) = x^3$

EXERCISE 7.3.1

1. Write in exponential notation.
(a) $2 \cdot 2 \cdot 2$
(b) $(-2)(-2)(-2)$
(c) $g \cdot g \cdot g \cdot g$
(d) $h \cdot h$
(e) $i \cdot i \cdot i \cdot i \cdot i$
(f) $j \cdot j \cdot j$
(g) $k \cdot k \cdot k \cdot k \cdot k \cdot k$
(h) $(-m)(-m)(-m)(-m)$
(i) $-t \cdot t \cdot t$
(j) $-(-d)(-d)(-d)(-d)$

2. Write without parenthesis:
   (a) $(-2)^2$
   (b) $(-5)^3$
   (c) $-(-5)^3$
   (d) $(-q)^4$
   (e) $(-r)^5$
   (f) $-(p)^3$
   (g) $-(-p)^3$
   (h) $-(-x)^4$
   (i) $(-2)^2(-3)^3$
   (j) $(-x)^3(x)x^2$

Multiplying Numbers in Exponential Notation

To multiply terms with the *same base*, add the exponents.

EXAMPLE

Solve: (a) $2^3 \cdot 2^4$ (b) a^2a^5 (c) $x^4 \cdot x \cdot x^2$

Solution
(a) $2^3 \cdot 2^4 = (2 \cdot 2 \cdot 2)(2 \cdot 2 \cdot 2 \cdot 2) = 2^7$
(b) $a^2a^5 = (a \cdot a)(a \cdot a \cdot a \cdot a \cdot a) = a^7$
(c) $x^4 \cdot x \cdot x^2 = (x \cdot x \cdot x \cdot x)(x)(x \cdot x) = x^7$

RULE

Multiplication of Exponential Terms
If a, m, and n stand for any numbers, then

$$a^m a^n = a^{(m+n)} \qquad a^n \text{ or } a^m \text{ cannot be } 0°.$$

To multiply terms that have the same base, add the exponents.

EXAMPLE

Multiply: (a) $2^5 \cdot 2^9$ (b) $5 \cdot 5^3$ (c) $x \cdot x^3$ (d) $y^2 \cdot y^5$

Solution
(a) $2^5 \cdot 2^9 = 2^{(5+9)} = 2^{14}$
(b) $5 \cdot 5^3 = 5^{(1+3)} = 5^4$
(c) $x \cdot x^3 = x^{(1+3)} = x^4$
(d) $y^2 \cdot y^5 = y^{(2+5)} = y^7$

EXERCISE 7.3.2

Multiply.

1. $(m^3)(m^6)$
2. $q \cdot q^5$
3. $x^5 \cdot x$
4. $t \cdot t^5 \cdot t$

5. $y^4 \cdot y \cdot y^3$

6. $p^2(p^3)$

7. $a(a^2)$

8. $z^2 \cdot z^6$

To multiply two terms with coefficients other than 1, we can use the commutative property and write

$$3x^2 \cdot 5x^3 = 3 \cdot 5 \cdot x^2 \cdot x^3 = 15x^5$$

EXAMPLE

Multiply: (a) $(-2a^4)(3a)$ (b) $(-5c^2)(3c^3)(4c)$ (c) $(3ab^2)(-4a^3b^2)$

Solution

(a) $(-2a^4)(3a) = -2 \cdot 3 \cdot a^4 \cdot a = -6a^5$

(b) $(-5c^2)(3c^3)(4c) = -5 \cdot 3 \cdot 4 \cdot c^2 \cdot c^3 \cdot c = -60c^6$

(c) $(3ab^2)(-4a^3b^2) = (3)(-4)a \cdot a^3 \cdot b^2 \cdot b^2 = -12a^4b^4$

RULE

Multiplication of Algebraic Terms

1. *Multiply* coefficients.

2. *Add* exponents of like bases.

(Always make sure the *bases* are the same *before* you add exponents.)

EXAMPLE

Multiply: (a) $2ab^2 \cdot 4a^2b$ (b) $(-5cd^2)(-2d^7)(-cde^3)$

Solution

(a) $2ab^2 \cdot 4a^2b = 2 \cdot 4 \cdot a^1 \cdot a^2 \cdot b^2 \cdot b^1 = 8a^3b^3$

(b) $(-5cd^2)(-2d^7)(-cde^3) = (-5)(-2)(-1)c \cdot c \cdot d^2 \cdot d^7 \cdot d \cdot e^3$
$$= -10c^2d^{10}e^3$$

EXERCISE 7.3.3

Multiply.

1. $3x \cdot 2x^2$

2. $t^2 \cdot 3t^4$

3. $-4p^4 \cdot 3p^6$

4. $5ax \cdot 3 \cdot ax^3$

5. $5s^2 \cdot 2t^3 \cdot 3s^4$

6. $7abc^4 \cdot 9a^5b^3c$

7. $-6x^3y^2(-4x^2y^6)$

8. $10s^5t^6(-10s^7t^8)$

9. $-2a^3 \cdot 5ab^2(-3a^4b)$

10. $(-3x^4y)(-2x^2y^2)(-5xy)$

Power to a Power

The product $a^2 \cdot a^2 \cdot a^2$ can be written as $(a^2)^3$, where a^2 is the base and 3 is the exponent. Since $a^2 \cdot a^2 \cdot a^2 = a^6, (a^2)^3$ is also equal to a^6. To raise a power to a power, multiply the exponents.

EXAMPLE

Simplify: (a) $(b^6)^2$ (b) $(c^4)^3$ (c) $(x^5)^{10}$

Solution
(a) $(b^6)^2 = b^{6 \times 2} = b^{12}$
(b) $(c^4)^3 = c^{4 \times 3} = c^{12}$
(c) $(x^5)^{10} = x^{5(10)} = x^{50}$

RULE

Raising a Power to a Power

If a, b, and c represent any numbers, then

$$(a^b)^c = a^{bc}$$

When a number in exponential form is raised to a power, multiply the exponents.

EXERCISE 7.3.4

Simplify.

1. $(t^4)^2$

2. $(c^2)^3$

3. $(s^3)^8$

4. $(w^2)^5$

5. $(b^1)^6(b^3)^2$

6. $(p^2)^4(p^4)^3$

Product to a Power

Suppose the term we are raising to a power begins with a coefficient or has two variables? Here are two examples.

$$(2a)^3 = (2a)(2a)(2a) = 2 \cdot 2 \cdot 2 \cdot a \cdot a \cdot a = 2^3 a^3 = 8a^3$$

and

$$(ab^2)^3 = (ab^2)(ab^2)(ab^2) = a \cdot a \cdot a \cdot b^2 \cdot b^2 \cdot b^2 = a^3 b^6$$

Here again there is a shortcut. We can raise *each* factor to the outside power:

$$(2a)^3 = 2^{(1 \times 3)} a^{(1 \times 3)} = 2^3 a^3 = 8a^3$$

EXAMPLE

Simplify:
(a) $(-5a^2b^3)^2$
(b) $(3x^2y^3)^7$
(c) $(-3x^3y^4)^2$

Solution
(a) $(-5a^2b^3)^2 = (-5)^2 \cdot (a^2)^2 \cdot (b^3)^2 = 25a^4b^6$
(b) $(3x^2y^3)^7 = 3^7(x^2)^7(y^3)^7 = 3^7x^{14}y^{21}$ or $2187x^{14}y^{21}$
(c) $(-3x^3y^4)^2 = (-3)^2(x^3)^2(y^4)^2 = 9x^6y^8$

RULE

Raising a Product to a Power

If *a*, *b*, and *c* represent any numbers, then

$$(ab)^c = a^c b^c$$

When a product is raised to a power, each factor is raised to that power.

EXERCISE 7.3.5

Simplify.
1. $(-2s)^3$
2. $(-3q)^4$
3. $(-1.5n)^2$
4. $(a^3bc^2)^2$
5. $(5s^2t)^2$
6. $3(-2xy^2z^3)^3$

7.4 DIVISION WITH EXPONENTS

In algebra we usually write division examples as fractions:
(Note: The variables in the denominator can never equal 0.)

$$j^5 \div j^2 = \frac{j^5}{j^2} = \frac{j \cdot j \cdot j \cdot j \cdot j}{j \cdot j} = j \cdot j \cdot j = j^3$$

$$k^4 \div k^3 = \frac{k^4}{k^3} = \frac{k \cdot k \cdot k \cdot k}{k \cdot k \cdot k} = k^1, \text{ or } k$$

$p^{25} \div p^{17} = ?$ If you said p^8, you would be correct.

To divide exponential terms with the same base, subtract the exponents.

EXAMPLE

Divide: (a) $3^8 \div 3^5$ (b) $5^2 \div 5$ (c) $m^7 \div m^2$ (d) $(-x)^5 \div (-x)^3$

Solution
(a) $3^8 \div 3^5 = 3^{8-5} = 3^3 = 27$
(b) $5^2 \div 5 = 5^{2-1} = 5$
(c) $m^7 \div m^2 = m^{7-2} = m^5$
(d) $(-x)^5 \div (-x)^3 = (-x)^{5-3} = (-x)^2 = x^2$

RULE

Division of Exponential Terms

If a, m, and n stand for any numbers and $a \neq 0$, then

$$a^m \div a^n = \frac{a^m}{a^n} = a^{m-n}$$

To divide exponential terms that have the same base, subtract the exponents.

EXERCISE 7.4.1

Divide.

1. $\dfrac{x^5}{x^2}$

2. $\dfrac{a^6}{a^5}$

3. $\dfrac{x^7 y^5}{x^2 y^3}$

4. $\dfrac{(-5)^4}{(-5)^2}$

5. $\dfrac{(-a)^6}{(-a)^3}$

6. $\dfrac{(x)^{11}}{(x)^8}$

7. $\dfrac{x^4 y^3 z^2}{xyz}$

8. $\dfrac{a^2 b^2 c^2}{ab^2 c}$

What difference do coefficients make?

EXAMPLE

Divide: (a) $8x^3 \div 4x^2$ (b) $\dfrac{-18y^7}{3y^4}$

Solution

(a) $8x^3 \div 4x^2 = \dfrac{8 \cdot x^3}{4 \cdot x^2} = \dfrac{8}{4}x^{3-2} = 2x$

(b) $\dfrac{-18y^7}{3y^4} = \dfrac{-18}{3}y^{7-4} = -6y^3$

In multiplication we multiply coefficients and add exponents; in division we *divide* coefficients and *subtract* exponents in terms where bases are alike.

RULE

Division of Algebraic Terms

1. *Divide the coefficients.*

2. *Subtract exponents* of like bases.

EXERCISE 7.4.2

Divide.

1. $\dfrac{30p^4}{5p}$

2. $\dfrac{25x^6}{5x^4}$

3. $\dfrac{8y^4}{2y^3}$

4. $\dfrac{-15z^6}{5z^2}$

5. $\dfrac{24x^3y^6}{6xy^4}$

6. $\dfrac{12q^3r^5}{-2qr^4}$

7. $\dfrac{-12a^3b^2c}{-3abc}$

8. $\dfrac{15a^2bc^2}{-5abc}$

Zero as a Power

Thus far we have only worked with problems where the exponents of the numerators are greater than those of the denominators.

$$x^6 \div x^4 = x^2$$

$$y^9 \div y^8 = y^1, \text{ or } y$$

What about $z^7 \div z^7$? By our rule it equals z^{7-7} or z^0.

When we divide a number by itself, the answer is 1.

$$7 \div 7 = 1, \qquad 19 \div 19 = 1, \qquad 2981 \div 2981 = 1$$

What about $q^5 \div q^5$?

$$\frac{q^5}{q^5} \begin{cases} = q^{5-5} = q^0 \\ = 1 \end{cases} \qquad \text{So } q^0 = 1.$$

We have already seen in Chapter 6 that $10^0 = 1$; by the same rule,

$$7^0 = 1, \qquad 15^0 = 1, \qquad (xy)^0 = 1, \qquad (-17x^2yz^3)^0 = 1$$

(An exception to this rule is 0^0; it has an undefined value.)

RULE

Any expression to the zero power equals 1.
The only exception is 0^0, which is undefined.

EXERCISE 7.4.3

Find the value.

1. 1^0
2. a^0
3. $2^3 + 2^2 + 2^1 + 2^0$
4. $2^3 \cdot 2^2 \cdot 2^1 \cdot 2^0$
5. $(-15)^0$
6. $5^0 + 9^0$
7. $2^0 - 3^0$
8. $(2^3 + 2^2 + 2^1 + 2^0)^0$

Negative Exponents

$$t^3 \div t^5 = \frac{t^3}{t^5} = \frac{t \cdot t \cdot t}{t \cdot t \cdot t \cdot t \cdot t} = \frac{1}{t^2}$$

But when we follow the rule for division, we have

$$t^3 \div t^5 = t^{3-5} = t^{-2}$$

Therefore,

$$t^{-2} = \frac{1}{t^2}$$

EXAMPLE

Simplify: $2^4 \div 2^7$

Solution $\dfrac{2^4}{2^7} = \dfrac{1}{2^{7-4}} = \dfrac{1}{2^3}$

In this example, we could have said

$$\frac{2^4}{2^7} = 2^{4-7} = 2^{-3}$$

Thus we see that

$$\frac{1}{2^3} = 2^{-3}$$

A number with a negative exponent equals its reciprocal written with a positive exponent. For example,

$$2^{-1} = \frac{1}{2^1} = \frac{1}{2} \qquad 2^{-2} = \frac{1}{2^2} = \frac{1}{4} \qquad 2^{-3} = \frac{1}{2^3} = \frac{1}{8}$$

$$\frac{1}{2^{-1}} = 2^1 = 2 \qquad \frac{1}{2^{-2}} = 2^2 = 4 \qquad \frac{1}{2^{-3}} = 2^3 = 8$$

When simplifying terms with negative exponents, be careful not to confuse the negative sign in front of the exponent with any that are in front of the coefficient or the entire term.

For example,

$$-2^{-3} = -\frac{1}{2^3} = -\frac{1}{8} \qquad \text{and} \qquad (-2)^{-3} = \frac{1}{(-2)^3} = \frac{1}{-8} = -\frac{1}{8}$$

However, with an even power, there's an important difference:

$$(-2)^{-2} = \frac{1}{(-2)^2} = \frac{1}{4} \qquad \text{but} \qquad -2^{-2} = -\frac{1}{2^2} = -\frac{1}{4}$$

EXAMPLE

Write the following expressions with positive exponents.

(a) $v^2 \div v^6$ (b) $\dfrac{1}{w^{-5}}$ (c) $\dfrac{x^{-2}}{y^{-3}}$

Solution

(a) $v^2 \div v^6 = v^{2-6} = v^{-4} = \dfrac{1}{v^4}$

(b) $\dfrac{1}{w^{-5}} = \dfrac{w^5}{1} = w^5$

(c) $x^{-2} = \dfrac{1}{x^2}$ and $y^{-3} = \dfrac{1}{y^3}$

So

$$\frac{x^{-2}}{y^{-3}} = \frac{1}{x^2} \div \frac{1}{y^3} = \frac{1}{x^2} \cdot \frac{y^3}{1} = \frac{y^3}{x^2}$$

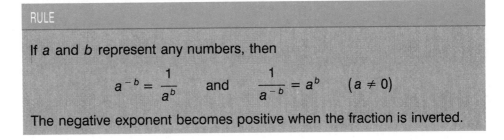

RULE

If *a* and *b* represent any numbers, then

$$a^{-b} = \frac{1}{a^b} \qquad \text{and} \qquad \frac{1}{a^{-b}} = a^b \qquad (a \neq 0)$$

The negative exponent becomes positive when the fraction is inverted.

EXAMPLE

Write with positive exponents and simplify.

(a) 3^{-3} (b) $5a^{-2}b^3$ (c) $\dfrac{8a^2b^{-5}}{2a^{-3}b^4}$

Solution

(a) $3^{-3} = \dfrac{1}{3^3} = \dfrac{1}{27}$

(b) $5a^{-2}b^3 = 5 \cdot \dfrac{1}{a^2} \cdot b^3 = \dfrac{5b^3}{a^2}$

(c) $\dfrac{8a^2b^{-5}}{2a^{-3}b^4} = \dfrac{8a^2a^3}{2b^5b^4} = \dfrac{4a^5}{b^9}$

EXERCISE 7.4.4

1. Simplify (write with positive exponents only).
    (a) 3^{-1}
    (b) 3^{-2}
    (c) 3^{-3}
    (d) x^{-1}
    (e) y^{-2}
    (f) $\dfrac{1}{3^{-1}}$
    (g) $\dfrac{1}{3^{-2}}$
    (h) 1^{-4}
    (i) $\dfrac{1}{p^{-2}}$
    (j) $-a^{-2}$
    (k) $-(-a)^{-2}$

2. Reduce to lowest terms. Write with positive exponents only.

    (a) $\dfrac{x^4}{x^6}$

    (b) $\dfrac{4a^4b^9}{8a^8b^5}$

    (c) $\dfrac{x^2y^4}{x^3y}$

(d) $\dfrac{-5rs^3t^2}{25r^2s^7t^7}$

(e) $\dfrac{36x^5y^2}{12x^3y^5}$

7.5 THE DISTRIBUTIVE PRINCIPLE

We know from our discussion about multiplication that

$$3x(5x) = 15x^2$$

But what about $3(2a + 3b)$?

Let's look first at an example with only numbers.

EXAMPLE

Simplify: $3(2 + 5)$

Solution $3(2 + 5) = 3(7)$ or 21 (by observing the rules of the order of operations). But it can also equal

$$3(2) + 3(5) = 6 + 15 = 21$$

Similarly,

$$3(5 + 2) = 3(5) + 3(2) = 15 + 6 = 21$$

In simplifying $3(2a + 3b)$ we cannot combine $2a + 3b$. We must use the second method of multiplication and multiply each term separately:

$$3(2a + 3b) = 3(2a) + 3(3b) = 6a + 9b$$

We do not need to multiply horizontally; we can set up the multiplication vertically.

$$
\begin{array}{r}
2a + 3b \\
\times \qquad 3 \\
\hline
6a + 9b
\end{array}
$$

EXAMPLE

Simplify the following both horizontally and vertically.
(a) $2ab(3a + 5b)$ (b) $x(x^2 - 5x + 7)$

Solution

(a) $2ab(3a + 5b) = 6a^2b + 10ab^2$ or $\begin{array}{r} 3a + 5b \\ \times \qquad 2ab \\ \hline 6a^2b + 10ab^2 \end{array}$

(b) $x(x^2 - 5x + 7) = x^3 - 5x^2 + 7x$ or $\begin{array}{r} x^2 - 5x + 7 \\ \times \qquad x \\ \hline x^3 - 5x^2 + 7x \end{array}$

In these examples we have *distributed* the multiplication *over* any addition or subtraction.

THE DISTRIBUTIVE PRINCIPLE FOR MULTIPLICATION

$$a(b + c) = ab + ac$$

In the same way we *distribute* division over addition and subtraction. We can solve $\dfrac{12 + 21}{2}$ in two different ways.

$$\frac{12 + 21}{3} = \frac{33}{3} = 11 \quad \text{or} \quad \frac{12 + 21}{3} = \frac{12}{3} + \frac{21}{3} = 4 + 7 = 11$$

THE DISTRIBUTIVE PRINCIPLE FOR DIVISION

$$\frac{ab + ac}{a} = \frac{ab}{a} + \frac{ac}{a} = b + c$$

EXAMPLE

Divide: (a) $\dfrac{9a + 6b}{3}$ (b) $\dfrac{4a^2b - 12ab^2}{4ab}$ (c) $\dfrac{15x^2y - 25xy + 10xy^2}{5xy}$

Solution

(a) We cannot combine the two terms $9a$ and $6b$, so we use the distributive principle.

$$\frac{9a + 6b}{3} = \frac{9a}{3} + \frac{6b}{3} = 3a + 2b$$

(b) $\dfrac{4a^2b - 12ab^2}{4ab} = \dfrac{4a^2b}{4ab} - \dfrac{12ab^2}{4ab} = a - 3b$

(c) $\dfrac{15x^2y - 25xy + 10xy^2}{5xy} = \dfrac{15x^2y}{5xy} - \dfrac{25xy}{5xy} + \dfrac{10xy^2}{5xy} = 3x - 5 + 2y$

EXAMPLE

Solve: (a) $2(5 + x) + x(x - 3)$ (b) $\dfrac{x^2 + 5x}{x} + x$

Solution

(a) As in arithmetic, we multiply and divide first, then add and subtract. Remember to combine like terms when you can. We usually write the terms in the answer in order starting with the highest exponent and listing the constant last.

$$2(5 + x) + x(x - 3) = 10 + 2x + x^2 - 3x = x^2 - x + 10$$

(b) $\dfrac{x^2 + 5x}{x} + x = \dfrac{x^2}{x} + \dfrac{5x}{x} + x = x + 5 + x = 2x + 5$

184

EXAMPLE

Solve: (a) $2(3 + x) - x(4 + x)$ (b) $10 - 4(x - 6)$ (c) $10 + \dfrac{4x - 24}{4}$

(d) $\dfrac{9x^2 - 3x}{3x} - \dfrac{10x^2 + 20x}{2x}$

Solution

(a) $2(3 + x) - x(4 + x) = 6 + 2x - 4x - x^2$

$$= 6 - 2x - x^2 = -x^2 - 2x + 6$$

(b) $10 - 4(x - 6) = 10 - 4x + 24 = 34 - 4x,$ or $-4x + 34$
Remember the order of operations: First we multiply by -4. Make sure to distribute the minus sign over the subtraction.)

(c) $10 + \dfrac{4x - 24}{4} = 10 + \left(\dfrac{4x}{4} - \dfrac{24}{4}\right) = 10 + (x - 6)$

$$= 10 + x - 6 = 4 + x = x + 4$$

(d) $\dfrac{9x^2 - 3x}{3x} - \dfrac{10x^2 + 20x}{2x} = \left(\dfrac{9x^2}{3x} - \dfrac{3x}{3x}\right) - \left(\dfrac{10x^2}{2x} + \dfrac{20x}{2x}\right)$

$$= (3x - 1) - (5x + 10)$$

$$= 3x - 1 - 5x - 10 = -2x - 11$$

EXERCISE 7.5.1

Distribute the multiplication and division over the addition or subtraction and simplify where possible.

1. $4(x - 3)$
2. $x(y^2 - 4y)$
3. $5a(3a + 6b)$
4. $3a^2(b^2 - abc)$
5. $-6xy(4x^2y - 6x)$
6. $5(x + 2) - 3(2x - 1)$
7. $-3(x^2 + 1) + 2(5x - 3)$
8. $4(xy - z) - (xy + z^2)$
9. $\dfrac{25x - 10}{5x}$
10. $\dfrac{2x^2 + 10x}{2x}$
11. $\dfrac{3a^2b^2 - 3a^3bc}{3a^2}$
12. $\dfrac{100x^2 + 10x}{10x}$
13. $\dfrac{16x^2 + 8x + 8}{8x}$

14. $\dfrac{5t - 10t^2}{5t} - 25t$

15. $\dfrac{s^2 - 15s + 8}{s^2}$

16. $\dfrac{6z - 15}{z} - \dfrac{3z^2 + 21z}{3z^2}$

17. $4x - 3(x - 3)$

18. $5 - 3(x - 4)$

19. $4(5 - 2x) + 3(3x - 4)$

20. $4(3x - 2) - 2(x + 1)$

21. $6 - 2(4x + 3) - 8$

22. $3(x - y) - 2(x + y)$

23. $4x + \dfrac{x^2 - 2x}{x}$

24. $7a - \dfrac{9a + 6a^2}{3a}$

7.6 FACTORING

When we write $24 = 2 \cdot 3 \cdot 4$ or $5x^2 = 5 \cdot x \cdot x$ or $36x^3y^4 = 36 \cdot x^3 \cdot y^4$ or $36x^3y^4 = 2 \cdot 2 \cdot 3 \cdot 3 \cdot x \cdot x \cdot x \cdot y \cdot y \cdot y \cdot y$, we are *factoring* the given expression. We are finding the terms that when multiplied together will give us the original expression. In fact, we *check* a factoring example by multiplying.

$$15a^2b = 3 \cdot 5 \cdot a^2 \cdot b$$

Check:

$$3 \cdot 5 \cdot a^2 \cdot b = 15a^2b$$

If the directions say "factor *completely*," we factor as far as possible, that is, all the way to the prime factors:

$$15a^2b = 5 \cdot 3 \cdot a \cdot a \cdot b$$

Suppose we have an expression with two terms: $3a + 9c$. If we want to factor the expression, we must look for a factor common to both terms. In this case 3 divides each, so we write

$$3(a) + 3(3c) \qquad \text{or} \qquad 3(a + 3c)$$

Check:

$$3(a + 3c) = 3a + 9c \qquad \text{(Remember the distributive process.)}$$

EXAMPLE

Factor completely: (a) $2 - 8x$ (b) $-6x - 15y$ (c) $x^2y + xy^2$

Solution

(a) Each term is divisible by 2.

$$2 \cdot 1 = 2 \quad\text{and}\quad 2(-4x) = -8x$$

$$2 - 8x = 2(1) + 2(-4x) = 2(1 - 4x)$$

Check:

$$2(1 - 4x) = 2 - 8x$$

(b) Factor (-3) from each term in $-6x - 15y$:

$$-3(2x) = -6x \quad\text{and}\quad -3(5y) = -15y$$

$$-6x - 15y = -3(2x) + (-3)(5y) = -3(2x + 5y)$$

Check:

$$-3(2x + 5y) = -6x - 15y$$

(c) In $x^2y + xy^2$, each term has an x and a y.

$$x^2y = (xy)x \quad\text{and}\quad xy^2 = (xy)y$$

$$x^2y + xy = xy(x + y)$$

Check:

$$xy(x + y) = x^2y + xy^2$$

EXAMPLE

Factor completely:

(a) $12x^2 - 36xy^3$ (b) $3a^2b - 6ab + 12ab^2$ (c) $5a^3b^2 + 10a^2b^3$

Solution

(a) 12 is a common factor: $12x^2 - 36xy^3 = 12(x^2 - 3xy^3)$. This is not yet factored completely because there is still a common factor of x.

$$x^2 - 3xy^3 = x(x - 3y^3)$$

Therefore,

$$12x^2 - 36xy^3 = 12x(x - 3y^3)$$

Check:

$$12x(x - 3y^3) = 12x^2 - 36xy^3$$

(b) $3a^2b - 6ab + 12ab^2 = 3ab(a - 2 + 4b)$ or $3ab(a + 4b - 2)$
Check the answer.

(c) $5a^3b^2 + 10a^2b^3 = 5a^2b^2(a + 2b)$
Check the answer.

EXERCISE 7.6.1

Factor completely.

1. 24

2. 72

3. $-18a^2b^3c$

4. $5p^2 - 35pq$

5. $8x - 12x^2$

6. $15x^2y - 3xy + 30y$

7. $-4a^2b + 8ab - 6ab^2$

8. $15p^2r^3 - 27p^3r^2$

9. $\frac{1}{5}t^3u^2v + \frac{3}{5}tuv^2$

10. $8c^2 - 4c$

7.7 EVALUATING EXPRESSIONS

A variable expression such as $a + b$ has no numerical value, but if we replace the letters with numbers, we can *evaluate* the expression. For example, if $a = 1$ and $b = 6$, then $a + b = 1 + 6 = 7$. The expression is evaluated as 7.

EXAMPLE

Evaluate $a + b$ (a) when $a = 2$ and $b = 3$, (b) when $a = 2$ and $b = -3$.

Solution
(a) $a + b = 2 + 3 = 5$
(b) $a + b = 2 + (-3) = 2 - 3 = -1$

EXAMPLE

Evaluate ab when $a = 5$ and $b = 2$.

Solution $ab = (5)(2) = 10$

It is a good habit to place parentheses around numbers when you substitute them for letters. If you forget when you evaluate ab for $a = 5$ and $b = 2$ that there is an invisible multiplication symbol between a and b, then you might get $ab = 52 = $ "fifty-two" instead of $(5)(2) = 10$ or evaluate ab for $a = 5$ and $b = -2$ as $5 - 2 = 3$ instead of $(5)(-2) = -10$.

EXAMPLE

Evaluate $3a^2b - 2ab^2$ for $a = -2$ and $b = -1$.

Solution

$$3(-2)^2(-1) - 2(-2)(-1)^2 = 3(4)(-1) - 2(-2)(1) = -12 + 4$$
$$= -8$$

EXAMPLE

Evaluate $x(x + y) - y(x - y)$ for $x = 2$ and $y = -2$.

Solution

$$2(2 + (-2)) - (-2)(2 - (-2)) = 2(0) - (-2)(4)$$
$$= 0 + 8 = 8$$

EXAMPLE

Simplify $x(x + y) - y(x - y)$ and then evaluate for $x = 2$, $y = -2$.

Solution

$$x(x + y) - y(x - y) = x^2 + xy - xy + y^2$$

$$= x^2 + y^2$$

$$(2)^2 + (-2)^2 = 4 + 4 = 8$$

EXERCISE 7.7.1

Evaluate for $x = 1$, $y = 3$, and $z = -2$.

1. $x + 2y$
2. $-xy$
3. $-2yz$
4. $x - yz$
5. $3xyz$
6. $x^2 + z^2$
7. $x + 2y - 3z$
8. $x^2y - 2y^2z$

When a problem can be simplified, simplify it before you substitute numbers for the letters. This makes the substitution easier.

EXAMPLE

(a) Evaluate $(2x^3)^2$ for $x = -2$.
(b) Evaluate $2(x + y) - 3(x - y)$ for $x = 1$, $y = -1$.

Solution
(a) First simplify the expression. $(2x^3)^2 = 4x^6$. Now substitute -2 for x:
$4(-2)^6 = 4(64) = 256$

Alternative Solution   Substitute without first simplifying.

$$\left[2(-2)^3\right]^2 = \left[2(-8)\right]^2 = (-16)^2 = 256$$

(b) $2(x + y) - 3(x - y) = 2x + 2y - 3x + 3y = -x + 5y$
Substitute 1 for x and -1 for y: $-(1) + 5(-1) = -1 - 5 = -6$
If you substitute first, you get

$$2[(1) + (-1)] - 3[(1) - (-1)] = 2(0) - 3(2) = -6$$

EXERCISE 7.7.2

Evaluate in two ways: First substitute the given values for each variable and simplify. Then simplify as far as you can before you substitute the given values. (You should get the same answer in both cases.)

1. $(-3xy^2)^3$ for $x = 2$ and $y = -1$
2. $-(-x^2)^2$ for $x = -2$
3. $\dfrac{xy^3}{x^2y}$ when $x = 3$ and $y = 5$
4. $\dfrac{15a^2b^3}{a^3b^4}$ when $a = -3$ and $b = 5$
5. $4x - 3x(2 - x)$ for $x = 3$
6. $4x + \dfrac{x^2 - 2x}{x}$ for $x = -3$
7. $2(x^2 - 3x - 4) - 3(x^2 + 5x + 4)$ for $x = 2$
8. Do Problem 7 for $x = -2$.

7.8 APPLICATIONS

Formulas

A *formula* is a statement of the rule connecting different variables. For example, there are formulas for geometric shapes, for physical relationships, for business and banking, and for many other situations. We will introduce some of the common formulas here.

Geometry

Rectangles The formula for the *area* of a rectangle is

$$\text{Area} = \text{length} \times \text{width}$$

This can also be written as $A = l \cdot w$ or $A = lw$. If the length of a rectangle is 5 cm and its width is 3 cm (Figure 7.1), then its area is $A = 5 \text{ cm} \times 3 \text{ cm} = 15 \text{ cm}^2$.

The *perimeter* is the distance around the rectangle,

$$P = l + w + l + w \qquad \text{or} \qquad P = 2l + 2w \qquad \text{or} \qquad P = 2(l + w)$$

In the rectangle in Figure 7.1, the perimeter is

$$P = 2(3 + 5) \text{ cm} = 2(8) \text{ cm} = 16 \text{ cm}$$

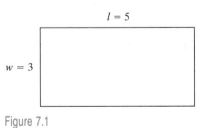

Figure 7.1

Triangles The formula for the area of a triangle is $A = \frac{1}{2}bh$, where b is the base and h is the height (altitude) of the triangle (see Figure 7.2).

For example, if the base of a triangle is 10 cm and its height is 5 cm, the area of the triangle is

$$A = \tfrac{1}{2}(10)(5) \text{ cm}^2 = \tfrac{1}{2}(50) \text{ cm}^2 = 25 \text{ cm}^2$$

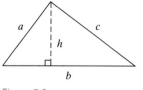

Figure 7.2

EXAMPLE

Find the area of a triangle with $b = 4$ and $h = 6$.

Solution   Formula: $A = \frac{1}{2}bh$

$$A = \tfrac{1}{2}(4)(6), \qquad A = 12$$

The perimeter of a triangle is $P = a + b + c$.

EXAMPLE

Find the perimeter of a triangle with sides 10 cm, 13 cm, and 18 cm.

Solution   The perimeter is the sum of the sides: 10 cm + 13 cm + 18 cm = 41 cm.

Circles Remember the symbol π (pi) ≈ 3.14.

$$\text{Area} = \pi r^2 \qquad \text{and} \qquad \text{Circumference (perimeter of a circle)} = 2\pi r$$

EXAMPLE

Find (a) the area and (b) the circumference of a circle with a radius of 4 cm.

Solution
(a) $A = \pi r^2 = (3.14)(4)^2 = (3.14)(16) = 50.24 \text{ cm}^2$
(b) $C = 2\pi r = 2(3.14)(4) \text{ cm} = 25.12 \text{ cm}$

Geometric Sequences Take the number 3. Triple it, triple it again, and triple it once more. What is the sum of $3 + 9 + 27 + 81$? (This is the same as asking for the sum of $3 + 3^2 + 3^3 + 3^4$.) By hand we find that the sum is equal to 120. There is a formula in mathematics for the sum S:

$$S = \frac{a(r^n - 1)}{r - 1}$$

a is the starting number, and r is the quotient of any number in the sequence and the number preceding it. n is the number of terms you add.

In the sum $3 + 9 + 27 + 81$, $a = 3$, $r = 3$, and $n = 4$. Substituting these into the formula, we have

$$S = \frac{3(3^4 - 1)}{3 - 1} = \frac{3(81 - 1)}{2} = \frac{3(80)}{2} = 120$$

EXAMPLE

Find the sum of $1 + 2 + 2^2 + 2^3 + 2^4 + \cdots + 2^{23}$.

Solution Formula: $S = \dfrac{a(r^n - 1)}{r - 1}$ $a = 1,$ $r = 2$ $(8 \div 4 = 2; 4 \div 2 = 2;$ etc.), and $n = 24$. (The first number is 1, the second is 2, the third is 4, ..., the nth is 2^{n-1}.) Thus,

$$S = \frac{1(2^{24} - 1)}{2 - 1} = 2^{24} - 1 = 16,777,216 - 1 = 16,777,215$$

In the introduction to Chapter 5 this was a sum of pennies. In dollars that would be $167,772 or approximately $170,000.

Physics

Distance, Rate, and Time Problems The formula $d = rt$ is used to show the relationship between distance (d), rate (r), and time (t).

EXAMPLE

How far can you drive in 3 hours if you drive at a rate of 50 miles per hour (mph)?

Solution Here $r = 50$ mph and $t = 3$ hr, so the distance is $50 \times 3 = 150$. The answer is 150 miles.

Temperature Conversions The formula $C = \frac{5}{9}(F - 32)$ translates temperature from degrees Fahrenheit (° F), which are used in the United States, to degrees Celsius (° C), which most other countries use. A corresponding formula $F = \frac{9}{5}C + 32$ translates from Celsius to Fahrenheit degrees.

EXAMPLE

Find the temperature in degrees Celsius if the temperature is 41° F. In other words, find C when F = 41.

Solution Formula: $C = \frac{5}{9}(F - 32)$

$$C = \frac{5}{9}(41 - 32) \qquad \text{Substitute 41 for F.}$$

$$C = \frac{5(9)}{9} = 5 \qquad \text{Reduce.}$$

41° F is equivalent to 5° C.

EXAMPLE

Find the temperature in degrees Fahrenheit when it is $-15°$ C.

Solution $F = \frac{9}{5}C + 32 = \dfrac{(9)(-15)}{5} + 32 = (-27) + 32 = 5$

$$-15° \text{ C} = 5° \text{ F}$$

Business

The formula $I = Prt$ is useful in solving problems that involve borrowing, lending, or investing money. I represents the interest that is paid or earned; P

is principal—the money we borrow or invest; r is the rate of interest earned or paid as a percent of the principal; and t is the time. This interest formula is for *simple interest*. It is useful for a quick estimate but not used by banks. They use *compound interest*.

EXAMPLE

What is the total simple interest if you borrow $1000 for 6 months at an interest rate of 10% per year?

Solution   Formula: $I = Prt$
$P = 1000$, $r = 10\% = 0.1$, $t = 0.5$ year

$$I = 1000(0.1)(0.5) = 50$$

The interest is $50.00.

EXERCISE 7.8.1

1. Find the area of a rectangle that is 8 ft long and 4 ft wide.
2. Find the perimeter of the rectangle in Problem 1.
3. Find the area of a rectangle whose length is 17 cm and whose width is 5 cm.
4. When the sides of a triangle are 3.1 ft, 4.2 ft, and 1.9 ft, what is its perimeter?
5. Find the area of a circle when the radius is $\frac{3}{5}$ ft.
6. Find the circumference of a circle when the radius is 3.4 m.
7. A trucker drives at 62 mph for $3\frac{1}{2}$ hr. How far does he travel?
8. How many miles less would the trucker in Problem 7 cover if he drove at 58 mph for the same length of time?
9. What is the temperature in degrees Fahrenheit when it is $0°$ C?
10. What is the temperature in degrees Celsius when it is $212°$ F?
11. What is the interest if you borrow $2000 for 2 years and the rate is 12%?
12. How much money do you make if you loan a friend $1500 for 8 months at $6\frac{1}{2}\%$ per year?

SUMMARY

Definitions

Like terms are terms in which variables and exponents match.

$3a^n b^m$ and $7a^n b^m$ are like terms.

$3a^n b^m$ and $7a^m b^n$ are not like terms.

Rules

If a, m, and n stand for any numbers and a^m or $a^n \neq 0^0$, then

$$a^m a^n = a^{m+n}$$

To multiply terms that have the same base, add the exponents. When multiplying algebraic terms, *multiply* coefficients and *add* exponents. (Always make sure the *bases* are the same *before* you add exponents.)

If a, b, and c represent any numbers, then

$$\left(a^b\right)^c = a^{bc}$$

When a number in exponential form is raised to a power, multiply the exponents.

If a, b, and c represent any numbers, then

$$\left(ab\right)^c = a^c b^c$$

When a product is raised to a power, each factor is raised to that power.

If a, m, and n stand for any numbers, then

$$a^m \div a^n = \frac{a^m}{a^n} = a^{m-n} \qquad (a \neq 0)$$

To divide terms that have the same base, subtract the exponents.

When dividing algebraic terms, *divide coefficients* and *subtract exponents* when bases are alike.

Any expression to the zero power equals 1. The only exception is 0^0, which is undetermined.

If a and b represent any numbers, then

$$a^{-b} = \frac{1}{a^b} \qquad \text{and} \qquad \frac{1}{a^{-b}} = a^b \qquad (a \neq 0)$$

A negative exponent becomes positive when the fraction is inverted.

The Distributive Principle for Multiplication:

$$a(b + c) = ab + ac.$$

The Distributive Principle for Division:

$$\frac{ab + ac}{a} = b + c. \qquad (a \neq 0)$$

Formulas

Rectangle: $A = lw$ $\qquad$ $P = 2(l + w)$

Triangle: $\quad A = \frac{1}{2}bh$ $\quad$ $P = a + b + c$

Circle: $\qquad A = \pi r^2$ $\quad$ $C = 2\pi r$

$Distance = rate \times time$ $\qquad D = rt$

$Interest = Principal \times rate \times time$ $\qquad I = Prt$

Temperature: $C = \frac{5}{9}(F - 32)$ $\qquad F = \frac{9}{5}C + 32$

VOCABULARY

Coefficient: The number before a letter.

Constant: A number or letter that remains fixed in the discussion.

Distributive principle: Distribute multiplication or division over addition and subtraction.

Evaluate: Find the numerical answer.

Factoring: Expressing a number or an expression as a product.

Like terms: Terms that have the same variables and exponents.

Numerical coefficient: Same as coefficient.

Substitute: Replace a variable with a number.

Term: A constant multiplied or divided by variables. Terms are separated by addition and/or subtraction.

Variable: A letter that can take on different number values.

CHECK LIST

Check the box for each topic you feel you have mastered. If you are unsure, go back and review.

☐ Recognizing like terms
☐ Adding and subtracting like terms
　Exponential notation
　　☐ Operations
　　☐ Zero as a power
　　☐ Negative exponents
☐ The distributive principle
☐ Factoring
☐ Evaluating expressions
☐ Simplifying expressions
☐ Using formulas

REVIEW EXERCISES

1. Simplify.
   (a) $3t + 2t$
   (b) $2c + c$
   (c) $8s - 5s$
   (d) $-2w - 8w$
   (e) $a - 7a + 9a$
   (f) $3q + 2q - 5q$

2. Combine like terms.
   (a) $a + 3a + 5b + 6b$
   (b) $8z - 2 - 3z + 6$
   (c) $2x^2 + 5 - 2x + x^2$
   (d) $3ps + 7p + 8s - 5p$
   (e) $8cd + 9cd^2 - 5cd - 11cd^2$
   (f) $2 - 3c^2 + c + 5 - 6c^2 - 7$

3. Write in exponential notation.
   (a) $t \cdot t \cdot t$
   (b) $v \cdot v \cdot v \cdot v \cdot v$
   (c) $(-m)(-m)(-m)(-m)$
   (d) $-p \cdot p \cdot p \cdot p$
   (e) $x \cdot x \cdot x \cdot x \cdot x$
   (f) $(-y)(-y)(-y)$

4. Rewrite without parentheses.
   (a) $(-2)^2$
   (b) $(-x)^3$
   (c) $-(-y)^3$
   (d) $(-3)^4$
   (e) $(-1)^7$
   (f) $(-1)^{12}$

5. Multiply.
   (a) $c \cdot c^3$
   (b) $d^2 \cdot d^3$
   (c) $2x^2 \cdot 3x^5$
   (d) $-2a^5 \cdot 3a$
   (e) $x^2 \cdot x \cdot x^5$
   (f) $-5t^3 \cdot t^4(6t^2)$

6. Simplify.

   (a) $\dfrac{p^5}{p^2}$

   (b) $\dfrac{-s^6}{s^5}$

   (c) $\dfrac{(-a)^3}{a^3}$

   (d) $-\dfrac{54p^4}{-6p}$

   (e) $\dfrac{-12a^3b^2}{-6a^2b^2}$

   (f) $\dfrac{7x^2y^3z^5}{7x^2y^3z^5}$

7. Find the value.

   (a) 5^0

   (b) $(-5)^0$

   (c) -5^0

   (d) $3^0 + 2^0$

   (e) $a^0 \cdot a^1 \cdot a^2$

   (f) $(a + 3a + 5)^0$

8. Write with positive exponents only.

   (a) 5^{-1}

   (b) 2^{-3}

   (c) x^{-2}

   (d) $\dfrac{1}{3^{-1}}$

   (e) $\dfrac{1}{4^{-2}}$

   (f) $\dfrac{1}{(2)^{-3}}$

   (g) $\dfrac{3a^2b^3c^4}{-3abc}$

   (h) $\dfrac{4r^{-2}s^5}{-4r^5s^{-3}}$

9. Distribute multiplication over addition or subtraction. Combine like terms.

   (a) $3(t - 5)$

   (b) $x(2x + 3)$

   (c) $-3r(r^2 - s^2)$

   (d) $3ab(b^2 - a^2)$

   (e) $3(c - 2) - 5(4c - 11)$

   (f) $-2xy(3x - y) + 6(x^2 - 4)$

10. Distribute division over addition or subtraction. Combine like terms.

    (a) $\dfrac{18x^2 + 27x}{3x}$

    (b) $\dfrac{15t^2v^2 - 5tv}{5t}$

    (c) $\dfrac{2r + 3s - 4r^2}{-r^2}$

    (d) $\dfrac{21s^2 - 7s + 35}{s^2}$

    (e) $\dfrac{15a - 9}{3} - \dfrac{14a^2 + 35a}{a^2}$

    (f) $\dfrac{x^2 - 6x - 9}{3} - \dfrac{6x^2 + x}{-x}$

11. Factor completely.

    (a) 36

    (b) 120

    (c) $-15a^3b^2$

    (d) $23a^2b^0c$

    (e) $3t^2 - 15ts$

    (f) $16y - 18y^2$

    (g) $x^2y - 2xy$

    (h) $-\frac{1}{2}p^2q + \frac{1}{6}pq^2$

    (i) $5a^2bc + 10ab^2c - 100abc^2$

    (j) $0.01x^2y^3 + 0.1x^3y^2 - 0.02x^2y^2$

12. Evaluate for $x = 2$, $y = -3$, $z = 4$.

    (a) $3x^2$

    (b) $2x - y$

    (c) $x^2 + 2y^2 - z^2$

    (d) $3(x + 1) - 2(y + 3) - z$

    (e) $x(x + 1) - 3x(x + 2) - x^2$

    (f) $xyz - 2xy + 3xz$

13. Find the area of a triangle if its base is 10 m and its height is 6 m.

14. Find the perimeter of a rectangle when its length is 16 feet and its width is 5 feet.

15. Find the area of a circle with a radius of 7.5 cm.

16. When the rate is 70 miles per hour and the time is $3\frac{1}{2}$ hours, what is the distance traveled?

17. What is the temperature in Fahrenheit when the temperature in Celsius is 100°?

18. What is the temperature in Celsius when the temperature in Fahrenheit is 32°?

READINESS CHECK

Solve the problems to satisfy yourself that you have mastered Chapter 7.

   1. Simplify: $3x + y - 5x + 4y$

   2. Simplify: $5x - 4(x - 5)$

3. Simplify: $(2x^5)(3x^4)$

4. Simplify: $\dfrac{25x^6}{5x^4}$

5. Simplify: $(3x^2)^4$

6. Write with positive exponents: x^{-5}

7. Evaluate: $4^0 + 1^0$

8. Factor completely: $5x^2 + 10x$

9. Evaluate: $a - b$ for $a = 1$, $b = -1$

10. Convert $5°$ F to Celsius degrees. Use the formula $C = \frac{5}{9}(F - 32)$.

EQUATIONS AND INEQUALITIES

Mathematical sentences

In Robert Recorde's *The Whetstone of Witte*, published in 1557, we find the first use of the *equals sign*, = . Recorde said

> And to avoide the tediouse repetition of these woordes: is equalle to: I will sette as I doe often in woorte use, a pair of paralleles or Gemowe (twin) lines of one lengthe, thees: = , bicause noe two thyngs con be moare equalle.

The first known record of a solved equation is contained in an old papyrus manuscript copied by Ahmes, an Egyptian, in the 17th century B.C. from an earlier manuscript written in the 19th century B.C. Its title was *"Rules for enquiering into nature, and for knowing all that exists, every mystery, every secret."* In this document, the oldest "book" on mathematics, the unknown quantity is called *hau* (heap), and the first problem reads, "Hau, its seventh, its whole makes nineteen," which will translate in this chapter into $\frac{1}{7}x + x = 19$.

In the Egyptian solution, the quantity "hau" is $16\frac{1}{2} + \frac{1}{8}$, and one-seventh of the quantity equals $2\frac{1}{4} + \frac{1}{8}$. (Remember, the Egyptians only used fractions with a numerator of 1.) In modern terms,

$$16\frac{1}{2} + \frac{1}{8} = 16\frac{5}{8} \qquad \text{and} \qquad \frac{1}{7} \text{ of } 16\frac{5}{8} = \frac{1}{7} \cdot \frac{133}{8} = \frac{133}{56} = \frac{19}{8} = 2\frac{3}{8}$$

Furthermore, $2\frac{3}{8} + 16\frac{5}{8} = 18\frac{8}{8} = 19$.

After this it took approximately 3400 years for the symbols we use in algebra to be developed. The equals sign is an example of the new symbols.

8.1 SOLVING EQUATIONS

Equations in One Variable

You may have seen examples of *equation* problems that look like the following:

    Put the right number in the box.

$$\square + 2 = 7$$
$$\square - 4 = 9$$
$$3(\square) = 15$$
$$\frac{\square}{2} = 8$$
$$3(\square) + 2 = 14$$

The correct answers are 5, 13, 5, 16, and 4.

    In algebra we use letters instead of boxes, but the problems are the same. Find the *number* to replace the letter:

$$p + 2 = 7; \qquad p =$$
$$s - 4 = 9; \qquad s =$$
$$3x = 15; \qquad x =$$
$$\frac{y}{2} = 8; \qquad y =$$
$$3z + 2 = 14; \qquad z =$$

The correct answers are again 5, 13, 5, 16, and 4.

    Did you recognize these equations as the same?

    When thinking about equations, consider an old-fashioned balance scale. To keep the scale balanced, whatever you do to one side must be done to the other. If you add 2 pounds to one side, you must add 2 pounds to the other. If you take away 2 pounds from one side, you must also take 2 pounds from the other (see Figure 8.1).

Figure 8.1

We must do the same with an equation.

    In solving an equation, our task is to find the value for the variable, whether it is n, x, y, or some other letter. But in doing that we must remember to keep the equation balanced.

EXAMPLE

Solve for n:

(a) $n + 3 = 12$ (b) $n - 3 = 12$ (c) $3n = 12$ (d) $\dfrac{n}{3} = 12$

Solution

(a)
$$n + 3 = 12$$
$$\; -3 \quad -3$$
$$\overline{n = \quad 9}$$

To find the value of n, we must isolate it on one side of the equation. To do this we subtract 3 from *both* sides.

To make sure our answer is right, we *check* it. This means we replace the n in the original equation with 9.

Check: $9 + 3 = 12\sqrt{}$ $n = 9$ is correct.

(b)
$$n - 3 = 12$$
$$\; +3 \quad +3$$
$$\overline{n = \quad 15}$$

This time to isolate the n, we add 3 to both sides.

Check: $15 - 3 = 12\sqrt{}$ $n = 15$ is the correct answer.

(c)
$$3n = 12$$

$$\frac{3n}{3} = \frac{12}{3}$$

This time we divide both sides by 3.

$$n = 4$$

Check: $3(4) = 12\sqrt{}$ $n = 4$ is the correct answer.

(d)
$$\frac{n}{3} = 12$$

$$\frac{3}{1} \cdot \frac{n}{3} = \frac{3}{1} \cdot 12$$

Multiply by 3.
(Remember: 3 is the same as 3/1.)

$$n = 36$$

Check: $\dfrac{36}{3} = 12\sqrt{}$ $n = 36$ is correct.

Now let's look at examples that require more steps to isolate the variable.

EXAMPLE

Solve for n: $2n + 3 = 11$.

Solution   Here we have two steps; we must remove both the 3 and the 2. To remove the 3 we must subtract, and to remove the 2 we must divide. We can do either first, but let's take the easier way and subtract the 3 from each side.

$$2n + 3 = 11$$
$$\; -3 \quad -3$$
$$\overline{2n = \quad 8}$$

Now we divide each side by 2:

$$\frac{2n}{2} = \frac{8}{2}$$
$$n = 4$$

Check: $2(4) + 3 = 11\sqrt{}$ The answer $n = 4$ is correct.

Suppose we had solved the equation in the other order, first dividing, then subtracting.

Alternative Solution

$$2n + 3 = 11$$

$$\frac{2n + 3}{2} = \frac{11}{2}$$

$$\frac{2n}{2} + \frac{3}{2} = \frac{11}{2} \qquad$$ Remember the distributive property, and divide each term on each side by 2.

$$n + \frac{3}{2} = \frac{11}{2}$$

$$-\frac{3}{2} \quad -\frac{3}{2} \qquad$$ Now subtract $\frac{3}{2}$ from both sides.

$$n = \frac{8}{2}$$

$$n = 4$$

We get the same answer, but the process is more complicated. As you gain more experience with equations, you will find the easiest way for you.

EXAMPLE

Solve for x:
(a) $3x + 5 = 17$ (b) $5x - 7 = 3$ (c) $-2x - 3 = 9$
(d) $-x + 2 = 4$ (e) $\frac{2}{3}x = 16$ (f) $5 = 4x - 3$

Solution

(a)

$$3x + 5 = 17 \qquad$$ Subtract 5 from both sides.
$$\underline{-5 \qquad -5}$$
$$3x = 12$$

$$\frac{3x}{3} = \frac{12}{3} \qquad$$ Divide each side by 3.

$$x = 4$$

Check: $3(4) + 5 = 17\sqrt{} \qquad x = 4$ is correct.

(b)

$$5x - 7 = 3$$
$$\underline{+7 \qquad +7} \qquad$$ Add 7 to both sides.
$$5x = 10$$

$$\frac{5x}{5} = \frac{10}{5} \qquad$$ Divide both sides by 5.

$$x = 2$$

Check: $5(2) - 7 = 3\sqrt{} \qquad x = 2$ is correct.

(c)

$$-2x - 3 = 9$$
$$\underline{+3 \qquad +3} \qquad$$ Add 3 to both sides.
$$-2x = 12$$

$$\frac{-2x}{-2} = \frac{12}{-2} \qquad$$ Divide both sides by -2.

$$x = -6$$

Check: $-2(-6) - 3 = 12 - 3 = 9\sqrt{} \qquad x = -6$ is correct.

(d)
$$-x + 2 = 4$$
$$\underline{\quad -2 \quad -2}$$ Subtract 2 from both sides.
$$-x \quad = \quad 2$$

We are not done yet. Our goal is to solve for x, not $-x$.

$$\frac{-x}{-1} = \frac{2}{-1}$$ Divide both sides by -1.

$$x = -2$$

Check: $-(-2) + 2 = 2 + 2 = 4\sqrt{}$ $x = -2$ is correct.

(e)
$$\frac{2}{3}x = 16$$

$$\frac{3}{2} \cdot \frac{2}{3}x = \frac{3}{2} \cdot 16$$ Multiply both sides by $\frac{3}{2}$ (the reciprocal of $\frac{2}{3}$) or multiply by 3 and then divide by 2.

$$x = 24$$

Check: $\frac{2}{3}(24) = 16\sqrt{}$ $x = 24$ is correct.

(f)
$$5 = 4x - 3$$
$$\underline{+3 \qquad +3}$$ Add 3 to both sides.
$$8 \quad 4x$$

$$\frac{8}{4} = \frac{4x}{4}$$

$$2 = x, \text{ or } x = 2 \ (x = 2 \text{ is the same as } 2 = x)$$

Check: $4(2) - 3 = 8 - 3 = 5\sqrt{}$ $x = 2$ is correct.

EXERCISE 8.1.1

1. Solve for x and check.
    (a) $x - 4 = 6$ (b) $3x = 9$

    (c) $-4x = 12$ (d) $\dfrac{x}{2} = 4$

    (e) $-\dfrac{x}{3} = 6$ (f) $\dfrac{2x}{5} = 4$

    (g) $15 = \dfrac{5x}{3}$ (h) $72 = \dfrac{9}{8}x$

2. Solve in two steps and check.
    (a) $2x + 5 = 7$ (b) $4 + 8m = 52$
    (c) $4x + 7 = 31$ (d) $2x - 1 = 6$
    (e) $2 - y = 5$ (f) $-9 = 4y - 1$
    (g) $\dfrac{1}{3}x + 6 = 15$ (h) $\dfrac{5x}{8} - 3 = 7$

Equations with Variables on Both Sides

Although the examples that follow are more complicated, the mathematical process is the same. In these examples there are variable terms on both sides of the equation. You have a choice about the order in which you simplify the equation. Remember, though, that whatever you do to one side of the equation, you must do to the other.

For example, to solve for x in $5x + 3 = 3x + 7$, we do the following:

$$
\begin{aligned}
5x + 3 &= 3x + 7 \\
-3x & -3x \qquad \text{Subtract } 3x \text{ from both sides.} \\
\hline
2x + 3 &= 7 \\
-3 & -3 \qquad \text{Subtract 3 from both sides.} \\
\hline
2x &= 4 \\
\end{aligned}
$$

$$\frac{2x}{2} = \frac{4}{2} \qquad \text{Divide both sides by 2.}$$

$$x = 2$$

To check this type of equation, we go back to the original equation and substitute the solution on both sides. It is common practice to have a question mark over the equals sign until we know that both sides of the equation are equal.

Check: $5(2) + 3 \overset{?}{=} 3(2) + 7$
$ 13 = 13\surd \qquad x = 2$ is correct.

EXAMPLE

Solve for x.
(a) $2x - 5 + 3x = 11 + 4x - 6$
(b) $3(x - 5) = x + 5$
(c) $2x - (3 - x) = 6x$
(d) $3 - x = 7 - 2x$

Solution   In these examples, simplify each side first.

(a)
$$
\begin{aligned}
2x - 5 + 3x &= 11 + 4x - 6 \\
5x - 5 &= 5 + 4x \qquad \text{Simplify.} \\
+5 & +5 \qquad \text{Add 5.} \\
\hline
5x &= 10 + 4x \\
-4x & -4x \qquad \text{Subtract } 4x. \\
\hline
x &= 10 \\
\end{aligned}
$$

Check: $2(10) - 5 + 3(10) \overset{?}{=} 11 + 4(10) - 6$

$ 20 - 5 + 30 \overset{?}{=} 11 + 40 - 6$

$ 45 = 45\surd \qquad x = 10$ is correct.

(b)
$$
\begin{aligned}
3(x - 5) &= x + 5 \\
3x - 15 &= x + 5 \qquad \text{Simplify.} \\
+15 & +15 \qquad \text{Add 15 to both sides.} \\
\hline
3x &= x + 20 \qquad \text{Subtract } x \text{ from both sides.} \\
-x & -x \\
\hline
2x &= 20 \\
\end{aligned}
$$

$$\frac{2x}{2} = \frac{20}{2} \qquad \text{Divide both sides by 2.}$$

$$x = 10$$

Check: $3(10 - 5) \overset{?}{=} 10 + 5$
$$3(5) = 15\surd \quad x = 10 \text{ is correct.}$$

(c)

$$2x - (3 - x) = 6x$$
$$2x - 3 + x = 6x$$
$$3x - 3 = 6x$$
$$\underline{-3x \qquad -3x}$$
$$\frac{-3}{3} = \frac{3x}{3}$$
$$-1 = x$$

Simplify. Remember that the minus before the parentheses tells you to change the sign of each term inside the parentheses.
Subtract $3x$ from both sides.

Divide both sides by 3.

Check: $2(-1) - (3 - (-1)) \overset{?}{=} 6(-1)$
$$-2 - (3 + 1) \overset{?}{=} -6$$
$$-2 - 4 = -6\surd \quad x = -1 \text{ is correct.}$$

(d)

$$3 - x = 7 - 2x$$
$$\underline{+2x \qquad +2x}$$
$$3 + x = 7$$
$$\underline{-3 \qquad -3}$$
$$x = 4$$

Add $2x$ to both sides,

Subtract 3 from both sides.

Check: $3 - 4 \overset{?}{=} 7 - 2(4)$
$$-1 = 7 - 8\surd \quad x = 4 \text{ is correct.}$$

EXERCISE 8.1.2

Solve. Check each answer to be sure it is correct.

1. $4x + 17 = 10x + 5$
2. $x - 2x + 2 = 3x + 4$
3. $5a + 2 = 3a - 7$
4. $2(x + 1) = x - 11$
5. $1 - (t - 1) = 2t$
6. $a - 2(a - 2) = 2(1 - 2a)$
7. $3x + 2(x + 2) = 13 - (2x - 5)$
8. $4x + 5 - 3(2x - 5) = 9 - (8x - 1) + x$

8.2 SOLVING EQUATIONS WITH FRACTIONS AND DECIMALS

When fractions occur in an equation, we can draw on the skills we learned when we worked with fractions and common denominators in Chapter 4 to simplify our work.
For example,

$$\frac{x}{2} + \frac{x}{3} = 5$$

This may look complicated. But if we multiply both sides by the least common denominator, we can clear the equation of fractions. Here, 6 is the LCD.

$$6\left(\frac{x}{2} + \frac{x}{3}\right) = 6(5) \quad \text{Multiply both sides by 6.}$$

$$6\left(\frac{x}{2}\right) + 6\left(\frac{x}{3}\right) = 30 \quad \text{Reduce and simplify.}$$

$$3x + 2x = 30$$

$$5x = 30$$

$$\frac{5x}{5} = \frac{30}{5}$$

$$x = 6$$

Check: $\dfrac{6}{2} + \dfrac{6}{3} = 3 + 2 = 5\sqrt{}$ $x = 6$ is correct.

EXAMPLE

Solve: $\dfrac{x}{3} + \dfrac{x}{4} = 14$

Solution   Again, let's clear the fractions.

$$12\left(\frac{x}{3} + \frac{x}{4}\right) = 12(14) \quad \text{Multiply both sides by the LCD, in this case 12.}$$

$$12\left(\frac{x}{3}\right) + 12\left(\frac{x}{4}\right) = 12(14)$$

$$4x + 3x = 168 \quad \text{Simplify.}$$

$$7x = 168 \quad \text{Divide both sides by 7.}$$

$$x = 24$$

Check: $\dfrac{24}{3} + \dfrac{24}{4} = 8 + 6 = 14\sqrt{}$ $x = 24$ is correct.

Equations with decimals can be handled in a similar way. Multiplying both sides by a power of 10 will clear the decimals from the equation.

EXAMPLE

Solve: $0.1x + 0.01x = 1.21$

Solution   If we multiply by 10, we will still have a decimal in the second term. So let's multiply by 100.

$$100(0.1x + 0.01x) = 100(1.21)$$

$$10x + x = 121$$

$$11x = 121$$

$$x = 11$$

Check: $0.1(11) + 0.01(11) \stackrel{?}{=} 1.21$

$1.1 + 0.11 = 1.21\sqrt{}$ $x = 11$ is correct.

EXAMPLE

Solve for x. (a) $4\% \, x = 15$ (b) $x\% \cdot 45 = 9$ (c) $\dfrac{1}{2} = \dfrac{1}{x} + \dfrac{1}{6}$

Solution

(a) $4\% x = 15$ First rewrite the equation in decimal form.

$0.04x = 15$ $4\% = \dfrac{4}{100}$ or 0.04.

$4x = 1500$ Multiply both sides by 100.

$x = 375$ Divide both sides by 4.

Check: 4% of $375 \overset{?}{=} 15$
$(0.04)(375) \overset{?}{=} 15$
$15 = 15\sqrt{}$ $x = 375$ is correct.

(b) $x\% \cdot 45 = 9$

$\dfrac{x}{100} \cdot 45 = 9$

$x \cdot 45 = 900$ Multiply both sides by 100.

$x = 20$ Divide both sides by 45.

Check: 20% of $45 \overset{?}{=} 9$
$(0.20)(45) \overset{?}{=} 9$
$9 = 9\sqrt{}$ $x = 20$ is correct.

(c) $\dfrac{1}{2} = \dfrac{1}{x} + \dfrac{1}{6}$ The least common denominator is $6x$.

$6x\left(\dfrac{1}{2}\right) = 6x\left(\dfrac{1}{x}\right) + 6x\left(\dfrac{1}{6}\right)$ Multiply both sides by $6x$.

$3x = 6 + x$ Simplify and solve.

$\underline{\quad -x \qquad\quad -x\quad}$

$2x = 6$

$\dfrac{2x}{2} = \dfrac{6}{2}$

$x = 3$

Check: $\dfrac{1}{2} \overset{?}{=} \dfrac{1}{3} + \dfrac{1}{6}$

$\dfrac{1}{2} \overset{?}{=} \dfrac{2}{6} + \dfrac{1}{6}$

$\dfrac{1}{2} = \dfrac{3}{6}\sqrt{}$ $x = 3$ is correct.

EXERCISE 8.2.1

1. Solve.

(a) $\dfrac{2}{5}x + \dfrac{4}{5}x = 36$

(b) $\dfrac{1}{4}t + \dfrac{1}{2}t = 24$

(c) $\dfrac{3a}{4} + 2 = 17$

(d) $\dfrac{c}{3} - \dfrac{2}{5} = 0$

(e) $\dfrac{2(x-3)}{3} = -4$

(f) $\dfrac{3(2-2x)}{2} = x$

(g) $\dfrac{2(x-1)}{3} = \dfrac{4}{5}$

(h) $\dfrac{2x+3}{6} - \dfrac{x}{6} = 0$

(i) $\dfrac{x}{3} - \dfrac{x}{5} = 2$

(j) $\dfrac{2}{3}x + \dfrac{1}{7}x = 34$

(k) $1 + \dfrac{x}{3} = 2 - \dfrac{x}{3}$

(l) $\dfrac{1}{4} = \dfrac{1}{x} + \dfrac{1}{3}$

2. (a) $0.2x + 0.7 = 1.7 - 0.3x$
    (b) $0.15 = 0.05(1 + x)$
    (c) $x \cdot 12\% + (12 - x) \cdot 8\% = 12 \cdot 9\%$
    (d) $8\%x + (10{,}000 - x) \cdot 10\% = 870$

8.3 FORMULAS

In Chapter 7 we introduced formulas and used them to demonstrate some applications of evaluating expressions. For example, in $A = lw$, if we know the values of l and w we can find A. Here, we have the same formula $A = lw$, but now we know the values of A and either l or w and we solve the equation to find the value of the unknown variable.

Geometry

Rectangles If we know the area or perimeter of a rectangle and one of the dimensions, we can find the other dimension.

EXAMPLE

If the area of a rectangle is 36 square inches and the length is 9 in., find the width.

Solution   Formula: $A = lw$. The rectangle is shown in Figure 8.2.

$36 = 9w$ Replace A and l with the known values.

$\dfrac{36}{9} = \dfrac{9w}{9}$ Divide both sides by 9.

$w = 4$ The width of the rectangle is 4 in.

Figure 8.2

EXAMPLE

Find the width of the rectangle in Figure 8.3. It has a length of 6 cm and a perimeter of 20 cm.

Solution   Formula: $P = 2l + 2w$

$$20 = 2(6) + 2w$$
$$20 = 12 + 2w \qquad \text{Replace } l + w \text{ with the known values.}$$
$$8 = 2w \qquad \text{Subtract 12 from both sides.}$$
$$4 = w \qquad \text{Divide both sides by 2.}$$

Check: $20 \overset{?}{=} 2(6) + 2(4)$

$20 = 12 + 8\surd$ The width of the rectangle is 4 cm.

Figure 8.3

Triangles The formula for the area of a triangle is $A = \frac{1}{2}bh$, where the area is A, the base is b, and the height (altitude) is h.

EXAMPLE

Find the base of a triangle with area $= 12$ in.^{2} and $h = 6$ in.

Solution $12 = \frac{1}{2}b(6)$

$12 = 3b \qquad$ Simplify.

$4 = b \qquad$ Divide both sides by 3.

The base of the triangle is 4 in.

Check: $12 \overset{?}{=} \frac{1}{2}4(6) \qquad\qquad 12 = 12\surd$

Trapezoids A trapezoid is a geometric figure with two sides that are parallel and two sides that are not. See Figure 8.4.

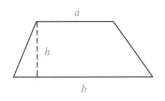

Figure 8.4

The formula for the area of a trapezoid is

$$A = \tfrac{1}{2}h(a + b)$$

EXAMPLE

Find the height of a trapezoid when the area is 60 cm^2 and the bases (the parallel sides—see the figure) are 8 cm and 12 cm.

Solution $60 = \tfrac{1}{2}h(8 + 12)$

$$60 = \tfrac{1}{2}h(20)$$

$$60 = 10h$$

$$6 = h$$

The height is 6 cm.
Check: $\tfrac{1}{2}(6)(8 + 12) = 3(20) = 60\surd$

Physics

Density If you fill a glass bottle with water, close it, and put it in the freezer, the glass will crack. Why?

When water freezes, the molecules spread out to form an open structure with holes in it. The same number of frozen water molecules take up more space than water molecules at higher temperatures. In other words, the density of ice is less than the density of water. The relationship between density, mass (weight), and volume is

$$\text{Density} = \frac{\text{mass}}{\text{volume}} \qquad \text{or} \qquad d = \frac{m}{v}$$

EXAMPLE

Ice has a density of 0.9 kg/L. Find the volume that 1 kilogram of ice occupies.

Solution   Formula: $d = m/v$

$$0.9 = \frac{1}{v} \qquad \text{or} \qquad v = \frac{1}{0.9} = 1.11$$

Although 1 kg of water has a volume of 1 L (see Chapter 6), 1 kg of ice has a volume of 1.11 L. The water expands as it freezes, and the bottle cracks!

EXAMPLE

Find the volume of a substance with density 7.13 g/cm^3 and mass 321 g.

Solution   Use the formula $d = \dfrac{m}{v}$.

$$7.13 = \frac{321}{v} \qquad \text{or} \qquad v = \frac{321}{7.13} = 45.0$$

Answer: 45.0 cm^3

Distance, Rate, and Time Problems

EXAMPLE

How long does it take (in hours) to drive 600 miles at a rate of 50 miles per hour (mph)?

Solution   Formula: $d = rt$. Here $d = 600$ miles and $r = 50$ mph. We are looking for the time t.

$$600 = 50t$$
$$12 = t \qquad \text{It takes 12 hours to drive the 600 miles.}$$

Temperature Conversions The following examples illustrate the conversion of temperature between Fahrenheit and Celsius values.

EXAMPLE

Find the temperature in degrees Fahrenheit if it is $10°$ C. In other words, find F when C = 10.

Solution   Formula: $C = \frac{5}{9}(F - 32)$

$$10 = \frac{5}{9}(F - 32) \qquad \text{Substitute 10 for C.}$$
$$9(10) = 9\left[\frac{5}{9}(F - 32)\right] \qquad \text{Multiply both sides by 9 to clear the fraction.}$$
$$90 = 5F - 160 \qquad \text{Simplify.}$$
$$250 = 5F \qquad \text{Add 160 to both sides.}$$
$$50 = F \qquad \text{Divide both sides by 5.}$$

The temperature is $50°$ F.

You could have used the formula that translates directly from Celsius to Fahrenheit degrees. These two formulas are actually the same.

EXAMPLE

Solve $F = \frac{9}{5}C + 32$ for C.

Solution   Multiply both sides by 5.

$$5F = 9C + 160$$
$$5F - 160 = 9C \qquad \text{Subtract 160 from both sides.}$$
$$5(F - 32) = 9C \qquad \text{Factor out 5.}$$
$$\frac{5}{9}(F - 32) = C \qquad \text{Divide by 9.}$$

The result is the formula we used earlier.

Business

EXAMPLE

What is the interest rate if you borrow $1000 for 6 months and have to pay $50 interest?

Solution   Formula: $I = Prt$. We know $I = 50$, $P = 1000$, $t = 0.5$, and we need to solve for r.

$$50 = 1000(r)(0.5)$$

$$50 = 500r \qquad \text{Simplify.}$$

$$\frac{50}{500} = r \qquad \text{Divide both sides by 500.}$$

$$0.1 = r$$

$r = 0.1 = 10\%$ (We express rate as % when we are dealing with interest.)

EXERCISE 8.3.1

1. Use the geometry formulas to find
   (a) the length of a rectangle when the area is 36 square inches and the width is 4 inches.
   (b) the height of a triangle when the area is 48 square centimeters and the base is 12 centimeters.
   (c) the height of a trapezoid when the area is 18 cm² and the parallel sides 1 cm and 3 cm, respectively.
   (d) the other parallel side of a trapezoid when the area is 36 cm², the height is 8 cm, and one parallel side is 5 cm.

2. Use the interest formula $I = Prt$ to find
   (a) t when $I = \$600$, $P = \$7500$, and $r = 2\%$.
   (b) P when $I = \$400$, $r = 5\%$, and $t = 2$ years.

3. A is the accumulated value of money in the bank when P is the principal, r is the interest rate, and t is the time. Use the formula $A = P + Prt$ to find
   (a) t when $A = \$4400$, $P = \$4000$, and $r = 5\%$.
   (b) r when $A = \$104$, $P = \$80$, and $t = 5$ years.

4. The mathematical formula for a straight line is $y = mx + b$.
   (a) Find b when $y = 0$, $m = -2$, and $x = 1$.
   (b) Find m when $y = 5$, $x = 2$, and $b = -3$.

5. Use the formula for temperature conversion $C = \frac{5}{9}(F - 32)$ to find
   (a) F when C = 0
   (b) F when C = -40

6. Use the physics formula $d = \dfrac{m}{v}$ to find
   (a) The mass m when the density $d = 2$ g/cm³ and the volume $v = 5$ cm³.
   (b) v when $d = 2$ g/cm³ and $m = 16$ g.

7. Use the formula that connects distance, rate, and time, $d = r \cdot t$, to find the time when the distance is 240 miles and the rate is 50 miles per hour.

8. Find the rate in miles per hour when Jean drives 115 miles in $2\frac{1}{2}$ hours.

9. How far does a cyclist go in 9 hours if he travels at a speed of 250 m/min? (*Hint*: Change the hours to minutes before you proceed.)

10. How long does it take a snail to move 34 m if its speed is 4 dm/min? (*Hint*: Change meters to decimeters.)

When we solve linear equations, we get a single number as the solution. For example, $x = 7$ or $y = -35$.

But many times a relation is not known exactly. Rather than using an equals sign ($=$), we may use one of these:

$>$ (is greater than)

$<$ (is less than)

$\leq$ (is less than or equal to)

$\geq$ (is greater than or equal to)

If the distance x is more than 100 miles to Boston, then we write

$$x > 100$$

Similarly, if it is less than 100 miles to Hartford, then

$$x < 100$$

If I have *at most* $15, the money I have is less than or equal to $15:

$$x \leq 15$$

If I have *at least* $10, the money I have is greater than or equal to 10:

$$x \geq 10$$

In many cases we can solve an inequality the same way we solve an equation: what we do to one side we do to the other. The exception, as we will see, is if we multiply or divide by a negative number.

EXAMPLE

Solve: (a) $x + 3 < 7$ (b) $3x < 12$ (c) $\frac{1}{2}x > 9$

Solution

(a)
$$\begin{array}{rl} x + 3 < & 7 \\ -3 & -3 \\ \hline x \quad < & 4 \end{array}$$
Subtract 3 from both sides.

(b)
$$3x < 12$$
$$\frac{3x}{3} < \frac{12}{3}$$
Divide both sides by 3.
$$x < 4$$

(c)
$$\frac{1}{2}x > 9$$
$$2\left(\tfrac{1}{2}x\right) > 2(9)$$
Multiply both sides by 2.
$$x > 18$$

There can be variable terms on both sides of an equality just as in an equation.

EXAMPLE

Solve: $5x + 3 < 2x + 15$

Solution

$$
\begin{array}{rl}
5x + 3 < & 2x + 15 \qquad \text{Subtract 3 from both sides.} \\
\underline{-3} & \underline{-3} \\
5x < & 2x + 12 \qquad \text{Subtract } 2x \text{ from both sides.} \\
\underline{-2x} & \underline{-2x} \\
3x < & 12 \qquad \text{Divide both sides by 3.} \\
x < & 4
\end{array}
$$

Let's now consider -2 and 18. Certainly $-2 < 18$. What happens if we divide -2 and 18 by negative 2?

$$\frac{-2}{-2} = 1 \qquad \text{and} \qquad \frac{18}{-2} = -9$$

but since $1 > -9$ we now have

$$\frac{-2}{-2} > \frac{18}{-2}$$

In other words, *when we divide an inequality by a negative number, the inequality sign gets reversed.*

EXAMPLE

Multiply: $7 < 9$ by -3

Solution $7(-3) = -21 \qquad$ and $\qquad 9(-3) = -27$

$-21 > -27 \qquad$ (Refer to the number line; the number to the left
$7(-3) > 9(-3) \qquad$ is the smaller number.)

When we multiplied the inequality by a negative number, the direction of the inequality sign was reversed. The same is true, of course, when variables are included in the inequality.

EXAMPLE

Solve for x. (a) $-2x < 14 \qquad$ (b) $\dfrac{x}{-3} > 5$

Solution
(a) $-2x < 14$

$$\frac{-2x}{-2} > \frac{14}{-2} \qquad \text{Divide both sides by } -2 \text{ and reverse the inequality sign.}$$

$$x > -7$$

Check: Let x be a number > -7, say 0. $-2(0) = 0 < 14\sqrt{}$

(b) $\qquad \dfrac{x}{-3} > 5$

$$-3\left(\frac{x}{-3}\right) < -3(5) \qquad \text{Multiply both sides by } -3 \text{ and reverse the inequality sign.}$$

$$x < -15$$

Check: Let $x = -18$. $\dfrac{-18}{-3} = 6 > 5\sqrt{}$

> **RULE**
>
> **Operations on Inequalities**
>
> You may add any positive or negative number to both sides of an inequality.
>
>   You may multiply or divide both sides by any *positive* number.
>
>   If you multiply or divide both sides of an inequality by a negative number, reverse the direction of the inequality sign.

EXERCISE 8.4.1

1. Solve in one step.
   (a) $3x < 27$
   (b) $-3x < 27$
   (c) $x + 3 > 8$
   (d) $x - 4 < 4$
   (e) $\dfrac{x}{3} < 3$
   (f) $-\dfrac{x}{2} > 4$
   (g) $x - 6 < -4$
   (h) $-4 + x > 12$

2. Solve in two steps.
   (a) $3x + 5 < 17$
   (b) $8x - 12 > 36$
   (c) $9x - x < 16$
   (d) $2x - 7x < 4$

3. Solve.
   (a) $5x + 3 < 3x + 8$
   (b) $9x - 5 > 4x + 5$
   (c) $5x + 10(12 - x) < 85$
   (d) $1 + x - 2x > x - 1$
   (e) $\dfrac{x - 2}{2} < \dfrac{x + 4}{4}$
   (f) $\dfrac{7 - x}{6} > \dfrac{x - 3}{2}$

8.5 SYSTEMS OF LINEAR EQUATIONS

The equation $x + y = 5$ has an infinite number of solutions. For example:

$$
\begin{array}{lll}
x = 2, & y = 3 & 2 + 3 = 5 \\
x = -1, & y = 6 & -1 + 6 = 5 \\
x = 0.5, & y = 4.5 & 0.5 + 4.5 = 5 \\
x = \tfrac{1}{2}, & y = \tfrac{9}{2} & \tfrac{1}{2} + \tfrac{9}{2} = 5 \\
x = 3, & y = 2 & 3 + 2 = 5
\end{array}
$$

and so on.

The equation $x - y = 1$ also has an infinite number of solutions, but there are only two numbers that *satisfy* (are solutions to) *both* equations. We say that the *system of equations*

$$x + y = 5$$
$$x - y = 1$$

has only one solution: $x = 3$, $y = 2$.

Check: $3 + 2 = 5\sqrt{}$ and $3 - 2 = 1\sqrt{}$

A system of equations can also be called *simultaneous equations*.

To find the value of the variables in simultaneous equations, we need two equations when we are working with two variables, three equations when we are working with three variables, and so on.

When we have more than one variable, we transform the different equations into one equation with one unknown by using either the *elimination* method or the *substitution* method. We can also solve equations with two variables by *graphing*, which will be discussed in Chapter 9.

The Elimination or Addition Method

To find the solutions to the following system of equations, we first *eliminate* one of the variables:

$$
\begin{array}{l}
x + y = 5 \\
\underline{x - y = 1} \\
2x \quad\;\; = 6
\end{array}
$$ Add the two equations to eliminate y.

$$x = 3$$

$$3 + y = 5$$ Substitute 3 for x in one of the original equations and solve for y.

$$y = 2$$

Check: $3 + 2 = 5$ and $3 - 2 = 1\sqrt{}$

EXAMPLE

Solve:

$$5x - 3y = 12$$
$$2x - 3y = 3$$

Solution

$$
\begin{array}{l}
5x - 3y = \quad 12 \\
\underline{-2x + 3y = \; -3} \\
3x \qquad = \quad\; 9
\end{array}
$$

To eliminate y, multiply the second equation by -1.

Add the equations.

$$x = 3$$ Divide both sides by 3.

$$5(3) - 3y = 12$$ Substitute 3 for x in one of the original equations and solve for y.

$$
\begin{array}{l}
15 - 3y = \quad 12 \\
\underline{-15 \qquad\quad -15} \\
\quad\; -3y = \; -3
\end{array}
$$

$$y = 1$$

Check: $5(3) - 3(1) = 15 - 3 = 12\sqrt{}$
$2(3) - 3(1) = 6 - 3 = 3\sqrt{}$

EXAMPLE

Solve the system of equations:

$$7x - 2y = 2$$
$$3x + 4y = 30$$

Solution

$$14x - 4y = 4$$
$$3x + 4y = 30$$

To eliminate y, multiply the first equation by 2.

$$17x = 34$$

Add the equations.

$$x = 2$$

Divide both sides by 2.

$$7(2) - 2y = 2$$
$$14 - 2y = 2$$

Substitute 2 for x in one of the original equations.

$$-14 -14$$
$$-2y = -12$$

Solve.

$$y = 6$$

Check: $7(2) - 2(6) = 14 - 12 = 2\sqrt{}$
$3(2) + 4(6) = 6 + 24 = 30\sqrt{}$

EXAMPLE

Solve the system:

$$2x + 3y = 6$$

$$3x + 4y = 7$$

Solution

$$6x + 9y = 18$$
$$-6x - 8y = -14$$
$$y = 4$$

Eliminate x by multiplying the first equation by 3 and the second equation by -2.
Add the equations.

$$2x + 3(4) = 6$$

Substitute 4 for y in one of the original equations.

$$2x + 12 = 6$$
$$2x = -6$$

Simplify and solve.

$$x = -3$$

Check: $2(-3) + 3(4) = -6 + 12 = 6\sqrt{}$
$3(-3) + 4(4) = -9 + 16 = 7\sqrt{}$

Alternative Method

$$8x + 12y = 24$$
$$\underline{-9x - 12y = -21}$$

Eliminate y by multiplying the first equation by 4 and the second equation by -3.

$$-x = 3$$

Add the equations.

$$x = -3$$

Divide by -1.

$$2(-3) + 3y = 6$$

Substitute -3 for x in one of the original

$$-6 + 3y = 6$$

equations. Simplify and solve.

$$3y = 12$$

$$y = 4$$

The answers are the same.

EXERCISE 8.5.1

Solve by the elimination method.

1. $x - 4y = 1$
    $x - 2y = 3$

2. $3x - y = 2$
    $x + y = 2$

3. $x - y = 0$
    $3x - 2y = 1$

4. $2x + 3y = 7$
    $x - y = 1$

5. $2x + 5y = 10$
    $5x + 2y = 4$

6. $6x - y = 9$
    $4x + 7y = 29$

The Substitution Method

There is another way to solve a system of two equations in two variables. Solve one of the equations for one variable (that is often done already). Substitute the expression into the other equation and solve for the only variable that's left. Then substitute the solution into one of the original equations to find the second variable.

For example, if $y = 2x$ and $3x + 2y = 21$, we substitute $2x$ for y in the second equation, getting $3x + 2(2x) = 21$, and solve for x.

$$3x + 4x = 21$$

$$7x = 21 \qquad \text{Simplify.}$$

$$x = 3 \qquad \text{Solve.}$$

$$y = 2(3) \qquad \text{Substitute 3 for } x \text{ in the other original}$$

$$y = 6 \qquad \text{equation.}$$

Check: $3(3) + 2(6) = 9 + 12 = 21\sqrt{}$

EXAMPLE

Solve the system:

$$y = x - 3$$
$$2x + 3y = 16$$

Solution

$$2x + 3(x - 3) = 16$$ Substitute $x - 3$ for y in the second equation. Simplify and solve.

$$2x + 3x - 9 = 16$$

$$5x - 9 = 16$$

$$5x = 25$$

$$x = 5$$

$$y = 5 - 3$$ Substitute 5 for x in the first original equation.

$$y = 2$$

Check: $2(5) + 3(2) = 10 + 6 = 16\sqrt{}$

EXAMPLE

Solve:

$$x + 2y = 10$$
$$2x - y = 5$$

Solution   Solve the first equation for x.

$$x = -2y + 10$$

$$2(-2y + 10) - y = 5$$ Substitute $x = -2y + 10$ in the second equation. Simplify and solve.

$$-4y + 20 - y = 5$$

$$-5y + 20 = 5$$

$$-5y = -15$$

$$y = 3$$

$$x = -2(3) + 10$$

$$x = -6 + 10$$

$$x = 4$$

Check: $4 + 2(3) = 4 + 6 = 10$
$2(4) - 3 = 8 - 3 = 5\sqrt{}$

Alternative Solution

$$2x - y = 5$$ Solve the second equation for y.
$$\underline{-2x \qquad\quad -2x}$$
$$-y = 5 - 2x$$
$$y = -5 + 2x$$

$$x + 2(-5 + 2x) = 10 \qquad \text{Substitute } -5 + 2x \text{ for } y \text{ in the first}$$
equation.

$$x - 10 + 4x = 10$$

$$5x - 10 = 10 \qquad \text{Simplify and solve.}$$

$$5x = 20$$

$$x = 4$$

$$y = -5 + 2(4) \qquad \text{Substitute 4 for } x \text{ in the second equation.}$$

$$y = -5 + 8$$

$$y = 3$$

EXERCISE 8.5.2

Solve by using the substitution method.

1. $x - y = 2$
   $2x + 2y = 8$

2. $y = 2x + 2$
   $x + y = 5$

3. $x - 3y = 0$
   $2x - y = 5$

4. $4x + y = -5$
   $2x - 5y = 3$

5. $2x - 3y = -2$
   $3x - y = 4$

6. $x + 2y = 2$
   $2x + y = 4$

7. $-3x + 2y = -1$
   $2x - y = 1$

8. $x + 3y = 7$
   $3x - y = 1$

SUMMARY

Formulas

(see also Chapter 7)

Geometry: Area of trapezoid $A = \frac{1}{2}h(a + b)$

Physics: Density $= \dfrac{\text{mass}}{\text{volume}}$ or $d = \dfrac{m}{v}$

Business: Interest = Principal × rate × time $I = Prt$
Accumulated value $A = P + Prt$
Mathematics: Straight line: $y = mx + b$

Rule

Operations on Inequalities

You may add any positive or negative number to both sides of an inequality.

You may multiply or divide both sides by any *positive* number.

If you multiply or divide both sides of an inequality by a negative number, reverse the direction of the inequality sign.

VOCABULARY

Addition method: Add two equations to eliminate one variable.

Elimination method: Same as addition method.

Equation: Two expressions that are equal.

Inequality: Two expressions that are not equal.

Satisfy an equation: A solution gives a true statement.

Simultaneous equations: Equations that are true at the same time.

Solving equations: Finding solutions.

Substitution method: Solve one equation for one variable. Substitute that expression into the other equation and solve for the second variable.

System of equations: Same as simultaneous equations.

CHECK LIST

Check the box for each topic you feel you have mastered. If you are unsure, go back and review.

Solving equations

☐ using one operation

☐ using more than one operation

☐ with variables on both sides of the equation

☐ with fractions

☐ with decimals

☐ with percents

☐ Checking equations

☐ Using formulas

☐ Solving inequalities

Solving systems of equations

☐ by the elimination method

☐ by the substitution method

REVIEW EXERCISES

1. Choose an operation, solve, and check to see if your answer is correct.
   (a) $x + 3 = 12$
   (b) $x - 3 = 12$
   (c) $3x = 12$
   (d) $\dfrac{x}{3} = 12$

2. Solve.
   (a) $n + 11 = 15$
   (b) $\dfrac{n}{7} = 3$

   (c) $n - 2 = 11$
   (d) $5n + 30 = 40$
   (e) $2n + 5 = 17$
   (f) $3n + 11 = 14$
   (g) $2n - 16 = 20$
   (h) $3n - 2 = 31$
   (i) $\dfrac{n}{5} + 3 = -1$
   (j) $\dfrac{q}{3} - 1 = 21$

(k) $\dfrac{2}{3}p - 15 = 9$

(l) $-n + 5 = -7$

(m) $-c + \dfrac{2}{3} = 4$

3. Solve.
   (a) $3x + 2 = x + 8$
   (b) $4x - 17 = x + 3$
   (c) $3d + 5 = 5d - 7$
   (d) $2 - 3(s + 1) = 2s$
   (e) $4(2 - p) = 3(2 - p)$

4. Solve.

   (a) $\dfrac{x}{2} - \dfrac{x}{3} = 11$

   (b) $\dfrac{x}{5} - \dfrac{x}{3} = -24$

   (c) $0.2x + 0.7 = 1.2 - 0.3x$

   (d) $1.3x - 0.02 = 1.9x + 0.16$

5. Use the given formula to find the missing value.
   (a) $P = 2(l + w);\ l = 10$ cm, $P = 30$ cm^2
   (b) $A = \frac{1}{2}bh;\ A = 25$ sq. in., $h = 5$ in.
   (c) $C = 2\pi r;\ C = 62.8$ cm, $\pi = 3.14$
   (d) $A = \dfrac{h(a + b)}{2};\ A = 50$ cm^2, $a = 8$ cm, $h = 5$ cm
   (e) $I = Prt;\ I = \$320,\ P = \$1000,\ r = 0.08$
   (f) $A = P + Prt;\ A = \$2720,\ P = \$2000,\ t = 6$ years
   (g) $F = \frac{9}{5}C + 32;\ F = -40°$
   (h) $C = \frac{5}{9}(F - 32);\ C = -40°$

6. Solve.
   (a) $x - 5 < 27$
   (b) $x - 5 < 18$
   (c) $3x + 2 > 14$
   (d) $\dfrac{x}{5} > 2$
   (e) $-3x < 15$
   (f) $3 - x > 5$
   (g) $2x + 3(5 - x) > 17$
   (h) $\dfrac{x - 3}{2} < \dfrac{3x - 5}{3}$

7. Solve by elimination.
   (a) $x + y = 5$
        $x - y = 1$
   (b) $3x + y = 4$
        $2x - 3y = -1$
   (c) $3x - 2y = 1$
        $2x - 3y = -1$
   (d) $x + 2y = 6$
        $3x + y = 8$

8. Solve by substitution.
   (a) $x = y + 2$
        $x + 2y = 5$
   (b) $y = 2x - 1$
        $2x + 3y = 5$
   (c) $x + y = 5$
        $2x + 3y = 12$
   (d) $2x + y = 5$
        $x - 2y = 10$

Solve the problems to satisfy yourself that you have mastered Chapter 8.

1. Solve: $8x + 4 = 36$
2. Solve: $\frac{5}{7}x = 35$
3. Solve: $x - 3(x - 5) = 2x - 11$
4. Solve: $\dfrac{x}{2} + \dfrac{x}{3} = 10$
5. Solve: $7.4x - 4 - 5.2x = 2.6$
6. Solve: $25\%(45) + 15x = 40\%(60)$
7. Find the base of a triangle when its area is 12 square centimeters and its height is 6 centimeters.
8. Solve: $-5x < 55$
9. Solve for x and y:

$$2x + 3y = 7$$
$$3x - 2y = 4$$

10. Solve for x and y:

$$y = 3x + 2$$
$$5x + 4y = 25$$

GRAPHING

The art in mathematics

Graphing was used in the ancient world by the Egyptians and the Romans in surveying and by the Greeks in mapmaking. However, it was not until the 14th century that graphing as we now know it had its origins in the works of the French mathematician Nicole Oresme. It was the one mathematical bright spot in that century of the Black Death and the Hundred Years War.

The idea of graphing with coordinate axes (pronounced ax-eeze) dates both to Descartes in the 17th century and before him to Apollonius in the 2nd century B.C. Graphing did not appear in the high school textbooks of the early 1900s. By the 1930s, however, even junior high teachers considered graphing to be an important part of algebra, one that could not easily be omitted. The rapid growth in its general importance was probably due to the use of graphic representation in statistics to show how one variable is dependent on another. Graphing is also an important part of navigation and of modern warfare.

A graph is a drawing that is intended to make a collection of data or a mathematical expression easier to understand. However, for many people the graph does nothing of the sort—it only serves to confuse. This chapter will try to change this.

9.1 DRAWING GRAPHS

To draw a graph yourself, start by drawing two axes, one horizontal and one vertical. You must decide what information should go along the horizontal axis and what should go along the vertical axis (Figure 9.1).

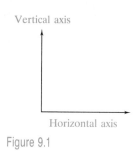

Figure 9.1

EXAMPLE

Draw a graph for the following data, which show the temperature for the week beginning January 1.

Date	Temperature (°F)
January 1	35
2	24
3	9
4	15
5	27
6	32
7	40

Solution   We usually place along the horizontal axis the numbers that change at a regular rate (the *independent variable*), in this case the days. Along the vertical axis we place the values that depend on the given value of the independent variable. These we call the *dependent variable.* In this case the temperature is the dependent variable because its value depends on which day the temperature was measured.

In this case the dependent numbers range from 9 to 40. Rather than filling in every number from 9 to 40, we could use intervals such as 5, 10, 15, . . . , 40 to represent this range. We write the dates along the horizontal axis, evenly spaced. Label both axes (Figure 9.2).

To mark the temperature for January 1 we go to 1 on the horizontal axis. The temperature on January 1 was 35° F, so we draw a point that is directly above 1 on the date axis and opposite 35 on the temperature axis (Figure 9.3). This point represents the temperature on January 1. For January 2, we mark a point above 2 and opposite where 24 would be on the

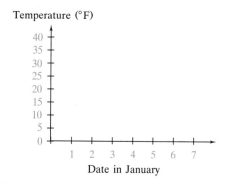

Figure 9.2

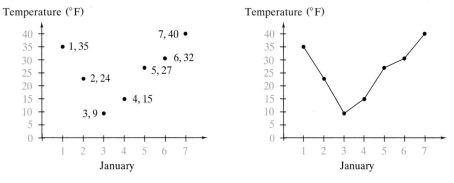

Figure 9.3

vertical axis. We continue marking points for the other dates until the graph looks like the one to the left in Figure 9.3. As a final step, we can connect the points to show how the temperature changed for the week of January 1 (right in the figure).

What conclusions can you draw about the temperature during the first week of January?

EXAMPLE

Plot the following student test scores.

Test	Average Score
1	72
2	81
3	96
4	43
5	92

Solution   Here the low score is 43 and the high score is 96; a good choice of intervals for the vertical scale would be $10, 20, 30, \ldots, 100$. The test numbers are the independent variable, so we list them across the horizontal axis. The average test score is the dependent variable—the score depends on which test we are looking at. We list the test scores along the vertical axis.

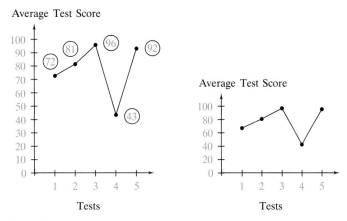

Figure 9.4

The average score for the first test is marked above the 1 (identifying the test) and across from where 72 would be on the vertical axis (identifying the score). Other scores are marked in a similar way. The points are then connected, showing how test scores varied. See Figure 9.4.

The choice of scales is up to you. The shape of the graph changes with different scales as you can see in the two graphs shown in Figure 9.4.

What conclusions can you draw from these graphs?

EXERCISE 9.1.1

1. A newspaper article reported on how the value of a certain stock fluctuated during the first 9 months of the year. The values are listed in Table 9.1. Use the values from the table to draw a graph. Label the horizontal axis "Month" and the vertical axis "Price per share ($)." What conclusions can you draw from the graph about how the stock fluctuated?

2. Experiments in science often give values that are plotted on a graph to clarify relationships between sets of numbers. For instance, the pressure of a confined gas increases when the temperature increases. Table 9.2 shows this relationship.

   Use a suitable scale and graph the pressure as a dependent variable. The temperature (the independent variable) should be plotted along the horizontal axis and the pressure along the vertical axis. Conclusion: If you heat a gas-filled bottle it may eventually explode!

Table 9.1

Month	Price per share ($)
January	26
February	30
March	28
April	26
May	30
June	24
July	28
August	26
September	18

Table 9.2

Temperature (°C)	Pressure (mm Hg)
−15	1
−10	2
−5	3
0	5
5	7
10	9
15	13
20	18
25	24

3. Given the data below, showing the yearly income of the ABC corporation, draw the corresponding graph.

Years	1978	1979	1980	1981	1982	1983
Income ($Million)	0.56	0.50	0.20	0.29	0.89	3.23

9.2 GRAPHING WITH ORDERED PAIRS

Martin and Nancy go shopping and spend exactly the same amount of money at each store. At the first store each spends $1; at the second, $4; and at the third, $10.

We can draw a graph showing this information. First we make a table of the given facts:

Martin	Nancy	(M, N)	
$1	$1	(1, 1)	(over 1, then up 1)
$4	$4	(4, 4)	(over 4, then up 4)
$10	$10	(10, 10)	(over 10, up 10)

Then we draw and label the axes and graph the points as in Figure 9.5.

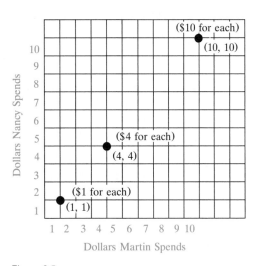

Figure 9.5

EXAMPLE

This time, at each store, Nancy spends a dollar more than Martin. For example, when Martin spends $1, Nancy spends $2; when Martin spends $2, Nancy spends $3. Revise your graph for this new information. See Figure 9.6.

Solution   Make a table of these facts before you draw the graph.

Martin	Nancy
$1	$2
2	3
3	4
4	5

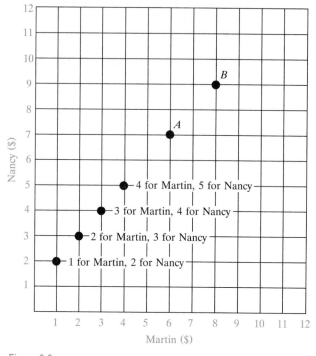

Figure 9.6

Do you see that these points lie in a straight line? Connect the points. Which values would you read for A? for B?

Point A: Martin $6, Nancy $7
Point B: Martin $8, Nancy $9

EXAMPLE

This time Nancy spends twice what Martin does. Make the new table and draw the graph. See Figure 9.7

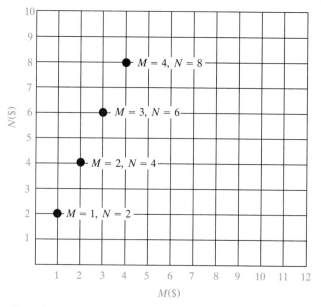

Figure 9.7

Solution Let *M* stand for what Martin spends and *N* for what Nancy spends.

M($)	N($)
1	2
2	4
3	6
4	8

$N = 2M$ expresses in symbols that Nancy's spending is twice Martins's.

EXAMPLE

Nancy spends $3 more than Martin each time. Prepare the table and draw the graph.

Solution Written in short form: $N = M + 3$.

M	N	(M, N)
1	4	(1, 4)
2	5	(2, 5)
3	6	(3, 6)
4	7	(4, 7)

Notice that this time beside each point on the graph instead of writing (1 for Martin, 4 for Nancy) or even $M = 1$, $N = 4$, etc., we write (1, 4), (2, 5), (3, 6), and (4, 7). These are called *ordered pairs* or *coordinates*. The first number (the independent variable) is always the number on the *horizontal* axis, and the second number (the dependent variable) is always the number on the *vertical* axis. See Figure 9.8.

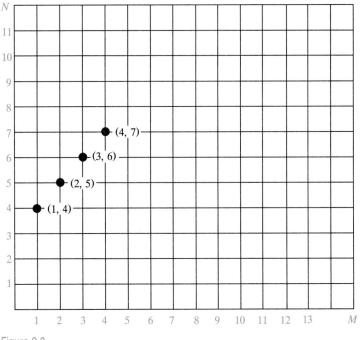

Figure 9.8

EXERCISE 9.2.1

Write an equation, construct a table, and draw the graph.

1. Nancy spends 3 times what Martin does.
2. Nancy spends $5 more than Martin does.
3. Nancy spends $1 less than Martin does.
4. Nancy spends $3 less than Martin does.
5. Nancy spends $3 more than twice what Martin spends.

Graphing Ordered Pairs

In order to generalize coordinate graphing, draw two axes, one vertical and one horizontal, so that they intersect at the point where both are zero. Positive numbers will be above and to the right of this intersection. Negative numbers will be below and to the left of the intersection. The intersection itself is called the *origin*. The number for the horizontal number line is listed first.

EXAMPLE

Plot the ordered pairs A $(1, 2)$, B $(3, 5)$, C $(-2, 3)$, D $(-3, -4)$, E $(2, -1)$, and F $(0, 5)$.

Solution

A $(1, 2)$: Start at the origin, move 1 step to the right and continue 2 steps up.

B $(3, 5)$: Start at the origin, move 3 steps to the right and 5 steps up.

C $(-2, 3)$: Start at the origin, move 2 steps to the left and 3 steps up.

D $(-3, -4)$: Start at the origin, move 3 steps to the left and 4 steps down.

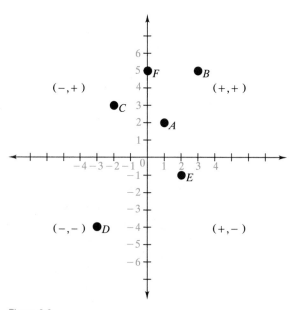

Figure 9.9

E $(2, -1)$: Start at the origin, move 2 steps to the right and 1 step down.

F $(0, 5)$: Start at the origin, move 0 steps to the right and 5 steps up.

See Figure 9.9.

EXERCISE 9.2.2

1. Draw two intersecting axes, as for Figure 9.9. Locate and label the following points:

A $(1, 1)$	D $(-1, 4)$	G $(2, 0)$	J $(0, 0)$
B $(2, 6)$	E $(1, -4)$	H $(3, -2)$	K $(-1, -3)$
C $(6, 2)$	F $(-3, -5)$	I $(0, 4)$	L $(-3, 0)$

2. Find the ordered pair that goes with each of the letters in Figure 9.10.

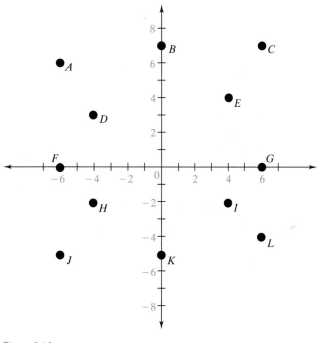

Figure 9.10

9.3 GRAPHING RELATED PAIRS

So far the graphs in this chapter have related years and income and two people's spending habits. Thus far we have used various labels for our number lines. Often, however, the letters x and y are used to stand for whatever pair of ideas we are relating. The horizontal axis shows the x-values, and the vertical axis shows the y-values. Now, instead of $N = M$, we have $y = x$; instead of $N = M + 3$, we have $y = x + 3$. Ordered pairs always list the x-value first. The x-value is the *independent* variable, and the y-value is the *dependent* variable (y depends on the value assigned to x).

Now, suppose we have the statement "y is 3 more than x" to graph. We rewrite the statement as an equation, $y = 3 + x$ or $y = x + 3$. Then we find ordered pairs (x, y) by choosing values for x and then solving the equation to find the values for y.

x	y	(x, y)	
1	4	$(1, 4)$	If $x = 1$, then $y = 1 + 3 = 4$.
2	5	$(2, 5)$	If $x = 2$, then $y = 2 + 3 = 5$.
3	6	$(3, 6)$	If $x = 3$, then $y = 3 + 3 = 6$.

Let's try some negative values for x:

x	y	(x, y)	
-1	2	$(-1, 2)$	If $x = -1$, then $y = -1 + 3 = 2$.
-3	0	$(-3, 0)$	If $x = -3$, then $y = -3 + 3 = 0$.
-6	-3	$(-6, -3)$	If $x = -6$, then $y = -6 + 3 = -3$.

All these points lie on one line. The equation $y = x + 3$ implies that for every x-value, there is only one y-value. In order for the graph to form a curve, there must be more than one y-value for some x-values or more x-values for some y-values.

When we graph a line, we locate at least three points. If they lie in a straight line we know that we have not made a mistake in our substitution. We could have used any three pairs. If we use $(2, 5)$, $(3, 6)$, and $(-6, -3)$, we will have *exactly* the same line as if we had used $(1, 4)$, $(3, 6)$, and $(-1, 2)$. (See Figure 9.11.)

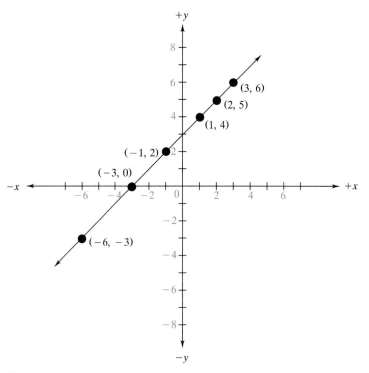

Figure 9.11

EXAMPLE

Use the process just described to graph the statement "y is 3 times as large as x."

Solution   We begin by translating the statement into an equation.

$$y = 3x$$

Next make the table of values for which $y = 3x$ is true.

x	y
1	3
2	6
3	9
0	0

Now we can write the ordered pairs (1, 3), (2, 6), (3, 9), and (0, 0) and locate them as points on a graph. Last, we draw a line connecting the points as at the right in Figure 9.12. This line is the graph of $y = 3x$.

Using zero for one of the values of x lets us find where the line crosses the y axis. We call this point the *y-intercept*. In Figure 9.12, $y = 0$ when $x = 0$. This is not always the case.

Look at the equation $y = \frac{1}{3}x$.
Again we can substitute any values we want for x. Let's try 1, 2, 3:

x	y
1	1/3
2	2/3
3	3/3 or 1

(1, 1/3) and (2, 2/3) are harder to graph because of the fractions that represent the y value. Let's try to find values for x that will lead to whole-number values for y.

x	y
0	0
3	1
6	2

For the equation $y = \frac{1}{3}x$, any multiple of 3 for the x value will lead to a whole-number value for y. Here are some more:

x	y
9	3
−3	−1
−6	−2

We can now plot three of the ordered pairs and draw the line connecting them. Figure 9.13 shows the result using (−3, −1), (3, 1), and (6, 2).

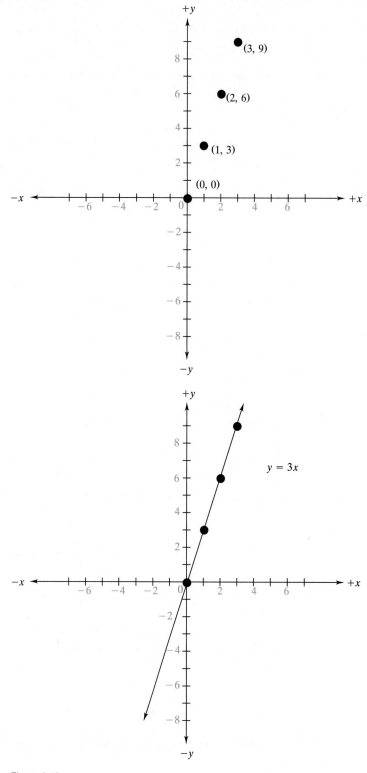

Figure 9.12

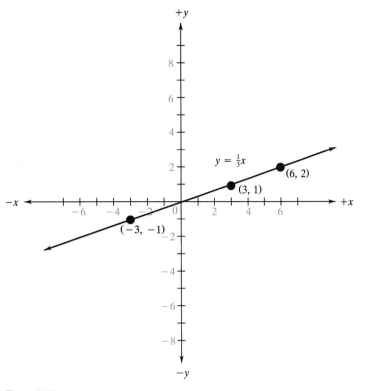

Figure 9.13

EXAMPLE

Graph $y = \frac{2}{5}x$.

Solution   This time, to avoid fractions we use multiples of 5 for the x value.

x	y
0	0
5	2
10	4

Plotting the points from the table and connecting them leads to the graph of $y = \frac{2}{5}x$ shown in Figure 9.14.

EXAMPLE

Graph $y = -x$.

Solution   First make the table with three values of your choice for x.

x	y	
0	0	
1	−1	
−1	1	[Remember, $-(-1) = 1$]

Then plot the points and connect them. (See Figure 9.15.)

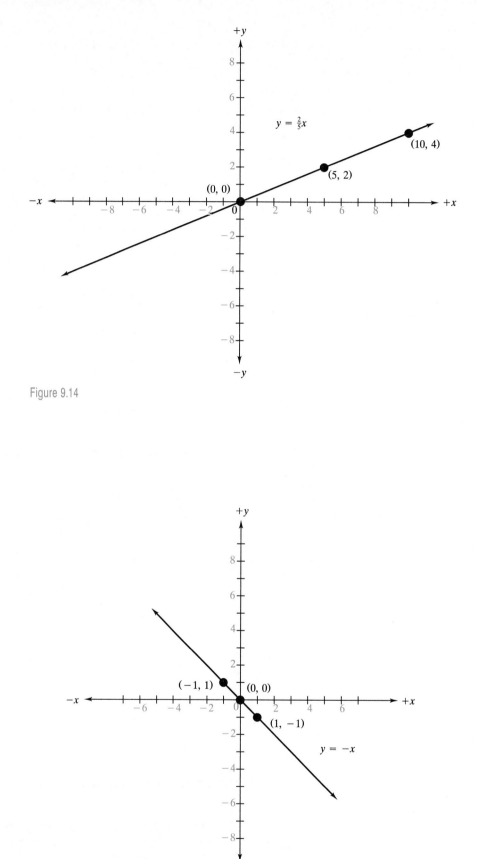

Figure 9.14

Figure 9.15

EXAMPLE

Draw the graph that corresponds to the statement "y is 3 more than 2 times x."

Solution   The equation is $y = 2x + 3$.
Substitute values for x to find corresponding values for y. Make a table.

x	y
0	3
1	5
2	7

If $x = 0$, then $y = 2(0) + 3 = 0 + 3 = 3$.
If $x = 1$, then $y = 2(1) + 3 = 5$.
If $x = 2$, then $y = 2(2) + 3 = 7$.

The ordered pairs (x, y) are $(0, 3)$, $(1, 5)$, $(2, 7)$. These points are used to graph the line $y = 2x + 3$ in Figure 9.16.

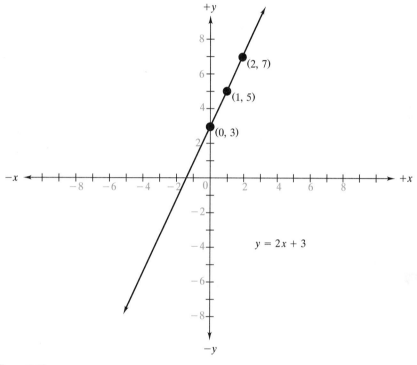

Figure 9.16

EXAMPLE

Graph $y = 3x - 5$.

Solution   Find three points, plot them, and connect the points.

x	y
0	-5
1	-2
2	1

If $x = 0$, then $y = 3(0) - 5 = -5$, giving the point $(0, -5)$.
If $x = 1$, then $y = 3(1) - 5 = -2$, giving the point $(1, -2)$.
If $x = 2$, then $y = 3(2) - 5 = 1$, giving the point $(2, 1)$.

These points are used in Figure 9.17 to graph $y = 3x - 5$.

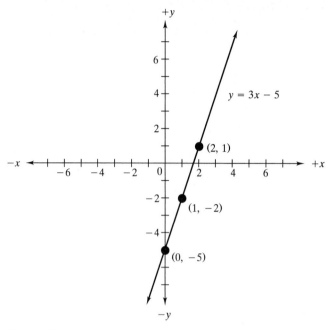

Figure 9.17

EXAMPLE

Graph $y = -\frac{1}{2}x + 1$.

Solution   Make a table with three ordered pairs. Plot the points and connect them.

x	y	(x, y)
0	1	$(0, 1)$
2	0	$(2, 0)$
4	-1	$(4, -1)$

The graph is shown in Figure 9.18.

Vertical and Horizontal Lines

Suppose the equation we want to graph is $y = 3$.
   If $x = 0$, what is y? $y = 3$
   If $x = 2$, what is y? $y = 3$
   If $x = -2$, what is y? $y = 3$
List these values in a table and then draw the graph.

x	y	(x, y)
0	3	$(0, 3)$
2	3	$(2, 3)$
-2	3	$(-2, 3)$

Figure 9.19 shows that the graph of $y = 3$ is a line through $(0, 3)$ parallel to the x axis.

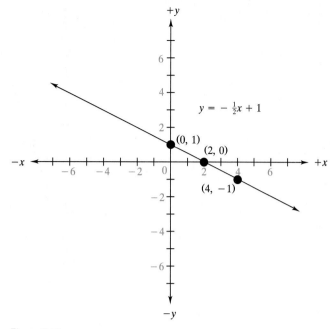

Figure 9.18

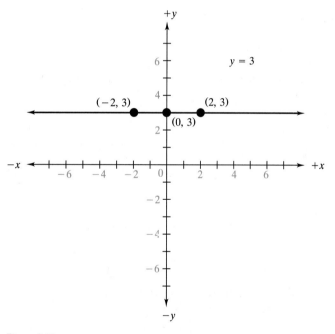

Figure 9.19

EXAMPLE

Graph $y = -3$.

    Solution Follow the same steps as before.

x	y	(x, y)
−1	−3	(−1, −3)
0	−3	(0, −3)
5	−3	(5, −3)

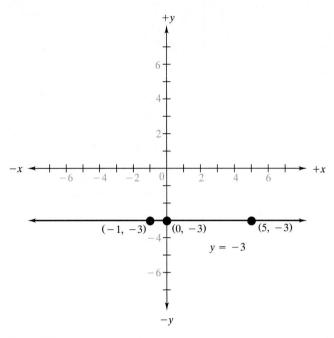

Figure 9.20

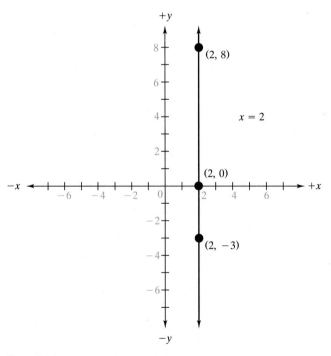

Figure 9.21

The line intercepts the y axis at $(0, -3)$ and is parallel to the x axis, as shown in Figure 9.20.

EXAMPLE

Graph $x = 2$.

> **Solution**   The equation states that x is always 2.
> Make the table with all x entries equal to 2. y can be anything.

x	y	(x, y)
2	0	$(2, 0)$
2	8	$(2, 8)$
2	-3	$(2, -3)$

The line intersects the x axis at $(2, 0)$ and is parallel to the y axis. (See Figure 9.21.)

All of the equations discussed above are called *linear* equations because their graphs are *straight* lines.

EXERCISE 9.3.1

Find three points that satisfy the given equation. Plot these points, and draw the line.

1. $y = 2x$
2. $y = x + 4$
3. $y = -\frac{1}{3}x$
4. $y = x - 4$
5. $y = 2x + 1$
6. $y = 3x - 4$
7. $y = \frac{2}{3}x + 1$
8. $y = -x + 5$
9. $y = -3x$
10. $y = -3x + 1$
11. $y = -x - 2$
12. $y = \frac{1}{2}x$
13. $y = -2x + 5$
14. $y = -2x + 3$
15. $y = -2x - 1$
16. $y = -\frac{1}{4}x - 2$
17. $y = 2$
18. $x = -3$
19. $y = -3$
20. $x = 7$

9.4 SLOPE AND y-INTERCEPT

Slope

Let's look at some more graphs of linear equations.

1. $y = x$

x	y
0	0
1	1
2	2

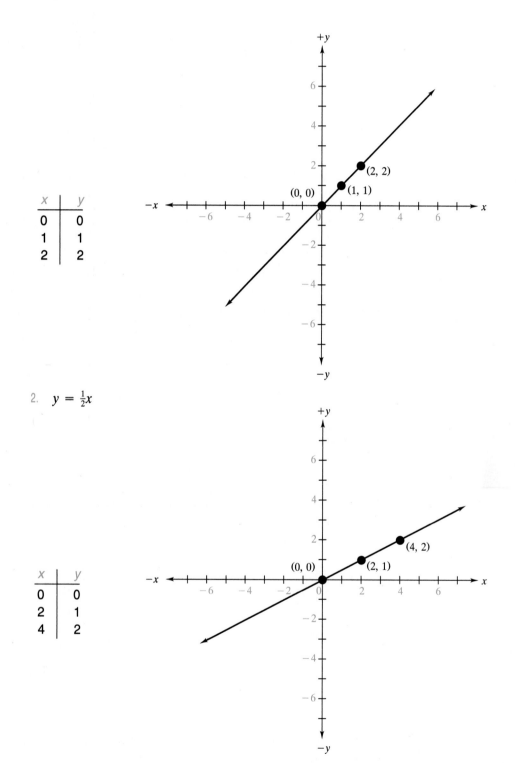

2. $y = \frac{1}{2}x$

x	y
0	0
2	1
4	2

3. $y = 4x$

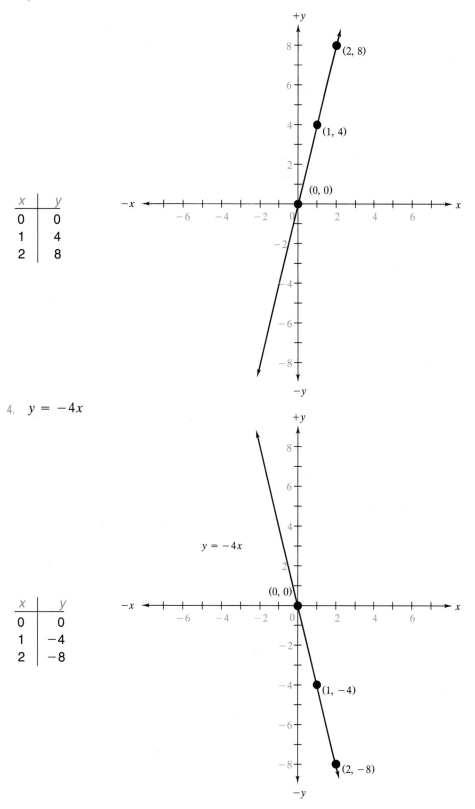

x	y
0	0
1	4
2	8

4. $y = -4x$

x	y
0	0
1	−4
2	−8

What do you see happening? Each of these lines goes through the point $(0, 0)$, the origin.

In the first three graphs, as we look from left to right, the line goes up; in the fourth graph, the line goes down. The rise or fall of a graph line as you travel along it from left to right is called its *slope*. Each of these lines has a different slope.

When the *coefficient* of x is positive $(1, \frac{1}{2}, 4)$, the slope is positive, and the line goes up from left to right. The larger the coefficient, the steeper the slope will be. We see that the graph of $y = 2x$ rises more quickly than the graph of $y = \frac{1}{2}x$.

Similarly, when the coefficient of x is negative, the slope is negative, and the line goes down from left to right. Its steepness depends on the size of the *absolute value* of the coefficient. Remember that the absolute value is the positive magnitude of the number. $|3| = 3, |-3| = 3$.

How do we define slope? We can use the carpenter's definition,

$$\text{Slope} = \frac{\text{rise}}{\text{run}}$$

Rise is the vertical change; *run* is the horizontal change.

The slope of the flight of stairs in Figure 9.22 is the rise (6 ft) divided by the run (10 ft). The steps rise from left to right, so the slope is positive.

$$\text{Slope} = \frac{6}{10} = \frac{3}{5}$$

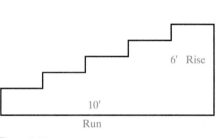

6′ Rise

10′

Run

Figure 9.22

The slope of a roof is positive or negative, depending on which part of the roof we are looking at. In part a of the roof in Figure 9.23, the vertical change is $+8$, and the horizontal change is 12.

$$\text{Slope} = \frac{8}{12} = \frac{2}{3}$$

In part b, the vertical change is -8, leading to a slope of $\frac{-8}{12} = -\frac{2}{3}$.

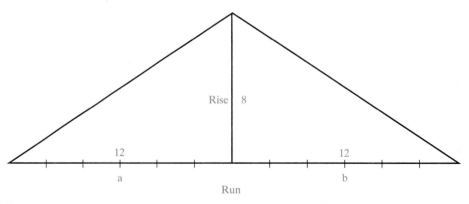

Rise   8

12   12

a   b

Run

Figure 9.23

EXAMPLE

Find the slope of the line connecting the two points $(0, 0)$ and $(2, 8)$.

> **Solution**   First we plot the two points (Figure 9.24).
> The rise is 8 and the run is 2. Therefore,
>
> $$\text{Slope} = \frac{8}{2} = 4$$

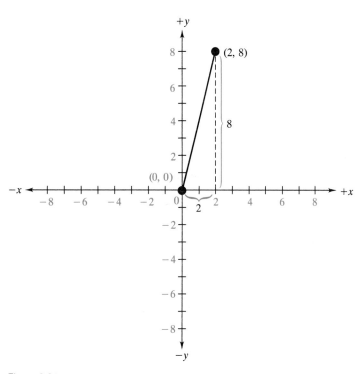

Figure 9.24

EXAMPLE

Find the slope of the line connecting $(0, 0)$ and $(3, -4)$.

> **Solution**   The two points are plotted and a line is drawn through them in
> Figure 9.25.
>
> $$\text{Rise} = -4, \quad \text{run} = 3, \quad \text{slope} = -\frac{4}{3}.$$

EXAMPLE

Find the slope of the line containing $(1, 4)$ and $(8, 7)$.

> **Solution**   See Figure 9.26.
> Rise (the change in the value of y) = 3
> Run (the change in the value of x) = 7
> $$\text{Slope} = \frac{3}{7}$$

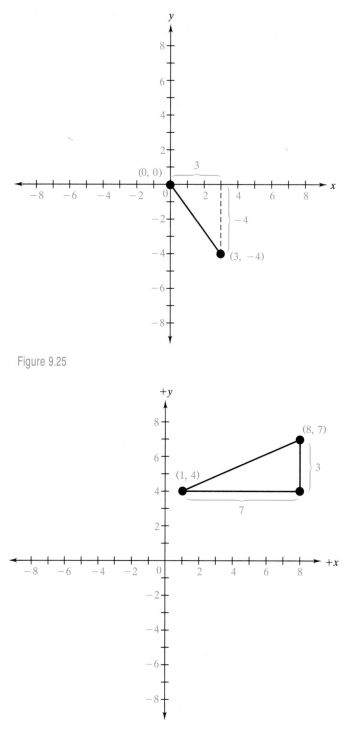

Figure 9.25

Figure 9.26

Suppose we measured from $(8, 7)$ to $(1, 4)$ in Figure 9.26. The change in y would be -3, while the change in x would be -7.

$$\text{Slope} = \frac{-3}{-7} \quad \text{or} \quad \frac{3}{7}$$

It makes no difference in which order we read the points; the slope will still measure the change in the line as it goes from left to right.

In definitions and formulas, we sometimes refer to a point as $P_1(x_1, y_1)$, which means that the point P_1 (pronounced "P sub one") has the coordinates x_1 and y_1.

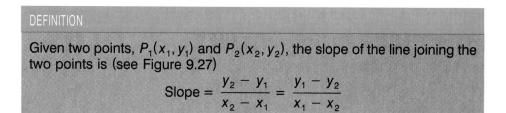

DEFINITION

Given two points, $P_1(x_1, y_1)$ and $P_2(x_2, y_2)$, the slope of the line joining the two points is (see Figure 9.27)

$$\text{Slope} = \frac{y_2 - y_1}{x_2 - x_1} = \frac{y_1 - y_2}{x_1 - x_2}$$

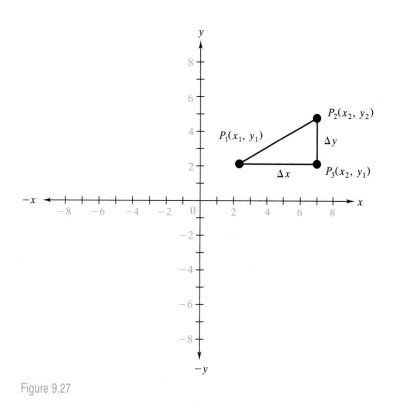

Figure 9.27

The letter m is used to represent the slope of a line. A capital Greek letter delta, Δ, is used to represent change. Thus Δy = change in y, and Δx = change in x. So the definition of slope can also be written

$$m = \frac{\Delta y}{\Delta x}$$

EXAMPLE

Given the equation $y = -3x + 1$. Find the slope.

Solution   First we find any two points on the line. Let's let $x = 0$ and 4.

x	y
0	1
4	-11

$P_1 = (x_1, y_1) = (0, 1)$
$P_2 = (x_2, y_2) = (4, -11)$

Then

$$m = \frac{y_2 - y_1}{x_2 - x_1} = \frac{-11 - 1}{4 - 0} = \frac{-12}{4} = -3$$

We could also have solved this by reversing the points:

$$m = \frac{y_1 - y_2}{x_1 - x_2} = \frac{1 - (-11)}{0 - 4} = \frac{12}{-4} = -3$$

The slope is the coefficient of x when the equation is in the form

$$y = mx + b$$

Here, m represents the slope of the line and b is a constant. In the equation, $y = -3x + 1$, -3 is m and 1 is b.

For the equation $y = 5x + 2$, the slope is $m = 5$, $b = 2$.

For the equation $y = -x - 4$, the slope is $m = -1$, $b = -4$.

The equation $y = 3$ can be thought of as $y = 0x + 3$. Here we can see that the slope is $m = 0$. A slope of 0 indicates that the line is horizontal.

To find the slope of the equation $x = 1$, take two points, let's say $(1, 5)$ and $(1, 3)$. By definition,

$$m = \frac{\Delta y}{\Delta x} = \frac{5 - 3}{1 - 1} = \frac{2}{0}$$

Since we cannot divide by zero, this slope is undefined. Such a slope indicates a vertical line.

Intercept

Let's look at the three equations

1. $y = 2x + 3$
2. $y = 2x$
3. $y = 2x - 3$

and plot them on the same axes, as in Figure 9.28.

All three lines have the same slope: $m = 2$. When two or more lines have the same slope, we say they are *parallel*. They will not intersect no matter how far we extend them in either direction. Each of these lines crosses the y-axis at a different point. We call these points the *y-intercepts*, the point where $x = 0$.

In each of the three equations above, we will set $x = 0$ to find the y-intercept.

1. $y = 2x + 3$
    $y = 2(0) + 3$
    $y = 3$ y-intercept is $(0, 3)$.
2. $y = 2x$
    $y = 2(0)$
    $y = 0$ y-intercept is $(0, 0)$.
3. $y = 2x - 3$
    $y = 2(0) - 3$
    $y = -3$ y-intercept is $(0, -3)$.

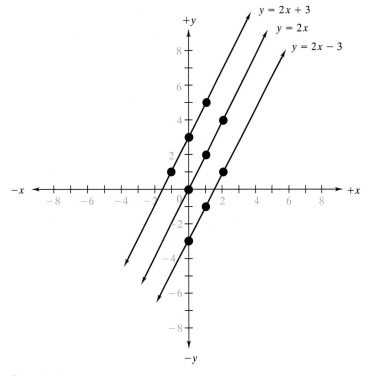

Figure 9.28

EXAMPLE

Find the y-intercept: (a) $y = 5x - 7$ (b) $y = x + 0.03$

Solution
(a) The y-intercept is $(0, -7)$.
(b) The y-intercept is $(0, 0.03)$.

To find the y-intercept of any line in the form $y = mx + b$, we set $x = 0$. Then

$$y = mx + b$$
$$y = 0x + b$$
$$y = b$$

Thus, in any line $y = mx + b$, m is the slope and $(0, b)$ is the y-intercept.

EXERCISE 9.4.1

1. Find the y-intercept.
    (a) $y = 2x + 3$
    (b) $y = \frac{2}{3}x - 4$
    (c) $y = -x + 6$
    (d) $y = -3x - 2$
    (e) $y = 2$
    (f) $y = -3$

2. Find the slope.
    (a) $y = 2x$
    (b) $y = -x + 3$

(c) $y = 2x - 1$
(d) $y = -\frac{1}{3}x - 6$
(e) $y = -5$
(f) $x = 2$

9.5 GRAPHING USING SLOPE AND y-INTERCEPT

We can use what we have learned about slope and intercept to graph linear equations in the form $y = mx + b$.

EXAMPLE

Graph $y = \frac{2}{3}x + 4$.

Solution   From the equation, we can see that $m = \frac{2}{3}$ and $b = 4$. Thus the slope is 2/3, and the y intercept is the point $(0, 4)$.
First, we locate the point $(0, 4)$ on the graph (Figure 9.29). From this point we use the slope to find a second point. With a slope of 2/3, the change in x is 3, and the change in y is 2. From $(0, 4)$ we move to the *right* 3 units and *up* 2 units to the point $(3, 6)$.

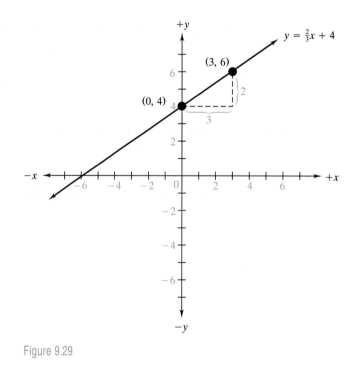

Figure 9.29

Now draw the line connecting the two points.

EXAMPLE

Graph $y = -3x + 5$.

Solution $y = -3x + 5$ is in the form $y = mx + b$. Here $m = -3 = \frac{-3}{1}$. So the slope has a rise of -3 and a run of 1.
$b = 5$; therefore the y intercept is $(0, 5)$. Locate $(0, 5)$ on the graph. From this point the change in x is 1, so go *right* 1 unit. The change in y is

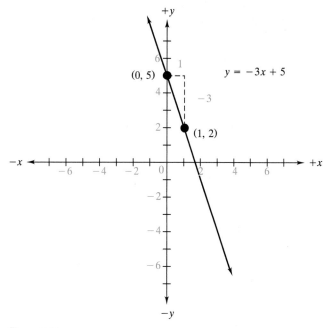

Figure 9.30

-3, so go *down* 3 units to point $(1, 2)$. Now connect the two points. The completed graph is shown in Figure 9.30.

EXAMPLE

Graph $y = -x - 2$.

Solution   This equation is also in the form $y = mx + b$. Think of it as $y = -1(x) + (-2)$. Here

$$m = -1 = \tfrac{-1}{1} \qquad \text{and} \qquad b = -2$$

Thus the slope has a rise of -1 and a run of 1, and the y intercept is $(0, -2)$. The graph is shown in Figure 9.31.

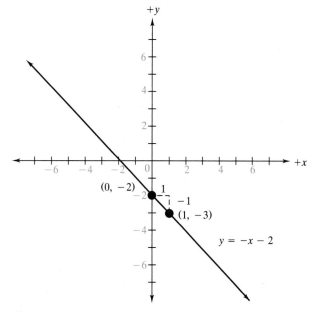

Figure 9.31

Suppose we are given the equation $2y - x = 6$ to graph. To rewrite this in the form $y = mx + b$, we must solve for y.

$$2y - x = 6$$
$$\underline{+x \qquad +x} \qquad \text{Add } x.$$
$$2y = x + 6$$
$$\frac{2y}{2} = \frac{x}{2} + \frac{6}{2} \qquad \text{Divide by 2.}$$
$$y = \frac{1}{2}x + 3 \qquad \text{Simplify.}$$

The equation is now in the form $y = mx + b$. $m = \frac{1}{2}$, $b = 3$.

EXAMPLE

Rewrite $\frac{1}{2}y + 3x - 4 = 0$ in the form $y = mx + b$.

Solution

$$\frac{1}{2}y + 3x - 4 = 0$$
$$\underline{+4 \qquad +4} \qquad \text{Add 4.}$$
$$\frac{1}{2}y + 3x = 4$$
$$\underline{-3x \qquad -3x} \qquad \text{Subtract } 3x.$$
$$\frac{1}{2}y = -3x + 4$$
$$y = -6x + 8 \qquad \text{Multiply by 2.}$$
$$m = -6 \quad \text{or} \quad \frac{-6}{1}, \qquad b = 8$$

EXERCISE 9.5.1

Find the slope and the y-intercept. Then graph the line.

1. $y = x - 2$
2. $y = 5x$
3. $y = \frac{1}{2}x - 5$
4. $y = -3x - 1$
5. $y = 3x + 2$
6. $y + \frac{2}{3}x = 1$
7. $2y - 3x + 4 = 0$
8. $\frac{1}{2}x - \frac{2}{3}y = 4$
9. $3x + 2y = 6$
10. $2x - 5y = 10$

SUMMARY

Definition

Slope $m = \dfrac{y_2 - y_1}{x_2 - x_1}$

Linear equation: $y = mx + b$, where m is the slope and b is the y-intercept.

VOCABULARY

Coordinates: The two numbers in an ordered pair (x, y). The first number is the x-coordinate; the second number the y-coordinate.

Delta (Δ)**:** A symbol used to signify a difference.

Dependent variable: The variable whose value depends on the value of another variable.

Graph: A picture signifying relationships between 2 sets of numbers.

Horizontal axis: The axis used for the independent variable.

Independent variable: A variable whose values can be chosen without regard to other variables.

Linear equation: An equation with two variables both to the first power. The slope form is $y = mx + b$.

Ordered pair: Two numbers (x, y) that are written in order. Commonly used to identify a point on a graph.

Origin: The intersection of the x-axis and the y-axis.

Parallel lines: Lines with the same slope.

Related pairs: Two numbers that are connected by a rule.

Rise: Change in y between two points.

Run: Change in x between two points.

Slope: The ratio of the change in y to the change in x.

Vertical axis: The axis used for the dependent variable.

x-value: The value of the variable x; the first number in an ordered pair.

y-intercept: The point where a graph line intersects the y axis.

y-value: The value of the variable y; the second number in an ordered pair.

CHECK LIST

Check the box for each topic you feel you have mastered. If you are unsure, go back and review.

☐ Ordered pairs and points
☐ Graphing ordered pairs
☐ Writing equations from rules
☐ Graphing related pairs
☐ y-intercept
☐ Slope
☐ Graphing vertical and horizontal lines
☐ Graphing lines using slope and y-intercept

REVIEW EXERCISES

1. Plot the following monthly average temperatures.

Month	Temperature (° F)
January	9
February	14
March	12
April	25
May	54
June	63
July	68
August	76
September	50
October	42
November	48
December	28

2. Write an equation, make a table of number facts, and plot the points.
   (a) Nancy spends 6 dollars more than Martin.
   (b) Nancy spends four times what Martin spends.
   (c) Nancy spends $3 less than Martin spends.
   (d) Nancy spends $1 more than two times what Martin spends.

3. Locate and label the ordered pairs.
   (a) $(2, 2)$ (b) $(2, -3)$ (c) $(-5, 1)$
   (d) $(-1, 0)$ (e) $(0, 0)$ (f) $(2, 0)$
   (g) $(0, -4)$ (h) $(-3, 0)$ (i) $(-1, -1)$

4. Express each point in Figure 9.32 as an ordered pair. For example, $A = (0, 0)$.

5. Find three points that satisfy the given equation. Plot the points, and draw the line.

(a) $y = x + 2$
(b) $y = x - 3$
(c) $y = 3x$
(d) $y = 3$
(e) $y = 2x + 1$
(f) $x = 4$
(g) $y = -x + 2$
(h) $y = 3x - 2$

6. Plot the two points and find the slope.
   (a) $(1, 3)\ (3, 6)$
   (b) $(2, -5)\ (8, 1)$
   (c) $(0, 6)\ (3, -1)$
   (d) $(4, 0)\ (0, 2)$

(e) $(4, 2)\ (8, 2)$
(f) $(2, 7)\ (7, 2)$

7. Find the slope and y-intercept. Graph the equation.
   (a) $y = x + 2$
   (b) $y = x - 5$
   (c) $y = 2x$
   (d) $y = -2x + 4$
   (e) $y = \frac{1}{2}x$
   (f) $y = -3$
   (g) $x + y = 6$
   (h) $2x - y = 2$

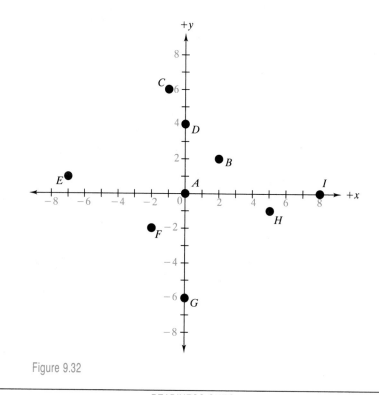

Figure 9.32

Solve the problems to satisfy yourself that you have mastered Chapter 9.
   1. Read point A.
   2. Graph the point $(2, -5)$.

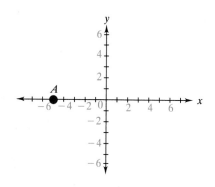

3. Graph the line $y = 2x + 4$.
4. Graph the line $2x - 3y = 12$.
5. Graph the line $x = -6$.
6. Graph the line $y = 4$.
7. What is the slope of the line $y = 6x$?
8. What is the y-intercept of the line $y = 3x + 4$?
9. Find the slope of the line that connects $(2, 3)$ and $(4, 9)$.
10. Plot the following relationship using temperature as the independent variable:

Temperature	$-10°$	$-5°$	$0°$	$5°$	$10°$	$15°$
Volume	1 L	2 L	3 L	4 L	5 L	6 L

WORD PROBLEMS

The agony or joy of mathematics

Here is a word problem about Diophantus from A.D. 300 from the *Greek Anthology* (dated about A.D. 500):

> Diophantus' youth lasted 1/6 of his life. He grew a beard after 1/12 more. After 1/7 more he married. 5 years later he had a son. The son lived exactly half as long as his father. Diophantus died just 4 years after his son. How old was he when he died?

The problem is solved at the end of the chapter.

10.1 TRANSLATING WORDS TO SYMBOLS

In order to be able to apply mathematics to real situations, we need to develop equations or inequalities that fit the problem. As a first step we must be able to replace words with symbols. There are certain words and phrases that occur frequently in mathematics; it helps to be aware of how they are translated.

WORDS AND SYMBOLS

English	Mathematics
Addition	
the sum of a and b	$a + b$
a plus b	
a increased by b	
b more than a	
add b to a	

Subtraction

the difference of a and b $a - b$
a minus b
a decreased by b
b subtracted from a
a less b
b less than a
take away b from a

Multiplication

the product of a and b ab
a times b $a \cdot b$
a multiplied by b $a * b$
                                              $(a)(b)$
                                              $a(b)$
                                              $(a)b$

Division

                                              $\dfrac{a}{b}$
the quotient of a and b

                                              a/b
a divided by b $a \div b$
b goes into a $b\overline{)a}$

Equality

is $=$
a is equal to b $a = b$
a equals b
a is the same as b
the result is

Inequality

a is greater than b $a > b$
a is less than b $a < b$
a is greater than or equal to b $a \geq b$
a is less than or equal to b $a \leq b$
a is at most b $a \leq b$
a is at least b $a \geq b$
a does not equal b $a \neq b$

According to the box, 5 more than a number n should be translated as $n + 5$. Since in addition the commutative property tells us that the order doesn't affect the sum, we could write either $n + 5$ or $5 + n$.

However, five less than a number *must* be written as $n - 5$. In subtraction the order *does* matter: $2 - 5$ does *not* equal $5 - 2$.

The commutative property also applies to multiplication; we can change the order and still have the same product. Thus $3 \cdot 2 = 2 \cdot 3$. But in division the order *does* matter: $6 \div 2$ does *not* equal $2 \div 6$.

In short, the sum $a + b$ can be written $b + a$, and the product ab can also be written as ba. But the difference and quotient *must* be written in the order in which the numbers are given. We interpret "difference" as the first number minus the second number, or $a - b$.

EXAMPLE

Write in mathematical symbols.
(a) The sum of six and a number
(b) The difference of six and a number
(c) The difference of a number and six
(d) The product of six and a number
(e) The quotient of six and a number
(f) The quotient of a number and six

Solution
(a) $n + 6$ or $6 + n$ (b) $6 - n$ (c) $n - 6$ (d) $6n$ or $n \cdot 6$
(e) $6 \div n$ or $\dfrac{6}{n}$ (f) $n \div 6$ or $\dfrac{n}{6}$

EXAMPLE

Write in symbols.
(a) Seven decreased by a number
(b) Take away a number from seven
(c) Seven less than a number
(d) Seven subtracted from a number

Solution   (a) $7 - n$ (b) $7 - n$ (c) $n - 7$ (d) $n - 7$

EXAMPLE

Write in symbols.
(a) The product of four and the sum of a number and three
(b) The difference of a number and five is subtracted from six

Solution
(a) The *sum* $(n + 3)$ should be enclosed in parentheses. The solution is $4(n + 3)$
(b) $6 - (n - 5)$

Mistakes made in solving word problems can often be traced to incorrectly translating words and phrases rather than to incorrectly carrying out the actual mathematical steps. Translate the following into symbols before you go to Exercise 10.1.1.

EXAMPLE

a and *b* are two numbers. Find:
(a) their sum
(b) their difference
(c) their product
(d) their quotient
(e) twice their product
(f) three times their sum
(g) the difference of three times the first number and three times the second number
(h) the quotient of their sum and their product

Solution (a) $a + b$ (b) $a - b$ (c) ab (d) $\dfrac{a}{b}$ (e) $2ab$

(f) $3(a + b)$ (g) $3a - 3b$ (h) $\dfrac{a + b}{ab}$

EXERCISE 10.1.1

For each word phrase write the mathematical phrase. Do not compute the answer.

 1. The sum of 3 and 25

 2. The sum of 12 and 5

 3. 5 more than 11

 4. 15 decreased by 3

 5. 3 less than 10

 6. 5 times the sum of 4 and 7

 7. 9 subtracted from 2

 8. The quotient of 5 and 6

 9. x less than 12

10. x decreased by 5

11. 5 times the sum of 6 and y

12. The sum of $3p$ and $5q$ is multiplied by 4

13. 20 divided by x

14. The quotient of $3t$ and x

15. $4x$ divided by the sum of a and b

16. $5c$ multiplied by the difference of $4x$ and $9y$

10.2 WRITING EQUATIONS AND INEQUALITIES

Equations

In Section 10.1 we translated phrases into symbols. Now we are going to translate sentences into symbols. In each of the following we translate a phrase and then use the phrase in a sentence.

1. The sum of a and b: $a + b$
    The sum of a and b is 4. $a + b = 4$

2. The product of a and b: ab
    The product of a and b is 15. $ab = 15$

3. 3 less than a number: $n - 3$
    3 less than a number is 7. $n - 3 = 7$

4. a divided by 4: $\dfrac{a}{4}$

    a divided by 4 is 9. $\dfrac{a}{4} = 9$

The difference between the phrase and the sentence is the presence of a verb (usually "is") in the sentence. As we said in the box at the beginning of this chapter, other verbs and phrases that are replaced by the equals sign (=) are "is equal to," "equals," "is the same as," etc. Remember that when you replace a sentence by an equation it is just like translating a foreign language: sum means add, difference means subtract, product means multiply, and quotient means divide.

A statement that two expressions are equal is an *equation*.

EXAMPLE

Translate the following statements into equations. (We can use any letter, but we usually use x.) The important words you should watch for are printed in *italics*.
(a) The *sum* of a number and 6 *is* 11.
(b) Two *less than* a number *equals* 6.
(c) The *difference* of a number and 11 *is* 6.
(d) Three *more than* 5 *times* an unknown number *is equal to* 23.
(e) When four *times* an unknown number is *subtracted from* 31, the result *is* the unknown number.

Solution
(a) $x + 6 = 11$ (b) $x - 2 = 6$ (c) $x - 11 = 6$
(d) $5x + 3 = 23$ (e) $31 - 4x = x$

EXERCISE 10.2.1

Write an equation, and solve for the unknown number.

1. Four more than a number is ten.

2. The sum of a number and twenty-five is thirty-six.

3. Three less than a number is twenty.

4. Seven minus a number is eleven.

5. Two times a number is eighteen.

6. The product of five and a number is forty.

7. Twice a number plus three is thirty-one.

8. Twenty-five more than three times a number is forty.

9. Five times a number decreased by eight is twenty.

10. Nine less than eight times a number is seven.

11. Twice an unknown number plus thirteen is twenty-five.

12. Twenty-five plus three times an unknown number is thirty-four.

13. Five times an unknown number, decreased by eight, is twenty-two.

14. Six times a number increased by seven is thirty-seven.

15. Twice an unknown number is equal to the sum of five and the unknown number.

16. The sum of seven and an unknown number is twice that unknown number.

Inequalities

Recall the symbols for inequalities:

$$< \text{ is less than}$$

$$> \text{ is greater than}$$

$$\leq \text{ is less than or equal to} \quad \text{or} \quad \text{is at most}$$

$$\geq \text{ is greater than or equal to} \quad \text{or} \quad \text{is at least}$$

EXAMPLE

Translate the following statements into inequalities and solve.
(a) The *sum* of a number and two is *less than* four.
(b) The *difference* of a number and 7 is *greater than* or *equal to* eleven.
(c) Five *times* the number is *at most* 10.

Solution
(a) $x + 2 < 4$; $x < 2$
(b) $x - 7 \geq 11$; $x \geq 18$
(c) $5x \leq 10$; $x \leq 2$

EXERCISE 10.2.2

Translate the sentence into an inequality and solve.

1. One-fifth of an unknown number is less than four.

2. Twice the sum of five and an unknown number is greater than or equal to twenty-six.

3. Two-thirds of an unknown number is at least 10.

4. Two times a number subtracted from 15 is less than or equal to 41.

5. If the product of 6 and a number is subtracted from 60, the result is at most 132.

6. Pete's diet requires no more than 1200 calories per day. For lunch he can have 200 more calories than for breakfast, and for dinner 3 times as many calories as for breakfast. What is the most calories he can have at each meal? (Hint: Call the breakfast calories x.)

7. The perimeter of a rectangle can be no greater than 84. The length is three times the width. Give examples of possible dimensions of the rectangle.

8. To get a B in his math class, Pat must have an average of at least 80 on 6 tests. His grades so far are 73, 85, 80, 79, 72. What grade does Pat need to get on the last test to have a B average?

10.3 STRATEGIES FOR SOLVING WORD PROBLEMS

In this section we introduce several types of standard word problems, those that often appear on school entrance tests. We also challenge you with some nonstandard word problems.

An orderly approach is extremely important for any kind of problem solving. The hints in the strategy box may be helpful to you.

STRATEGY FOR PROBLEM SOLVING

1. Read the problem *carefully*. Draw a picture, if possible.
2. Read the problem again, and list the quantities you are looking for.
3. If possible, make an estimate of what you expect the answer to be.
4. Which variable do the others depend on? That will be your independent variable.
5. Express each of your unknown quantities in terms of the independent variable.
6. Write as many equations as you need.
7. Solve the equations.
8. Make sure you've found values for all listed quantities.
9. Check your answer. Even if your answer solves the equation you wrote, *does it make sense*?
10. Reread the question. Have you answered the question, or do you have to continue with more calculations?

Number Problems

While number problems may seem to have little to do with "real-life" problems, they give you more practice translating words to symbols and developing your problem-solving skills. Many word problems involve consecutive numbers, consecutive even numbers, and consecutive odd numbers.

Consecutive numbers are numbers that follow each other in natural order, for example, $9, 10, 11, 12$; $100, 101, 102$; or $-7, -6, -5, \ldots$. If we let the first number be x, the next one will be $x + 1$, the next one $x + 2$, and so forth.

If, for example, the sum of three consecutive numbers is 33, we can express the sum as $x + (x + 1) + (x + 2) = 33$.

$x + (x + 1) + (x + 2) = 33$	The sum of the numbers equals 33.
$3x + 3 = 33$	Simplify the left side.
$3x = 30$	Divide each side by 3.
$x = 10$	This is the first number.
$x + 1 = 11$	This is the second number.
$x + 2 = 12$	This is the third number.

Check: $10 + 11 + 12 = 33\sqrt{}$

The three consecutive numbers are 10, 11, and 12.

Often algebra problems involve consecutive odd numbers (i.e., $3, 5, 7$) or even numbers (i.e., $2, 4, 6$). These odd numbers and even numbers are 2 apart. If x were the first number, then $x + 2$ would be the second number and $x + 4$ the third.

EXAMPLE

The sum of three consecutive odd integers is 21. Find the numbers.

Solution

x	First number
$x + 2$	Second number
$x + 4$	Third number
$x + (x + 2) + (x + 4) = 21$	The sum of the numbers is 21.
$3x + 6 = 21$	Simplify the equation.
$3x = 15$	Solve.
$x = 5$	First number
$x + 2 = 7$	Second number
$x + 4 = 9$	Third number

Check: $5 + 7 + 9 = 21\sqrt{}$

The three consecutive odd integers are 5, 7, and 9.

EXAMPLE

Find three consecutive even integers such that three times the first equals the sum of the other two.

Solution   Let the numbers be x, $x + 2$, and $x + 4$.

$$3x = \underbrace{(x + 2) + (x + 4)}$$

Three times the first equals   the sum of the other two

$3x = x + 2 + x + 4$	Write the equation.
$3x = 2x + 6$	Simplify the right-hand side.
$x = 6$	Subtract $2x$ from both sides. This gives the first number.
$x + 2 = 8$	Second number
$x + 4 = 10$	Third number

Check: $3(6) \overset{?}{=} 8 + 10$

$18 = 18\sqrt{}$

The three consecutive integers are 6, 8, and 10.

Yet another type of problem involves sums and differences of two numbers.

EXAMPLE

If the sum of two numbers is 10 and their difference is 2, find the numbers.

> **Solution**   Call the numbers x and y.
>
> $$x + y = 10 \qquad \text{Write the equations.}$$
> $$\underline{x - y = 2}$$
> $$2x = 12 \qquad \text{Add the equations,}$$
> $$x = 6 \qquad \text{Solve.}$$
> $$6 + y = 10 \qquad \text{Substitute 6 for } x.$$
> $$y = 4 \qquad \text{Solve.}$$

Check: $6 + 4 = 10\sqrt{}$

$6 - 4 = 2\sqrt{}$

The numbers are 6 and 4.

Other number problems deal with the value of digits. If x is the digit in the tens place, its value is $10x$. Remember that if 2 is in the tens place, its value is 20.

EXAMPLE

The sum of the digits of a two-digit number is 12. If the digits are reversed, the new number is 18 less than the original number. What is the original number?

> **Solution**   Let's label the digits x and y and make a table of what we know.

	First Digit (Tens Place)	Value	Second Digit (Ones Place)	Value	Value of Number
Original number	x	$10x$	y	$1 \cdot y$	$10x + y$
New number	y	$10y$	x	$1 \cdot x$	$10y + x$

The new number $(10y + x)$ is 18 less than the original $(10x + y)$. Thus,

$$10y + x = 10x + y - 18$$
$$9y = 9x - 18 \qquad \text{Subtract } x \text{ and } y \text{ from both sides.}$$
$$y = x - 2 \qquad \text{Divide both sides by 9.}$$
$$y + 2 = x \qquad \text{Add 2 to both sides.}$$

We also know that $x + y = 12$. We can substitute $y + 2$ for x. Thus,

$$x + y = 12 \qquad \text{The sum of the digits is 12.}$$
$$(y + 2) + y = 12 \qquad \text{Substitute } y + 2 \text{ for } x.$$
$$2y + 2 = 12 \qquad \text{Simplify.}$$
$$y = 5 \qquad \text{Solve.}$$
$$x + y = 12 \qquad \text{The sum of the digits is 12.}$$
$$x + 5 = 12 \qquad \text{Substitute.}$$
$$x = 7 \qquad \text{Solve.}$$

Since x is 7 and y is 5, the original number would be 75, the reversal would give 57. According to the original problem, $75 - 18 = 57$. This is true. It is also true that the sum of the digits $(7 + 5)$ is 12.

EXERCISE 10.3.1

1. Find three consecutive numbers whose sum is 24.

2. Find three consecutive numbers whose sum is 54.

3. Find three consecutive odd numbers whose sum is 45.

4. Find three consecutive even numbers whose sum is 66.

5. The sum of the digits of a two-digit number is 7. If the digits are reversed, the new number increased by 3 equals 4 times the original number. Find the original number.

6. The sum of two numbers is 32, and their difference is 4. Find the numbers.

7. The sum of two numbers is 22. Five times one number is equal to 6 times the second number. Find the numbers.

8. The sum of the digits of a two-digit number is 13. Twice the tens digit increased by two equals five times the units digit. Find the number.

Percent Problems

In Chapter 3 we solved percent problems of the type "Some percent of a number equals what number?" These problems can all be solved using equations. Since we know two numbers and are looking for a third number, we can call the unknown x (or some other letter) when we set up an equation.

EXAMPLE

(a) 7% of 200 is what?
(b) What percent of 54 is 27?
(c) 15% of what number is 30?

Solution
(a) Recall that 7% = 0.07.

$$(0.07)(200) = x \qquad \text{Write the equation.}$$

$$14 = x \qquad \text{Simplify.} \qquad \text{The number is 14.}$$

(b)

$$x\% = \frac{x}{100} \qquad \text{Express \% as hundredths.}$$

$$\frac{x}{100}(54) = 27 \qquad \text{Write the problem as an equation.}$$

$$x \cdot 54 = 2700 \qquad \text{Multiply both sides by 100.}$$

$$x = \frac{2700}{54} \qquad \text{Divide both sides by 54.}$$

$$x = 50 \qquad \text{Simplify.}$$

Check: 50% of 54 = 0.5(54) = 27√

The answer is 50%.

(c) 15% can be written as either 0.15 or $\dfrac{15}{100}$. Let's work with 0.15 in this problem.

$$0.15x = 30 \qquad \text{Write the equation.}$$

$$15x = 3000 \qquad \text{Multiply both sides by 100.}$$

$$x = \frac{3000}{15} \qquad \text{Divide by 15.}$$

$$x = 200 \qquad \text{Simplify.}$$

Check: 15% of 200 = 0.15(200) = 30√

The number is 200.

Percent problems often involve discount, tax, or interest.

EXAMPLE

A dress costs $75.33 after the tax is added. If the tax is 8%, how much is the dress?

Solution   Let x = the price of the dress.

$$x + 8\%x = 75.33 \qquad \text{Write the equation.}$$

$$x + 0.08x = 75.33 \qquad \text{Change the percent to decimal.}$$

$$1.08x = 75.33 \qquad \text{Simplify.}$$

$$x = \frac{75.33}{1.08} \qquad \text{Divide both sides by 1.08.}$$

$$x = 69.75 \qquad \text{Simplify.}$$

Check: 69.75 + 8%(69.75) $\overset{?}{=}$ 75.33
          69.75 + 5.58 = 75.33√

The original price for the dress was $69.75.

EXAMPLE

John invests $4500, some of it at 8% simple interest and the remainder at 9%. How much is invested at each rate if the total annual income from his investments is $380?

Solution   Make a table.

	Investment	Rate	Interest
Part 1	x	8%	8%(x)
Part 2	y	9%	9%(y)
Total	$4500		$380

Write the equations:

$$x + y = 4500$$

$$8\%x + 9\%y = 380$$

$x + y =$	4500	The first equation.
$8x + 9y =$	38,000	Second equation multiplied by 100.
$-8x - 8y =$	$-36,000$	First equation multiplied by -8.
$y =$	2,000	Add the last two equations.
$x + 2000 =$	4500	Substitute for y in the first equation.
$x =$	2500	Solve for x.

Thus, John invests $2500 at 8% and $2000 at 9%. We leave the checking to you.

EXERCISE 10.3.2

1. What percent of 85 is 17?

2. What percent of 25 is 4?

3. The price of eggs increased from 70¢ to 98¢. What was the percent increase?

4. The price of eggs decreased from $1 to 70¢. What was the percent decrease?

5. Of a total of 718,000 workers, 26,925 worked in mines. What percent were mine workers?

6. A painting was bought for $4000 and sold for $4560. What percent was the profit? (The base is the buying price.)

7. 16% of what number is 96?

8. 32 is 40% of what number?

9. $\frac{3}{4}$% of what number is 120?

10. $\frac{1}{5}$% of what number is 90?

11. One night 300 students attended a concert, and the auditorium was 75% full. How many people can the room hold?

12. I wrote a check for $127.20 to cover my weekly rent and a 6% tax. How much was my rent?

13. The price of a radio alarm clock was $28.09 when the 8.25% sales tax was added. Find the price of the clock itself. (You'll need to round your answer.)

14. A speculator bought stocks and later sold them for $4815, making a profit of 7%. How much did the stocks cost her?

15. A woman paid $100.80 for a dress after a 20% discount had been taken off. How much did the dress cost before the discount?

16. Jane has $12,000 invested, some at 10% and some at 14%. The yearly incomes from the two investments are equal. How much is invested at each rate?

17. A 6% investment brings an annual return of $28 more in interest than a 5% investment. The total amount invested is $1200. Find the two amounts invested.

18. John's weekly gross pay is $550 but 27% of his check is withheld. How much is John's take-home pay?

Ratio and Proportion Problems

In Chapter 4 (Fractions) we solved problems using equivalent fractions. A fraction is a comparison of two numbers. Another word for the comparison of two numbers is *ratio*. 2 out of 5, $\frac{2}{5}$, can be written as the ratio $2:5$. When two ratios are alike, such as 2 is to 5 as 4 is to 10 ($2:5 = 4:10$), or we have two equivalent fractions, we have a *proportion*. $\frac{2}{5} = \frac{4}{10}$ is a proportion. In proportion problems in mathematics, we usually know three of the numbers involved and are asked to find the fourth number.

If $\frac{2}{7} = \frac{8}{?}$, for example, we can replace the ? with x and have an equation:

$$\frac{2}{7} = \frac{8}{x}$$

A proportion can be solved just like any other fractional equation by multiplying both sides by a common denominator.

$$7x\left(\frac{2}{7}\right) = 7x\left(\frac{8}{x}\right) \qquad \text{Multiply both sides by } 7x.$$

$$2x = 56 \qquad \text{Simplify.}$$

$$x = 28 \qquad \text{Divide both sides by 2.}$$

Check: $\dfrac{2}{7} = \dfrac{8}{28}\,\checkmark$

Essentially we have multiplied the numerators by the opposite denominators and found the *cross products*. We call this *cross multiplication*.

$$\frac{2}{7} = \frac{8}{x}$$

$$2(x) = 7(8)$$

EXAMPLE

Glen earns $97 in 4 days. How many days will it take him to earn $485?

Solution

$$\$97 \text{ in } \quad 4 \text{ days}$$

$$\$485 \text{ in } \quad x \text{ days}$$

We ask ourselves, "$97 is to $485 as 4 days is to how many days?"

$$\frac{97}{485} = \frac{4}{x} \qquad \text{Express the problem as a proportion.}$$

$$97x = 485(4) \qquad \text{Cross multiply.}$$

$$97x = 1940 \qquad \text{Simplify.}$$

$$x = 20 \qquad \text{Divide both sides by 97.}$$

It will take Glen 20 days to earn $485.

EXAMPLE

A tree casts a shadow of 28 feet, while a 6-foot post nearby casts a shadow of 4 feet. What is the height of the tree?

Solution

Tree: x ft high   28 ft shadow

Pole: 6 ft high   4 ft shadow

$\dfrac{x}{6} = \dfrac{28}{4}$ Write the proportion.

$4x = 6(28)$ Cross multiply.

$4x = 168$ Simplify.

$x = 42$ Divide both sides by 4.

The height of the tree is 42 feet.

EXAMPLE

A 14-inch-long ribbon is cut into two pieces in a ratio of $3:4$. How long are the pieces?

Solution   Let the pieces be x and y inches long.

$x + y = 14$ so $y = 14 - x$

$\dfrac{x}{y} = \dfrac{3}{4}$ Write the proportion.

$4x = 3y$ Cross multiply.

$4x = 3(14 - x)$ Substitute.

$x = 6$ Solve for x.

$y = 8$ Solve for y.

The pieces are 6 in. and 8 in. long.

Alternative Solution: Call the pieces $3x$ and $4x$.

$3x + 4x = 14$

$x = 2$

The pieces are 3(2) in. = 6 in. and 4(2) in. = 8 in.

EXERCISE 10.3.3

1. A flagpole casts a shadow of 27 feet, while a boy 5 feet tall casts a shadow of 3 feet. Find the height of the flagpole.

2. A motorist uses 25 gallons of gasoline to travel 350 miles. How much gasoline will he use to go 462 miles?

3. A 12-oz can of frozen orange juice costs $1.92 and makes 48 oz of diluted juice. How much would a glass (6 oz) of orange juice cost?

4. Three gallons of fruit juice will serve 40 people. How much juice is needed to serve 100 people?

5. The length and width of a rectangle are in a ratio of 6 to 4. The perimeter is 360 ft. What are the dimensions?

6. A woman cuts a 60-inch-long board into two pieces. They are in a ratio of 2 to 3. How long is each piece?

7. A farmer plants 15 acres of corn and potatoes in a ratio of 2 : 1. How many acres does he plant of each?

8. The tax for a car that cost $10,000 was $350. What was the tax for a car that cost $12,000?

Geometry Problems

As we mentioned earlier in this text, the perimeter of a geometric figure is the sum of all sides. In particular, the perimeter of a rectangle is $P = 2l + 2w$, where l is the length and w is the width of the rectangle.

The area of a rectangle equals length times width. Areas of other geometric figures can be calculated by the use of appropriate formulas.

EXAMPLE

A rectangle has a perimeter of 76 inches. If its length is 8 inches more than its width, find the length and width.

Solution   Draw the rectangle.

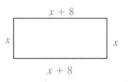

If x is the width, then $x + 8$ is the length.

$2x + 2(x + 8) = 76$	Perimeter
$4x + 16 = 76$	Simplify.
$4x = 60$	Subtract 16 from both sides.
$x = 15$	Divide both sides by 4.
$x + 8 = 15 + 8 = 23$	Substitute the x value to find length.

Check: $2(15) + 2(23) \stackrel{?}{=} 76$
$30 + 46 = 76\surd$

The length is 23 inches, and the width is 15 inches.

EXAMPLE

The perimeter of a rectangle is 18 cm. Its length is twice the width. Find the area.

Solution   Draw a diagram and list what you know.

$$x \qquad \text{Width}$$

$$2x \qquad \text{Length}$$

$$2(2x) + 2(x) = 18 \qquad \text{Perimeter}$$

$$4x + 2x = 18 \qquad \text{Simplify.}$$

$$6x = 18$$

$$x = 3 \qquad \text{Solve to get the width.}$$

$$2x = 6 \qquad \text{Length}$$

The area is $(3\,\text{cm})(6\,\text{cm}) = 18\ \text{cm}^2$.

EXAMPLE

The base of an isosceles triangle is 10 cm and the perimeter is 26 cm. Find the other sides.

Solution   An isosceles triangle has two equal sides. The third side is the base. Call each of the equal sides x. The perimeter is

$$x + x + 10 = 26$$

$$2x + 10 = 26$$

$$2x = 16$$

$$x = 8 \qquad \text{The sides are each 8 cm.}$$

EXERCISE 10.3.4

1. Find the dimension of a rectangle whose length is twice its width and whose perimeter is 54 inches.

2. The width of a rectangular dining room is three times the width of the table. The length of the room is three times the length of the table. If the dining room is twice as long as it is wide and its perimeter is 72 feet, what are the dimensions of the table?

3. A piece of wire 60 cm long is bent to form an isosceles triangle (a triangle with two equal sides). The base of the triangle is 8 cm longer than one of its sides. Find the length of the sides and the base of the triangle.

4. The length of a rectangle is 8 units more than twice its width. If the perimeter is 52 units, find the length and the width.

5. The base in an isosceles triangle is 4 ft shorter than one of its sides, and its perimeter is 14 ft. Find the length of the sides.

6. A triangular garden has a perimeter of 27 m. The longest side is 5 m longer than the shortest side, and the third side is 4 m less than the longest side. What are the lengths of the three sides?

7. A circular braided rug just fits a square room with sides of 16 feet (Figure 10.1). When the rug is in place, how much of the old floor will show?

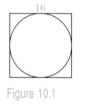

Figure 10.1

8. A new window installed in an old farmhouse is shaped like a semicircle (half a circle) on top of a rectangle (see Figure 10.2). The diameter of the semicircle is $2\frac{1}{2}$ ft. The perimeter of the rectangle is 13 ft. What is the total area of the window?

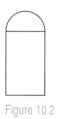

Figure 10.2

Motion Problems

In distance problems, as we saw in Chapter 7, we often use the formula Distance = rate $\times$ time, or $D = rt$.

EXAMPLE

A car and a motorcycle stop at a rest stop on the highway. The car then travels north at 60 mph, and the motorcycle travels south at 45 mph. If they leave the rest stop at the same time, how long will it take them to be 210 miles apart?

Solution   Let's draw a simple diagram.

Rest stop

South ←————————————|————————————————→ North
     45 mph   60 mph

We are looking for the time. We can start by making a table, calling the unknown time x.

	Rate (mph)	Time (hr)	Distance (mi)
Car	60	x	$60x$
Motorcycle	45	x	$45x$

Then,

$$60x + 45x = 210$$
$$105x = 210$$
$$x = 2 \quad \text{The time is 2 hours.}$$

EXAMPLE

Sandy can row 24 miles downstream in 3 hours. But when he rows upstream, the same distance takes 6 hours. What would Sandy's rate be in still water? What is the rate of the current?

Solution If Sandy can row x mph and the current is y mph, the rate downstream is $x + y$ and the rate upstream is $x - y$.

	Rate (mph)	Time (hr)	Distance (mi)
Downstream	$x + y$	3	24
Upstream	$x - y$	6	24

Use the formula rate $\times$ time = distance

$3(x + y) = 24$	Equation 1
$6(x - y) = 24$	Equation 2
$x + y = 8$	Divide both sides of equation 1 by 3.
$x - y = 4$	Divide both sides of equation 2 by 6.
$2x = 12$	Add the equations.
$x = 6$	Divide both sides by 2.
$y = 2$	Substitute and solve.

Check: $(3)(6 + 2) \stackrel{?}{=} 24$

$$(3)(8) = 24 \surd$$

$$(6)(6 - 2) \stackrel{?}{=} 24$$

$$(6)(4) = 24 \surd$$

Sandy rows at 6 mph, and the current is 2 mph.

EXERCISE 10.3.5

1. Two cars start at the same time from the same place. One travels north at 40 mph, the other south at 60 mph. In how many hours are they 300 miles apart?

2. A car is driven 3 hours at a certain speed, and then the speed is increased by 10 miles per hour for the next 2 hours. If the total distance is 250 miles, what are the two rates?

3. A boat can move at 8 mph in still water. It can travel downstream 20 miles in the same time it takes to travel 12 miles upstream. Determine the rate of the stream.

4. Two joggers start toward each other at the same time from schools 10 miles apart. Amy is jogging at 4 mph and Marni at 6 mph. (a) How long will it take them to meet? (b) How far will they be from Amy's school?

5. On my way to the cabin I drove at the rate of 55 mph. Coming home in a heavy rain, I only averaged 35 mph. It took me 2 hours less to get to the cabin than to get home. How many miles is it to the cabin?

6. On a canoe trip we stopped for lunch on a shady beach. After lunch we paddled downstream for $3\frac{1}{2}$ hours before Jim realized he had left his jacket with his billfold in it at the lunch stop. It took us 5 hours to get back to the beach. What would be our speed in still water if the rate of the stream is 3 miles per hour?

7. John's plane flies three times as fast as Jim's. Over a distance of 3000 miles Jim's plane takes 10 hours longer than John's. Find the speeds of the two planes.

8. Hiking up to Chimney Pond on Mt. Katahdin takes 5 hours, but hiking down takes only $1\frac{1}{2}$ hours. The distance is 3.3 miles. What is the difference between the two average hiking rates?

Age Problems

There is a time in life when a parent is twice the age of the child. To find out when this happens, we use equations.

EXAMPLE

A father is 24 years older than his son. In 8 years he will be twice as old as his son. Determine their present ages.

Solution   Set up a table.

	Current Age	Age in 8 Years
Father	$x + 24$	$x + 24 + 8 = x + 32$
Son	x	$x + 8$

$$x + 32 = 2(x + 8) \quad \text{Write the equation.}$$
$$x + 32 = 2x + 16 \quad \text{Simplify.}$$
$$16 = x \quad \text{Solve.}$$

Check: $16 + 24 + 8 \stackrel{?}{=} 2(16 + 8)$
$$48 = 48 \sqrt{}$$

The son is now 16 years old and the father $16 + 24 = 40$ years old. (In 8 years, the son will be 24 years old and the father twice that, 48 years old.)

EXERCISE 10.3.6

1. Mary is 15 years older than her sister Jane. Six years ago, Mary was six times as old as Jane. Find their present ages.

2. Jon is 6 years older than his wife Liz. In 4 years, twice his age plus 1 will be 3 times Liz's age 3 years ago. How old are they now?

3. The sum of Betsy's age and her father's age is 41, and the difference is 31. How old is each of them now?

4. Casey is 4 years older than Aaron. Three times Casey's age in 3 years will equal 2 less than 5 times Aaron's age in 3 years. How old are they now?

5. The sum of Fiori's and Omn's ages is 19, and the difference is 3. If Fiori is older than Omn, how old is each?

6. The difference in age between Deborah and her daughter is 25 years. The sum of their ages is 43. What are their ages?

7. Aaron is half as old as his sister, Cassandra. In 12 years he will be four-fifths as old. How old are they now?

8. The ratio of Jonathan's age to David's age is $6:5$. In seven years it will be $7:6$. What are their ages now?

Work Problems

Many problems ask us to find out how long it would take workers or machines to complete a job either working together or working alone.

EXAMPLE

Working alone, Anita can complete a job in 2 hours, while Betty can do the same job in 3 hours. How long does it take the two women to complete the job if they work together?

Solution   Anita can complete 1/2 the job in 1 hour while Betty completes 1/3 of the job in 1 hour. Let's set up a table.

	Time to Complete Job	Part of Job per Hour	Part of Job Done in x hr
Anita	2 hr	$\dfrac{1}{2}$	$\dfrac{x}{2}$
Betty	3 hr	$\dfrac{1}{3}$	$\dfrac{x}{3}$

In x hours Anita will complete $\dfrac{x}{2}$ of the job.

In x hours Betty will complete $\dfrac{x}{3}$ of the job.

Together, they will complete the total job, which is represented by 1.

$$\frac{x}{2} + \frac{x}{3} = 1 \qquad \text{Write the equation.}$$

$$6\left(\frac{x}{2}\right) + 6\left(\frac{x}{3}\right) = 6(1) \qquad \text{Multiply both sides by 6.}$$

$$3x + 2x = 6 \qquad \text{Reduce and simplify.}$$

$$5x = 6$$

$$x = \frac{6}{5} = 1\frac{1}{5}$$

The time is $1\frac{1}{5}$ hours, or 1 hr and 12 min.

Check: $\dfrac{6}{5} \cdot \dfrac{1}{2} + \dfrac{6}{5} \cdot \dfrac{1}{3} \overset{?}{=} 1$

$$\dfrac{6}{10} + \dfrac{6}{15} = \dfrac{18}{30} + \dfrac{12}{30} = \dfrac{30}{30} = 1\sqrt{}$$

It takes Anita and Betty 1 hr and 12 min to complete the job together.

EXAMPLE

One pipe alone will fill a tank in 20 hours. A second pipe alone will fill it in 15 hours. If the first pipe was open for 16 hours and then closed, how long would it take the second pipe to finish filling the tank?

Solution Let x be the hours needed to finish filling the tank. Make a table.

	Time to Fill	Job per hour	In 16 hr	In x hr
1st pipe	20 hr	1/ 20	16/ 20 = 4/ 5	—
2nd pipe	15 hr	1/ 15	—	x/ 15

$\dfrac{4}{5} + \dfrac{x}{15} = 1$ Write an equation.

$15\left(\dfrac{4}{5}\right) + 15\left(\dfrac{x}{15}\right) = 15(1)$ Multiply both sides by 15.

$12 + x = 15$ Simplify.

$x = 3$ Solve.

It would take the second pipe 3 hours to fill the tank.

EXAMPLE

If 6 people complete a job in 10 days, how many days would it take 12 people to complete the same job?

Solution Here the size of the job is $6(10) = 60$ "people-days." If it takes 12 people x days, we get

$$12x = 60 \qquad \text{Equation}$$
$$x = 5 \qquad \text{Solve.}$$

It would take 12 people 5 days to complete the job.

EXERCISE 10.3.7

1. Sylvia can complete a job in 45 minutes working alone. Carla takes 30 minutes to complete the same job. How long would it take if they worked together?

2. If 16 men finish a piece of work in $28\frac{1}{3}$ days, how long would it take 12 men to do the same work?

3. Assad and Abduhl can do a job together in 12 days. After Abduhl has worked alone 26 days, Assad finishes the job alone in 5 days. How long would the whole job take each man alone?

4. Sam could paint Sherry's house in 7 days. Sue could paint it by herself in 5 days. How long will it take if they work together?

5. It would take Judy 4 hours to cut the grass but it would take John 6 hours. How long will it take them to cut the grass if they work together?

6. It took Dave and his mother 4 hours to paper the bedroom. It would take Dave 6 hours to do it alone. How long would it take his mother to do it alone?

7. John and Peter, aged 11 and 13, were given a job to do at 9 A.M. Working alone, John could do the job in 5 hours, and Peter in 4 hours. At 10 A.M. they went swimming until 1:30, then came back and finished the job. At what time did they finish it?

Coin and Mixture Problems

Many word problems deal with mixtures; a grocer mixes coffee or candy, a person has a mixture of coins in his pocket and so on.

EXAMPLE

A merchant mixes 25 pounds of one kind of candy with 15 pounds of another kind to make a mixture that costs 60 cents per pound. If the first kind of candy costs 45 cents per pound, how much per pound does the second kind cost?

Solution   Make a table.

	Amount	Price per Pound	Total Value
Candy A	25	$0.45	$25(0.45) = \$11.25$
Candy B	15	x	$15x$
Mixture	40	$0.60	$40(0.60) = \$24.00$

$$11.25 + 15x = 24.00 \quad \text{Equation}$$
$$x = 0.85 \quad \text{Solve.}$$

The second kind of candy costs 85¢ per pound.

EXAMPLE

Eva has 10 coins consisting of nickels and dimes. The value of her money is 60¢. How many of each coin does she have?

Solution   Make a table.

Number of Coins	Value per Coin	Total Value
x	5¢	$5x$ ¢
y	10¢	$10y$ ¢
10		60¢

$x + y = 10$	Write the equation for the number of coins.
$5x + 10y = 60$	Write the equation for their value.
$-5x - 5y = -50$	Multiply the first equation by -5.
$5y = 10$	Add the equations.
$y = 2$	Solve for y.
$x + 2 = 10$	Substitute for y in first equation.
$x = 8$	Solve for x.

Eva has 8 nickels and 2 dimes. The total value is 60 cents.

EXAMPLE

Forty people signed up to take a bus trip through the White Mountains to see the fall foliage in October. The senior citizens received a 20% discount. $1164.80 was collected from the 40 people. How many senior citizens went on the trip, if the price of the regular ticket was $32?

Solution   Let x be the number of senior citizens taking the trip and y the number of other passengers.

Number of People	Price per Ticket ($)	Total Cost ($)
x	80% (32)	80% (32)(x)
y	32	$32y$
40		1164.80

$$x + y = 40$$
$$80\%(32)x + 32y = 1164.80$$
$$0.80x + y = 36.40 \qquad \text{Divide both sides by 32.}$$
$$0.20x = 3.60 \qquad \text{Subtract from Equation 1.}$$
$$x = 18 \qquad \text{Divide by 0.2.}$$

Answer: 18 senior citizens went on the trip.

EXAMPLE

For John's wedding, Jean decided to make a punch of passion fruit and grapefruit juice. She wanted 20 quarts of punch and had some 20% passion fruit juice to mix with. How much of each juice should she use to get a punch with 5% passion fruit juice?

Solution

	Amount	Percent Passion Fruit Juice	Pure Passion Fruit Juice
Juice mix	x qt	20%	20% x qt
Grapefruit	y qt	0%	0
Punch	20 qt	5%	5% (20) qt

$$x + y = 20$$
$$20\%x = 5\%(20)$$
$$20x = 100$$
$$x = 5$$
$$y = 20 - 5 = 15$$

Answer: Jean needs 5 quarts of the 20% passion fruit juice and 15 quarts of grapefruit juice.

EXERCISE 10.3.8

1. A sum of money amounting to $4.32 consists entirely of dimes and pennies, 108 coins in all. How many dimes and how many pennies are there?

2. A purse contains $3.05 in nickels and dimes, 19 more nickels than dimes. How many coins are there of each kind?

3. Lubricating oil worth 28¢ per quart is to be mixed with oil worth 33¢ per quart to make up 45 qt of a mixture to sell at 30¢ a quart. What volume of each grade should be taken?

4. There were 2500 persons watching a game. The adults paid 75¢ for admission, and children paid 25¢. If $1503 was collected in ticket sales, how many adults and how many children saw the game?

5. A clerk bought 150 stamps for the office. Some were 23¢ stamps, and the rest were 29¢ stamps. How many stamps of each kind did he buy if he spent $37.50?

6. Sarah and Jonathan Jr. (both eight-years-old) found Grandma's piggy bank. First they counted the coins. There were 146, all dimes and quarters. Then they counted the money. There was $26.60. How many of each coin were there?

7. Twenty ounces of a punch containing 30% grapefruit juice was added to 40 ounces of punch containing 20% grapefruit juice. Find the percent of grapefruit juice in the resulting mixture.

8. How many fluid ounces of pure acid (100%) must be added to 60 fluid ounces of a 20% acid solution to obtain a 40% acid solution?

10.4 ADDITIONAL WORD PROBLEMS

EXAMPLE

A bridge is divided into three sections. The last section is 120 ft longer than the first, and the middle section is 3 times as long as the last. If the bridge is 1000 ft long, how long is each section?

Solution   Draw a picture (Figure 10.3).

x	First section
$x + 120$	Last section
$3(x + 120)$	Middle section
$x + x + 120 + 3(x + 120) = 1000$	Equation
$x + x + 120 + 3x + 360 = 1000$	Distribute.
$5x + 480 = 1000$	Simplify.
$5x = 520$	Solve.
$x = 104$	First section
$104 + 120 = 224$	Last section
$3(104 + 120) = 672$	Middle section

Check: $104 + 224 + 672 = 1000\sqrt{}$
The sections are, in order, 104 feet, 672 feet, and 224 feet.

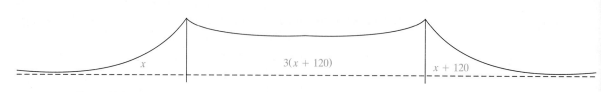

Figure 10.3

EXERCISE 10.4.1

1. The sum of the heights of Lars and Hans is equal to 312 cm. Lars is 18 cm taller than Hans. How tall is each of the boys?

2. In a mixture of wine and cider, $\frac{1}{2}$ of the whole plus 25 gallons was wine, and $\frac{1}{3}$ of the whole minus 5 gallons was cider. How many gallons of each were there?

3. Tom is 25 years old. His youngest brother is 15. How many years must elapse before their ages are in a ratio of 5 to 4?

4. The first digit of a number exceeds the second by 4, and if the number is divided by the sum of the digits, the quotient is 7. Find the number.

5. At the university the ratio of in-state to out-of-state students is 5 to 2. There are 3750 more in-state students than out-of-state. How many of each are enrolled?

6. A man is paid $20 for each day he works and has to pay $5 for each day he is idle. At the end of 25 days he nets $450. How many did he work?

7. Deborah, Dave, and Judy are practicing their piano lessons. Deb practices $\frac{2}{3}$ as long as Dave, and Judy practices twice as long as Dave. If their combined practice time is 110 minutes, how long does each practice?

Here are some very old problems written in the days when the only numerator allowed, except for the fraction $\frac{2}{3}$, was 1. Try solving a few of them. (Some problems have been edited a bit to make the language easier to read.)

EXAMPLE

Solve the problem about Diophantus in the introduction to this chapter.

   Solution   Let his age equal x. Then
$$\frac{x}{6} + \frac{x}{12} + \frac{x}{7} + 5 + \frac{x}{2} + 4 = x$$
The least common denominator of 6, 12, 7, and 2 is 84.
Multiply both sides of the equation by 84.
$$14x + 7x + 12x + 420 + 42x + 336 = 84x$$
$$75x + 756 = 84x$$
$$756 = 9x$$
$$84 = x$$

Diophantus lived until he was 84 years old.

EXERCISE 10.4.2

1. King Crossius dedicated six bowls having a total weight of 100 drachmas. Each bowl was 1 drachma heavier than another. Find the weight of the bowls. (From *Greek Anthology*.)

2. Demochares has lived $\frac{1}{4}$ of his life as a boy, $\frac{1}{5}$ as a youth, $\frac{1}{3}$ as a man, and has spent 13 years in his old age. How old is he? (From *Greek Anthology*.)

3. A traveler on a ship on the Adriatic asked the captain, "How much sea do we still have to cross?" The captain answered, "Voyager, our total trip between Cretan Ram's Head and Sicilian Peloria is 6000 stades [about 600 miles]. We still have to travel $\frac{2}{3}$ the distance we have already traveled." Find (a) the distance they have traveled and (b) the distance still to be traveled. (From *Greek Anthology*.)

4. The sum of a number, $\frac{2}{3}$ of the number, $\frac{1}{2}$ of the number, and $\frac{1}{7}$ of the number is 33. Find the number. (From the Rhind Papyrus, an Egyptian manuscript dated 1650 B.C.)

5. Silversmith, throw in the furnace and mix the bowl, $\frac{1}{3}$ of its weight, $\frac{1}{4}$ of its weight, and $\frac{1}{12}$ of its weight so that the total mass weighs 100 drachmas. Find the weight of the bowl. (From *Greek Anthology*.)

6. Mother, don't scold me about the walnuts. Pretty girls took them from me. Melission took $\frac{2}{7}$, and Titane took $\frac{1}{12}$. Playful Astyoche and Philinna have $\frac{1}{6}$ and $\frac{1}{3}$. Thetis seized 20, and Thiabe 12. Glauce, smiling sweetly there, has 11 in her hand. This 1 nut is all I have left. How many walnuts were there in all? (From *Greek Anthology*.)

7. Of a collection of mango fruits, the King took $\frac{1}{6}$, the Queen $\frac{1}{5}$ of the remainder, and the three chief princes $\frac{1}{4}$, $\frac{1}{3}$, and $\frac{1}{2}$ of the successive remainders. The youngest child took the remaining 3 mangoes. Of you who are clever in miscellaneous problems on fractions, give the measure of that collection of mangoes. (From the works of the Hindu Mahavira, about A.D. 850.)

8. A man was lying on his deathbed and wished to make out his will. His estate amounted to 1000 ducats. 200 ducats he left to the church, and 800 ducats remained. The man's wife was shortly to give birth to a child, and he wished to make specific provision for both his widow and his orphan. He therefore made this disposition: If the child was a girl, then she and her mother were to share the money equally, 400 ducats each; if, on the other hand, the child was a boy, he was to have 500 ducats and the widow only 300. Shortly afterwards the man died, and in due course his widow's time came. But she gave birth to twins, and, to make things more complicated, one of the twins was a boy and the other a girl. The problem is: If the proportions between mother, son, and daughter desired by the deceased are honored, how many ducats will mother, son, and daughter receive? (From *Greek Anthology*.)

SUMMARY

English	*Mathematics*
Addition	
the sum of *a* and *b*	$a + b$
a plus *b*	
a increased by *b*	
b more than *a*	
add *b* to *a*	

Subtraction

the difference of *a* and *b*	$a - b$
a minus *b*	
a decreased by *b*	
b subtracted from *a*	
a less *b*	
b less than *a*	
take away *b* from *a*	

Multiplication

the product of *a* and *b*	ab
a times *b*	$a \cdot b$
a multiplied by *b*	$a * b$
	$(a)(b)$
	$a(b)$
	$(a)b$

Division

the quotient of *a* and *b*	
a divided by *b*	$\dfrac{a}{b}$
b goes into *a*	$a \div b$
	$b\overline{)a}$

Equality

the result is	
is, is equal to,	$=$
equals	
is the same as	

Inequality

a is greater than *b*	$a > b$
a is less than *b*	$a < b$
a is greater than or equal to *b*	$a \geq b$
a is less than or equal to *b*	$a \leq b$
a is at least *b*	$a \geq b$
a is at most *b*	$a \leq b$
a does not equal *b*	$a \neq b$

Strategy for Solving Word Problems

1. Read the problem *carefully*. Draw a picture, if possible.
2. Read the problem again and list the quantities you are looking for.
3. If possible, make an estimate of what you expect the answer to be.
4. Which variable do the others depend on? That will be your independent variable.
5. Express each of the unknown quantities in terms of the independent variable.
6. Write as many equations as you need.

7. Solve the equations.

8. Make sure you've found values for all listed quantities.

9. Check your answer. Even if your answer solves the equation you wrote, *does it make sense*?

10. Reread the question. Have you answered the question or do you have to continue with more calculations?

VOCABULARY

Consecutive numbers: Integers that follow each other; for example, 2, 3, 4, 5 or 125, 126, 127.

Consecutive even numbers: Every other even integer; for example, 4, 6, 8.

Consecutive odd numbers: Every other odd integer; for example, 7, 9, 11.

Cross multiply: To multiply crosswise in a proportion.

Cross product: The product of the first and fourth numbers or the second and third numbers in a proportion.

Decrease: Minus.

Difference: The answer in a subtraction.

Product: The answer in a multiplication.

Proportion: An equation formed by two ratios.

Quotient: The answer in a division.

Ratio: Comparison of two numbers by division.

Sum: The answer in an addition.

CHECK LIST

Check the box for each topic you feel you have mastered. If you are unsure, go back and review.

☐ Translating words into symbols

☐ Remembering when to switch the order in translations

☐ Writing equations

☐ Writing inequalities

Solving

☐ Number problems

☐ Percent problems

☐ Ratio and proportion problems

☐ Geometry problems

☐ Motion problems

☐ Age

☐ Work

☐ Mixing problems

☐ Nonstandard problems

REVIEW EXERCISES

1. Translate from words to symbols.
   (a) The product of a and b
   (b) The sum of a and b
   (c) The difference of a and b
   (d) The quotient of a and b
   (e) a divided by b
   (f) Take away b from a
   (g) b subtracted from a
   (h) a decreased by b

2. Translate into equations and solve.
   (a) Four less than a number is eleven.
   (b) Eight into a number is two.

(c) Eight divided by a number is two.

(d) Twice an unknown number less five is eleven.

(e) If the sum of an unknown number and six is divided by seven, the result is three.

(f) Three times an unknown number increased by three is twenty-four.

3. The sum of three consecutive numbers is 66. Find the numbers.

4. The sum of four consecutive odd numbers is one more than five times the first. Find the numbers.

5. After lowering the thermostat from 70 to 64, the Smiths found that their heating bill decreased from $205.00 to $160.00. What was the percent of decrease?

6. A number decreased by 20% of itself is 85. What is the number?

7. Three out of every five voters vote in an election. How many would be expected to vote in a city of 60,000?

8. A person drives 800 miles in a day and a half. How long would it takes to drive 2000 miles?

9. The value of a personal computer is $2400. This is 80% of the computer's value last year. What was its value last year?

10. A construction company pays $12,000 for a truck. The truck depreciates at a constant rate of 20% yearly. How long will it take for the truck to be worth $7200?

11. Two men were talking about their ages. One said he was 94 years old. "Then," said the younger, "the sum of your age and mine multiplied by the difference will be 8512." How old was the younger man?

12. The sum of Tom's, Dick's, and Harry's ages is 31. Tom is twice as old as Dick, and Harry is 6 years older than Tom. How old is each?

13. Bob is 10 years older than his brother Pete. In 3 years he will be twice as old as Pete will be then. Find their present ages.

14. It would take Betsey 4 hours to clean her room. Her mother could do it in 2 hours. How long would it take them both working together?

15. If six people could paint a house in 3 days, how long would it take two people to paint it?

16. Liz found 17 coins (dimes and quarters), totaling $2.45, in the glove compartment of her car. How many of each coin does she have?

READINESS CHECK

Solve the problems to satisfy yourself that you have mastered Chapter 10.

1. Write an equation and solve: A number subtracted from eight is three.

2. Write an inequality and solve: Thirteen is less than four times a number.

3. The sum of three consecutive odd numbers is thirty-nine. Find the numbers.

4. 6% of what number is 36?

5. In a small town, two out of three cars are domestic. If there are 1500 cars in town, how many domestic cars are there?

6. A rectangle is 5 feet longer than it is wide. If its perimeter is 22 feet, find the dimensions of the rectangle.

7. Lina could travel 51 kilometers in 3 hours in her canoe when she paddled with the current. Against the current she traveled 39 kilometers in 3 hours. Find Lina's speed in calm water and the rate of the current.

8. Ellen is two years older than Eva. The sum of their ages is sixty-eight. How old is Ellen?

9. It takes one water tap 15 minutes to fill a small pool. Another water tap fills the pool in 30 minutes. How long does it take to fill the pool when both taps are open?

10. A purse contains $2.45 in nickels and dimes. If there are 32 coins all together, how many nickels are in the purse?

PRACTICAL
MATHEMATICS

MATHEMATICS IN BANKING

Pay interest and earn interest

By the year 2000 B.C. there already existed a banking system in Babylonia. Trusted men kept valuables for others for a fee. The Romans and the Chinese had banking systems by the years 500 B.C. and 1100 B.C., respectively. In Venice the *bancherius de scripta* accepted deposits and made loans from A.D. 1318 on. Loan offices and banks were formed in the 18th century in the United States. The Bank of North America was chartered in 1781.

In this chapter we discuss some of the mathematics necessary to understand loans and bank interest. Anybody should be able to go into a bank and ask questions about interest and loan rates, for example. Bank officers are often reluctant to give out this information but the public has a right to know. It is our money we are talking about!

11.1 LOANS AND MORTGAGES

Simple Interest

One of the most common uses we have for percent is with bank interest rates. If we have money in the bank we earn interest; if we borrow money, we pay interest. The easiest way to calculate interest is to use "simple interest." With simple interest, we pay (or receive) a percent of what we have borrowed (or have in the bank) for a certain time.

If the bank would use simple interest and if you keep $300 in the bank for 3 years and are paid 5% per year simple interest, you would have earned $300 \cdot 5\% \cdot 3 = 300 \cdot 0.05 \cdot 3 = \$45$ interest. You would then have \$300 + \$45 = \$345 in the bank.

The simple interest formula $I = Prt$ is never used in practical life. However, it is convenient to use to get an estimation of what you have to pay or what you can earn. Also the formula is used to calculate interest for a certain

Table 11.1

Time (Month)	Starting Principal	Payment	Interest	Loan Repayment	Remaining Principal
1	1000	88.65	9.65	79.00	921.00
2	921.00	88.65	8.89	79.76	841.24
3	841.24	88.65	8.12	80.53	760.71
4	760.70	88.65	7.34	81.31	679.40
5	679.40	88.65	6.56	82.09	597.31
6	597.31	88.65	5.76	82.89	514.42
7	514.42	88.65	4.96	83.69	430.73
8	430.73	88.65	4.16	84.49	346.24
9	346.23	88.65	3.34	85.31	260.93
10	260.92	88.65	2.52	86.13	174.80
11	174.78	88.65	1.69	86.96	87.81
12	87.81	88.65	0.85	87.80	0.01
		1063.80	63.84	999.96	

time such as a month. If the yearly rate is 5%, then the monthly rate is 5%/12, for example.

Loans

When you borrow money (take out a loan), you have to pay *interest*. If you borrow $1000 for a year at an *interest rate* of 11.58%, the bank states that the *finance charges* are $63.80 for the year and that you must pay $88.65 every month. Interest is paid on the unpaid balance, and banks use tables or computers to create a schedule of payments.

11.58% of $1000 for one year is $115.80, so your interest charges are not simple interest.

Your interest rate is 11.58%/12 or 0.965% of the principal remaining the beginning of each month. Table 11.1 presents a breakdown of what is going on. The 1 cent left over and the 4 cents extra interest charges stem from the rounding off using a calculator.

In all exercises in this chapter there might be differences between our answers and yours due to differences in rounding. A bank computer uses many more decimals than we do with calculators. Don't worry about the pennies in your calculations. As long as the millions and thousands are correct, so are you.

EXERCISE 11.1.1

1. Construct a table like Table 11.1 for a loan of $6000 for 12 months at an annual percentage rate of 16% if the monthly payments are $544.39. Calculate the total finance charges for the year. Here is a beginning:

Month	Principal	Payment	Interest	Repayment	New Principal
1	6000.00	544.39	80.00	464.39	5535.61
2	5535.61	544.39			

With some loans you *prepay* the interest. For instance, if you borrow $600 at an annual *percentage rate* of 8.00%, you have to pay interest of $25.21, so you receive only the difference of $574.79 from the bank. You then pay back the $600 in 12 installments of $50 each.

The interest after one month will be $(8\%/12)574.79 = \$3.83$. The 12 interest payments should add up to $25.21.

EXERCISE 11.1.2

1. Complete the table using the information above.

Month	Principal	Payment	Interest	Loan Repayment	New Principal
1	574.79	50	3.83	46.17	528.62
2	528.62	50	3.52	46.48	482.14
3	482.14	50			
4		50			
5		50			
6					
7					
8					
9					
10					
11					
12					
		Total:	600		

The Rule of 78

If you have borrowed money from a bank and prepaid your interest and decide to pay back a one-year bank loan early, perhaps after 6 months, you will get some interest back, but not, as you might have thought, half of the prepaid interest. The "Rule of 78" comes in here.

The Rule of 78 says that the first month you have a loan you can use the total amount of the loan $(\frac{12}{12})$ and pay interest accordingly. The next month you have only $\frac{11}{12}$ of the loan left, since you already paid back $\frac{1}{12}$ of it at the end of the first month. This goes on month after month until the end of the twelfth month, when you have paid back everything. Add $\frac{12}{12}$, $\frac{11}{12}$, $\frac{10}{12}$, and so forth to $\frac{1}{12}$ and you have $\frac{78}{12}$. Here is where we get the term "Rule of 78."

The interest to the bank is divided into 78 parts. The first month your interest is $\frac{12}{78}$ of the total interest, the second month it is $\frac{11}{78}$ of the total interest, and so on. In the first 6 months the interest portion is

$$\frac{12}{78} + \frac{11}{78} + \frac{10}{78} + \frac{9}{78} + \frac{8}{78} + \frac{7}{78} = \frac{57}{78} \text{ of total interest}$$

You get back $\frac{78}{78} - \frac{57}{78} = \frac{21}{78}$ of the interest you prepaid. This is about one-fourth of the interest, not one-half!

EXAMPLE

If you were able to pay back your loan after 3 months, how much of the interest payment would you get back?

Solution

$$\frac{12}{78} + \frac{11}{78} + \frac{10}{78} = \frac{33}{78} \quad \text{and} \quad \frac{78}{78} - \frac{33}{78} = \frac{45}{78}$$

You will get back $\frac{45}{78}$ of your interest payments.

EXERCISE 11.1.3

1. If you repay a one-year loan after 4 months, what part of your prepaid interest do you get back?

2. If you repay a one-year loan after 8 months, what part of your prepaid interest do you get back?

Mortgages

There are tables available to find how much money you have to pay back each month when you obtain a mortgage from a bank, for example. It might be of interest to you to find out how much of your monthly payment goes to the bank as interest and how much of your mortgage you are actually paying off.

EXAMPLE

Make a table showing the first 12 months of interest and repayments for a mortgage of $50,000 over 10 years. The mortgage rate is 10% and you have to pay $660.75 every month.

Solution   The yearly rate is 10%, and the monthly rate is $10\%/12 = 0.00833\ldots$. The first year's payments are shown in Table 11.2.

After 1 year there is still $46,932.99 left to pay. In the first year, $4862 is paid out in interest and only about $3067 actually goes toward paying off the $50,000. (This situation does improve, though. Note that the numbers

Table 11.2

Time (month)	Principal ($)	Rate (per month)	Interest ($)	Loan Repayment ($)
1	50,000.00	0.00833	416.67	244.08
2	49,755.90		414.63	246.12
3	49,509.80		412.58	248.17
4	49,261.60		410.51	250.24
5	49,011.40		408.43	252.32
6	48,759.10		406.33	254.42
7	48,504.70		404.21	256.55
8	48,248.10		402.07	258.68
9	47,989.40		399.91	260.84
10	47,728.60		397.74	263.01
11	47,465.60		395.55	265.20
12	47,200.40		393.34	267.41

are getting smaller in the interest column and larger in the loan repayment column.)

Mortgage tables can be produced with a computer. The results will be more accurate, since the computer programs use more decimals than is convenient in hand calculations.

EXERCISE 11.1.4

1. A $50,000 mortgage for 30 years at 10% requires payments of only $438.79 per month. How much of the mortgage is left to pay after 1 year?

2. A $90,000 mortgage for 25 years at 8% requires payments of $694.63/month. How much interest is paid out in the first 6 months?

11.2 BANK INTEREST

Banks never use the simple interest formula to determine the amount of money a person has in the bank. If, for example, a person keeps $300 in the bank for 3 years at 5%, the interest will be calculated at certain given periods and added to the principal. This is *compound interest*.

If your $300 were invested at 5% compounded yearly, you would have:
After 1 year: $300 + 5%($300) = $315
After 2 years: $315 + 5%($315) = $315 + $15.75 = $330.75
After 3 years: $330.75 + 5%($330.75) = $330.75 + $16.54 = $347.29

You would have a little more money than with simple interest if the interest was compounded yearly. The more frequently the interest is compounded, the more interest you earn.

There are formulas for these calculations. However, you can also set up a month-by-month table to calculate your interest and principal.

EXAMPLE

If you invest $1000 at 10% annual interest compounded monthly, how much will you have after 3 months? (Monthly interest of 10% is $\frac{1}{12} \times 10\%$, which rounds to 0.83% = 0.0083.)

Solution   After 1 month:

$$1000 + (10\%/12)(1000) = 1000 + 0.83\% \cdot 1000 = 1008.30$$

After 2 months: $1008.30 + 0.83\% \cdot 1008.30 = 1016.67$

Principal	Time (month)	New Principal
$1000.00	1	$1000 + 0.0083 × $1000 = $1008.30
$1008.30	2	$1008.30 + 0.0083 × $1008.30 = $1016.67
$1016.67	3	$1016.67 + 0.0083 × $1016.67 = $1025.11

The simple interest for one year would be 10% of 1000 = 100, so you would end up with $1100. As you will see from Problem 1 of the following

Exercise, compound interest gives you more interest. Actually, you would earn more than the table indicates; we rounded the monthly interest to 0.83%, which is 9.96% annually, not 10%.

EXERCISE 11.2.1

1. Complete the table in the example for 12 months.

2. Calculate the interest on $200 for one year at 8% compounded monthly. Compare the compounded interest with simple interest.

3. Calculate the interest earned on $5000 deposited for one year at 7.4% compounded monthly. Compare the compounded interest with simple interest.

Formula for Compound Interest

Interest is calculated at certain intervals such as every week or every day. Banks use the following formula for compound interest:

$$A = P(1 + r)^t$$

where P is the money invested, r the percent paid per compounding period, and t the number of times the interest is compounded. A is the accumulated principal (the new principal).

In Problem 1 of Exercise 11.2.1 we invested $1000 at 10% compounded monthly for 1 year. There are 12 months in a year, so $r = 10\%/12 = 0.83\%$ (rounded). $t = 12$, since the money is compounded 12 times in 1 year. P is the original principal, in this case $1000.

The compound interest formula lets us calculate that after one year we would have

$$\$1000(1 + 0.0083)^{12} = \$1000 \times 1.0083^{12}$$

Enter the following on your calculator:

$$1.0083, y^x, 12, =, \times, 1000, =$$

Some calculators allow you to go from left to right using the parentheses. You should get 1104.2749 or $1104.27, which is what we got earlier. This means that if we invest $1000, we earn $104.27 in interest in one year.

EXAMPLE

Find the new principal if $600 is invested at 3.2% for 5 years and the interest is compounded daily.

Solution $P = 600$, $r = 3.2\%/365 = 0.0000877$, $t = 5 \times 365$. So

$$A = 600(1.0000877)^{5 \times 365} = 600 \times 1.1736 = 704.14$$

The new principal is $704.14.

EXERCISE 11.2.2

1. Find the new principal if $1000 is invested at 8.5% for one year (a) compounded monthly; (b) compounded quarterly (every 3 months); (c) compounded daily.

Annual Yield

Banks often advertise the "annual yield" and "annual percentage rate" of money market certificates and similar items.

One ad states that the annual yield is 11.07% and the annual rate is 10.50%, interest compounded daily. An annual rate of 10.50% corresponds to a daily rate of 10.50%/365. Since the interest is compounded 365 times during the year, the $1000 would grow to

$$1000(1 + 10.50\%/365)^{365} = 1110.6937$$

(This result was obtained by going from left to right and introducing parentheses as well as an $\times$ for multiplication on the calculator.)

If you invest $1000 for one year and get $110.6937 in interest, what percent interest would you get if this was simple interest? In other words, what percent of $1000 is $110.6937? We have to divide here:

$$110.6937 \div 1000 = 0.1106937 = 11.06937\% \approx 11.07\%$$

This simple interest rate is the *annual yield*.

EXAMPLE

A bank advertises that money on deposit is compounded daily. A money market account has an annual rate of 10.10%, and the annual yield is advertised as 10.63%. Check if this is true.

Solution   Use the formula for compound interest.

$$A = (1 + 10.10\%/365)^{365} = 1.1062612.$$

This is how much money we would get after 1 year if we invested $1; we have earned $0.1062612, which is 10.63% of $1. This is the simple interest, and thus the annual yield is 10.63%.

EXERCISE 11.2.3

1. A bank advertises the following rates.

Today's Rate	Annual Yield
4.50%	4.60%
3.05%	3.10%
3.20%	3.25%
3.45%	3.51%
4.25%	4.34%
4.75%	4.86%

Verify the annual yields. Answers may vary slightly due to rounding.

The Rule of 70

If you deposit $1000 in the bank, how long will it be until your money has doubled? This depends, of course, on the interest you earn.

For simplicity, let's work with annual yields, so we'll say that our interest is compounded yearly.

EXAMPLE

If you deposit $1000 and the annual yield is 1%, how much money do you have after 70 years?

> **Solution**   Use the formula for compound interest.
> $$P = 1000, \qquad r = 1, \qquad t = 70$$
> $$A = P(1 + r)^t = 1000(1 + 1/100)^{70}$$
> $$= 1000(1.01)^{70}$$
> $$= 1000 \times 2.0067 = 2007 \approx 2000$$

$1000 at 1% compounded annually doubles after 70 years.

EXERCISE 11.2.4

Use the compound interest formula to determine the amount when $1000 is invested at the given rate for the stated time.

1.   2%,   35 yr
2.   3%,   23 yr
3.   4%,   18 yr
4.   5%,   14 yr
5.   6%,   12 yr
6.   7%,   10 yr
7.   8%,   9 yr
8.   9%,   8 yr
9. 10%,   7 yr

In all the problems of Exercise 11.2.4, your new principal should turn out to be approximately $2000. In other words, your money doubled after the number of years indicated. It is obvious that there is some kind of a "trick" in choosing these particular numbers of years. In constructing this exercise, we used the Rule of 70 to determine how many years money has to stay in the bank to double.

This Rule of 70 tells us to divide 70 by r (without the percent symbol %) to find the years to double your principal. In other words,

> if the rate is 1%, money doubles in 70/1 = 70 years
> 2%   70/2 = 35 years
> 3%   70/3 = 23 years

EXAMPLE

How long will it take to double your money if the annual yield is 3.5%?

> **Solution**   70 ÷ 3.5 = 20
> It will take 20 years to double your money if you invest at an annual yield of 3.5%.

EXAMPLE

The house in the country increased in value from $18,000 to $160,000 in 26 years.

What yearly interest does that correspond to?

Solution

$$18,000 \times 2 = 36,000$$
$$36,000 \times 2 = 72,000$$
$$72,000 \times 2 = 144,000$$

The value of the house has doubled more than three times in 26 years, or it doubled approximately every 9 years.

$$70 \div 9 = 7.7 \approx 8$$

The yearly interest is approximately 8%.

EXERCISE 11.2.5

1. How long will it take to double your money if you invest at an annual yield of (a) 12%? (b) 8.7%?
2. A property valued at $74,000 fifteen years ago is now worth $500,000. What yearly interest does this correspond to?

SUMMARY

Formula: Compound interest: $A = P(1 + r)^t$

Rule of 70: To find out how many years it takes to double your money in the bank, divide 70 by the interest rate (without the percent symbol).

Rule of 78: In a prepaid loan, the bank considers the interest the first month to be 12/78 of the prepaid interest, the second month 11/78 of the prepaid interest, and so on.

VOCABULARY

Accumulated principal: The original principal plus interest.

Annual percentage rate: The interest rate per year.

Annual yield: The interest rate per year recalculated as simple interest.

Compound interest: Interest calculated periodically on the amount in the account at the beginning of each period.

Finance charges: The total interest on a loan.

Interest: The money paid to the bank or earned from the bank for the use of money.

Loan: The money we borrow.

Mortgage loan: Money borrowed for a house.

Prepaid interest: Loan interest paid in advance.

Principal: Money in the bank.

Rate: Percent per time period.

Simple interest: Interest calculated as Principal × rate × time.

CHECK LIST

Check the box for each topic you feel you have mastered. If you are unsure, go back and review.

☐ Calculating simple interest
☐ Interest rates for certain periods

- ☐ Finance charges
- ☐ Setting up loan repayment tables
- ☐ Loans with prepaid interest
- ☐ Mortgage calculations
- ☐ The Rule of 78
- ☐ Bank interest
- ☐ Compound interest
- ☐ Annual yield
- ☐ The Rule of 70

REVIEW EXERCISES

1. A savings bank shows an illustration of an annuity savings plan in which contributions of $5000 are made the first day of each year for 5 years. The interest rate is 8% compounded daily. Show that the balance is approximately $32,000 at the end of 5 years.

2. According to another savings bank, if you borrow $3000 for 12 months at an annual rate of 8.00% and prepay the interest of $126.03, you pay the bank $250 every month. Make a schedule for the monthly interest and repayment.

3. Make a table for mortgage payments for the first year when the mortgage is $60,000 for 30 years and the annual percentage rate is 9%. The monthly payments are $539.84.

4. If you repay your loan in Exercise 2 after 2 months, how much of the prepaid interest do you get back?

5. How much interest do you earn if you deposit $500 for 1 year at 4.5% compounded daily?

6. What is the annual yield in Exercise 5?

7. How many years will it take for your money in Exercise 5 to double?

8. Chicken livers cost 90¢ a pound in 1960. If inflation was 6% each year, how much should chicken livers have cost in 1992 if their price went up at the same rate as inflation?

READINESS CHECK

Solve the problems to satisfy yourself that you have mastered Chapter 11.

1. If you borrow $1200 for one year and repay your loan in a lump sum at the end of the year with $1272, what rate of interest are you paying?

2. A 10-year 10% $30,000 mortgage costs $396.45 each month. What is the interest the first month?

3. What is the repayment of the loan the first month in Problem 2?

4. If you had invested $20,000 at 5% compounded annually 20 years ago, how much would your investment be worth today?

5. If you had invested $20,000 at 10% compounded annually 20 years ago, how much would your investment be worth today?

6. If you repay a loan after 9 months, what part of your prepaid interest do you get back?

7. What is the yearly interest on $1000 if the bank pays 4.5% interest compounded monthly?

8. What is the annual yield in Problem 7?

9. After how many years will your money double if you invest at an annual rate of 6.2%?

10. Jarlsberg cheese cost 99¢ per pound 30 years ago. Now it costs $5.00 per pound. Assuming that the rate of price increase has been constant over the years, what is the rate of increase?

CHAPTER 12

STATISTICS AND PROBABILITY

Organizing information

Governments have been collecting and analyzing data for centuries. For example, according to the New Testament, almost 2000 years ago the census drew Joseph and Mary to Bethlehem. Two thousand years before that, the Egyptians were already collecting and recording data about floods, crops, and buildings in progress.

The word *statistics* comes from the Latin *statisticus*, meaning "of the state." All the collecting and analyzing of data was the responsibility of the state in those early days.

Predicting from statistical data what might happen and the *probability* or likelihood of it happening developed from games and gambling in the 15th and 16th centuries in Italy. In France in the 17th century, a systematic study of the mathematics involved in these predictions was begun by two mathematicians, Fermat and Pascal.

12.1 DESCRIPTIVE STATISTICS

The branch of statistics that includes collecting and analyzing data to describe actual situations is called *descriptive statistics*. The other important branch of statistics, which predicts from these data, is called *inferential statistics*. (In this chapter we concentrate on descriptive statistics.)

Frequency Distribution

The numbers in Table 12.1 are the result of tossing four coins 30 times and recording the number of heads that came up with each toss.

The information we collected and listed as we observed it is called *raw data*. To make these raw data more meaningful, we put them in order in a

Table 12.1

1	2	4	2	1	1
3	2	2	2	3	4
0	3	2	3	1	2
2	1	2	2	2	4
3	2	3	1	0	3

Table 12.2

FREQUENCY DISTRIBUTION		
Number of Heads	Tally	Frequency (Number of Times)
0	||	2
1	ЖⅠ	6
2	Ж Ж||	12
3	Ж||	7
4	|||	3
	Total	30

table, which is called the *frequency distribution*. According to Table 12.1, zero heads (i.e., all tails) came up twice. The tally in Table 12.2 shows ||. One head came up six times; the tally shows ЖⅠ (5 + 1). It is practical to use tally marks, especially when the number of raw data is very large.

Graphs

Several types of graphs are used to illustrate statistical data. For example, if a bowl with 50 M & M candies contains 20 light brown, 10 green, 5 red, 10 yellow, and 5 dark brown M & M's, there are

$$\frac{20}{50} = 40\% \text{ light brown}$$

$$\frac{10}{50} = 20\% \text{ green}$$

$$\frac{5}{50} = 10\% \text{ red}$$

$$\frac{10}{50} = 20\% \text{ yellow}$$

$$\frac{5}{50} = 10\% \text{ dark brown}$$

This can be shown on a circle graph as in Figure 12.1.

The same information can be illustrated on a bar graph as in Figure 12.2.

A *frequency polygon* is a type of line graph that also shows the frequency distribution. Figure 12.3 is a frequency polygon.

We will now graph the data from Table 12.2. First we list our data on ordered pairs.

Number of Heads	Frequency	Ordered Pairs
0	2	(0, 2)
1	6	(1, 6)
2	12	(2, 12)
3	7	(3, 7)
4	3	(4, 3)

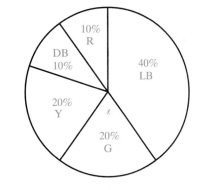

Figure 12.1

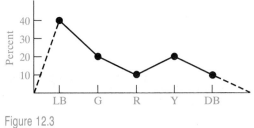

Figure 12.2

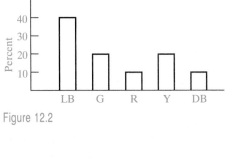

Figure 12.3

We will graph these data in three different ways:

1. Bar graph
2. Frequency polygon
3. Circle graph

Bar Graph In a bar graph, each frequency is represented as a vertical bar centered on the value of the independent variable. The height of each bar is the amount of the corresponding frequency. (See Figure 12.4.)

Frequency Polygon In Figure 12.5 we graph the ordered pairs and connect the points. It is common to complete this polygon by connecting the first and last points to the baseline (dashed lines in graph).

Circle Graph For a circle graph we need to find out what percent each frequency is of the whole. (The whole or sum of the frequencies in this case is 30.) In other words, "What percent of 30 is each of the frequencies?"

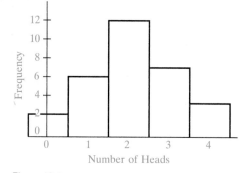

Figure 12.4

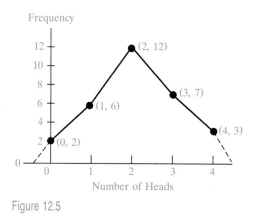

Figure 12.5

What % of 30 is 2?	$\frac{2}{30} = 0.07 = 7\%$
What % of 30 is 6?	$\frac{6}{30} = 0.20 = 20\%$
What % of 30 is 12?	$\frac{12}{30} = 0.40 = 40\%$
What % of 30 is 7?	$\frac{7}{30} = 0.23 = 23\%$
What % of 30 is 3?	$\frac{3}{30} = 0.10 = 10\%$
Total $\overline{30}$	Total $\overline{100\%}$

A circle consists of 360°, and to find the number of degrees in each percentage, we do the following:

7%	of	360°	is	25°
20%	of	360°	is	72°
40%	of	360°	is	144°
23%	of	360°	is	83°
10%	of	360°	is	36°
$\overline{100\%}$				$\overline{360°}$

To draw this graph, divide the circle into proportionate parts, like cutting a pie into pieces. (Another name for this graph is *pie chart*.) Start anywhere you want and then divide the circle into five slices. Approximate the size of the slices or use a protractor for a more accurate figure. Our circle graph for this coin toss problem is shown in Figure 12.6.

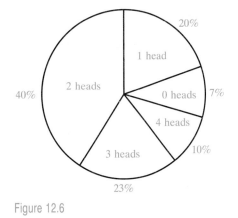

Figure 12.6

In all three graphs, the zero and four heads appear similar in size, the one and three heads are similar, while the two heads is much larger.

In statistical studies, bar graphs and frequency polygons are more commonly used than circle graphs.

Grouped Intervals

In a continuing education course in art, the ages of the students were as listed in Table 12.3. In the previous section we were dealing with a small set of numbers $(0, 1, 2, 3, 4)$ on the horizontal axis. Here the numbers range from 20 to 63. $63 - 20 = 43$, so there is a spread of 43 years. We couldn't fit 43 scores on the horizontal axis, so we group the given data into *intervals* or *classes*. The usual rule is to have 6 to 15 classes, making sure that no two classes overlap and that classes are of the same size. (The only exception to this last condition is when there is an extremely large empty space at one end or the other or both.) With grouped data a special bar graph, a *histogram*, is used. It has no spaces between the bars. The width of the bars is the class interval.

Table 12.3

20	47	40	22	37
30	63	55	32	42
42	49	59	27	31
47	25	33	34	34
34	27	30	44	29

EXAMPLE

Draw a histogram and a frequency polygon with the data in Table 12.3.

Solution We will use 9 classes of interval size 5 and start with 20. The raw data are the ages in intervals from 20–24 to 60–64. First we list the frequencies in Table 12.4.

Table 12.4

Ages	Tally	Frequency
20–24	||	2
25–29	||||	4
30–34	̶H̶H̶T̶ |||	8
35–39	|	1
40–44	||||	4
45–49	|||	3
50–54		0
55–59	||	2
60–64	|	1

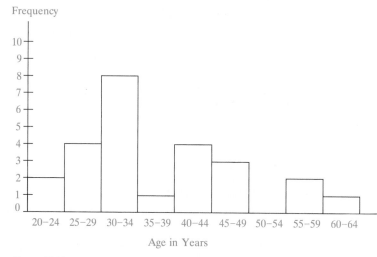

Figure 12.7

The histogram can start anywhere (i.e., not to scale). In this example only whole numbers are used. Most of the time the border between two rectangles would be a decimal number. In this case the borders are 24.5, 29.5, etc. (See Figure 12.7.)

Figure 12.8 is a frequency polygon constructed from the histogram by connecting the midpoints of the tops of the bars. The endpoints of the graph have been connected to the baseline to complete the polygon.

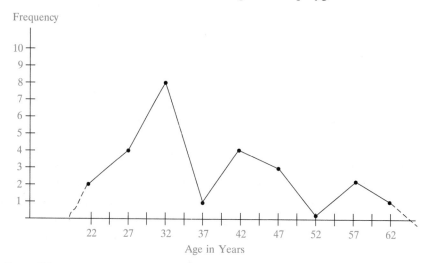

Figure 12.8

EXAMPLE

A teacher gives a test with 30 questions; 40 students take the test. The scores are the following.

1	2	24	29	23	15	7	5
8	18	12	6	28	4	19	18
17	9	16	19	20	27	12	14
19	3	9	21	16	13	16	27
3	27	28	16	19	23	28	8

Summarize the test scores using frequency intervals.
(a) Draw a histogram (a) with an interval size of 3.
(b) Draw a histogram with an interval size of 4.
(c) From the histogram in (b), construct a frequency polygon.

Solution
(a) Using an interval size of 3, the frequencies are as tabulated in Table 12.5. Figure 12.9 is the corresponding histogram.

Table 12.5

Interval	Tally	Frequency
1–3	IIII	4
4–6	III	3
7–9	HHT	5
10–12	II	2
13–15	III	3
16–18	HHT II	7
19–21	HHT I	6
22–24	III	3
25–27	III	3
28–30	IIII	4

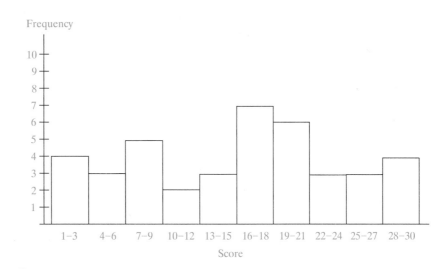

Figure 12.9

(b) Using an interval size of 4, we construct Table 12.6. The corresponding histogram is shown in Figure 12.10.

Table 12.6

Interval	Tally	Frequency
1–4	‖‖	5
5–8	‖‖	5
9–12	‖‖‖	4
13–16	‖‖ ‖	7
17–20	‖‖ ‖‖	8
21–24	‖‖‖	4
25–28	‖‖ ‖	6
29–30	‖	1

The graphs in Figures 12.9 and 12.10 look different, but they both represent the same data from the test scores.

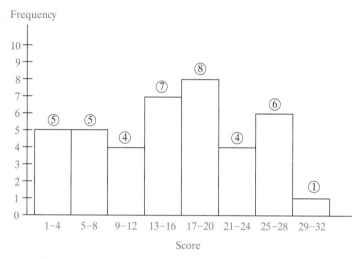

Figure 12.10

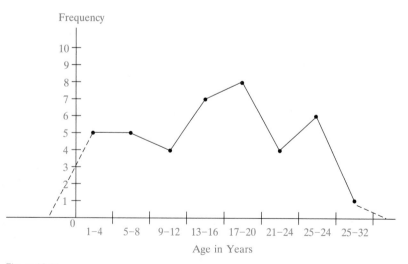

Figure 12.11

(c) The frequency polygon constructed from Figure 12.10 is shown in Figure 12.11.

When you group your data into intervals, remember the following rules.

RULE

Construction of a Histogram

1. Each score must occur in one and only one interval.

2. There should generally be at least 6 intervals but no more than 15.

3. All intervals must be of equal size (occasionally exceptions are allowed, usually for the highest or lowest interval).

EXAMPLE

Use the data in Table 12.7 from the Environmental Protection Agency (1990) to construct a circle graph.

Solution   First we calculate the percent of total waste represented by each type of waste. Then we use that to calculate the size of the "slice of pie" in degrees (see Table 12.8). The graph is presented in Figure 12.12.

Table 12.7

Municipal Solid Waste Generation (in millions of tons)	
Yard wastes	31.6
Food wastes	13.2
Paper	71.8
Metals	15.3
Glass	12.5
Plastics	14.4
Other	20.8

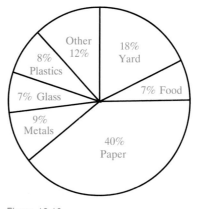

Figure 12.12

Table 12.8

Type of Waste	Million Tons	Percent of Total	Degrees (percent × 360°)
Yard wastes	31.6	17.6%	63°
Food wastes	13.2	7.3%	26°
Paper	71.8	40%	144°
Metals	15.3	8.5%	31°
Glass	12.5	7%	25°
Plastics	14.4	8%	29°
Other	20.8	11.6%	42°
Total	179.6	100%	360°

EXERCISE 12.1.1

1. Toss four coins and record the number of heads. (There can be 0, 1, 2, 3, or 4 heads in each toss.) Toss a total of 16 times.
    (a) Make a bar graph of these data.
    (b) Make a frequency polygon of the data.

2. The grades in two algebra classes were as follows:

72	93	93	84	52	98	84	94	91
85	91	89	79	88	62	91	71	46
88	62	64	62	71	73	77	68	88
79	78	71	73	78	91	68	75	76
91	84	80	98	93	75	74	83	93
62	99	81	70	62	91	85	76	98

Use the following intervals to make a histogram.

A^+	97–99
A	94–96
A^-	90–93
B^+	87–89
B	84–86
B^-	80–83
C^+	77–79
C	74–76
C^-	70–73
D^+	67–69
D	64–66
D^-	60–63
F	0–59

3. In Maine the average temperatures for the month of July were as follows:

78	82	83	82	88	91	82
85	76	70	83	93	95	81
84	80	73	75	81	88	96
94	91	86	88	80	72	78
93	88	85				

(a) Choose an interval size.
(b) Make a frequency distribution.
(c) Draw a frequency polygon.

4. Forty students take a 10-question quiz (1 point per question). The scores are

4	6	10	5	6	4	3	2
5	2	9	3	9	3	4	5
8	9	8	7	4	5	7	6
4	7	6	5	5	6	5	4
7	3	5	4	6	0	2	3

Construct (a) a bar graph and (b) a frequency polygon of the scores.

5. Make a circle graph from the following information obtained from the U.S. Department of Energy.

Energy Source	Percent of U.S. Energy Use
Coal	22.2
Natural gas	22.5
Petroleum	40.3
Nuclear	6.9
Other	8.1

12.2 MEASURES OF CENTRAL TENDENCY AND DISPERSION

In statistics there are three measures that tell how the data (here referred to as scores) cluster near the center of the data. These averages are called measures of *central tendency*.

Mean, Median, and Mode

Mean The mean is the numerical average of the scores. To find the mean, add the scores and then divide the sum by the number of scores.

EXAMPLE

Find the mean of the following scores: $81, 85, 82, 89, 83$.

Solution

$$\frac{81 + 85 + 82 + 89 + 83}{5} = \frac{420}{5} = 84$$

Median To find the median, arrange the scores in numerical order. The median is the number in the middle of the data—half of the data are lower than or equal to the median, and the other half are greater than or equal to the median.

EXAMPLE

Find the median of the scores in the example above.

Solution Sort the scores in order from smallest to largest: $81, 82, 83, 85, 89$. 83 is the middle score or the median.

If the number of scores is an *odd* number, the median is simply the middle score. For example, in 2, 5, 6, 8, 10, the middle score is 6, so 6 is the median.

If the number of scores is an *even* number, the median is the mean of the two middle scores. Consider the scores 2, 4, 9, 11, 14, 17. The two middle scores are 9 and 11. The median is

$$\frac{9 + 11}{2} = 10$$

Mode The mode is the score that occurs most often. If there are two modes, the data set is *bimodal*. For example, in 13, 8, 8, 7, the mode is 8.

EXAMPLE

Find the mode(s): 2, 2, 2, 4, 7, 8, 9, 9, 9, 10

Solution   This is a bimodal set. The modes are 2 and 9.

EXAMPLE

Find the mode(s): 1, 2, 3, 4, 5

Solution   If no score occurs more than once, then there is no mode.

EXAMPLE

The scores on a quiz are as follows:

$$6, 9, 8, 8, 7, 8, 10, 9, 7$$

Find the mean, median, and mode.

Solution   The *mean* (the numerical average) is usually written as $\bar{x}$ ("x bar").

$$\bar{x} = \frac{6 + 9 + 8 + 8 + 7 + 8 + 10 + 9 + 7}{9} = \frac{72}{9} = 8$$

The *median* (the middle score) is usually written $\tilde{x}$ ("x tilde"). First we must arrange the scores in order:

$$6, 7, 7, 8, \underline{8}, 8, 9, 9, 10$$

The middle score is 8. Note that four scores are lower than or equal to 8, and four scores are greater than or equal to 8. $\tilde{x} = 8$.

The *mode* is the score that occurs most often. Because 8 occurs three times, more than any other score, 8 is the mode.

EXAMPLE

Suppose the salaries in a department are $22,000, $23,000, $25,000, $27,000, $35,000, $38,000, $96,000. Find the mean, median, and mode. Decide which salary would be the best indication of the center of the salary range.

Solution   Mean:

$$\bar{x} = (22,000 + 23,000 + 25,000 + 27,000 + 35,000 + 38,000 + 96,000)/7$$
$$= (266,000)/7 = \$38,000$$

Median: $\tilde{x} = \$27,000$. There are three salaries higher and three salaries lower.

Mode: Since no salary appears more than once in the data set, there is no mode.

The mean is high and not very revealing because the $96,000 is an unusually high salary. There is no mode. The median of $27,000 with three numbers above and three below gives the best indication of the center of the salaries in this example.

Measures of Variance and Dispersion

While the mean, median, and mode help us focus on the center or middle of the set of data, the *range* and *standard deviation* measure the spread of the data, how the data vary from their mean—their *dispersion*.

The *range* is simply the high score minus the low score. In order to discuss the deviation we need to start with the *variance*, or how spread out the data are. The definition of variance is the mean (average) of the sum of the squares of the differences of the scores and the mean. In other words, if in your research you get the data $1, 2, 3, 4, 5$, the mean is $\dfrac{1 + 2 + 3 + 4 + 5}{5} = 3$.

Then $(1 - 3)^2 = 4$, $(2 - 3)^2 = 1$, $(3 - 3)^2 = 0$; $(4 - 3)^2 = 1$, and $(5 - 3)^2 = 4$.

The variance is $\dfrac{4 + 1 + 0 + 1 + 4}{5} = \dfrac{10}{5} = 2$.

Table 12.9 lists three sets of data with equal means but different ranges.

If these were three students' marks on five tests, the range would show us that the first student was consistent in his degree of understanding of the material while the third showed wide variety in the level of understanding, but also the greatest improvement.

Standard Deviation

To look specifically at the average deviations from the mean, we find the standard deviation. For the Set 1 data we construct Table 12.10.

Mean of squares = $10/5 = 2$ (variance)

The standard deviation (SD) of a set of data equals the square root of the mean of squares.

$$SD = \sqrt{2} = 1.41$$

Table 12.9

	Set 1	Set 2	Set 3
	78	70	60
	79	75	70
	80	80	80
	81	85	90
	82	90	100
Mean =	80	80	80
Range =	$82 - 78 = 4$	$90 - 70 = 20$	$100 - 60 = 40$

Table 12.10

Score x	Deviation from Mean $x - \bar{x}$	Deviation Squared $(x - \bar{x})^2$
78	−2	4
79	−1	1
80	0	0
81	1	1
82	2	4
400	0	10
$\bar{x} = 80$		

Set 2:

x	$x - \bar{x}$	$(x - \bar{x})^2$
70	−10	100
75	−5	25
80	0	0
85	5	25
90	10	100
$\bar{x} = 80$	0	250

Variance = Mean of squares = 250/5 = 50

$$SD = \sqrt{50} = 7.07$$

Set 3:

x	$x - \bar{x}$	$(x - \bar{x})^2$
60	−20	400
70	−10	100
80	0	0
90	10	100
100	20	400
$\bar{x} = 80$	0	1000

Variance = Mean of squares = 1000/5 = 200

$$SD = \sqrt{200} = 14.1$$

The *standard deviation* is the square root of the variance.

DEFINITION

Standard Deviation

The standard deviation is the square root of the mean of the squares of the differences between each score and the mean.

Find the mean, median, mode, range, and standard deviation.

1. $3, 8, 6, 4, 9, 4, 1$

2. $9, 8, 2, 7, 1, 1, 5, 7$

3. $4, 1, 8, 11, 5, 7, 15, 38, 3, 5, 2$

12.3 PROBABILITY AND ODDS

Probability

As we said at the beginning of this chapter, the origins of probability theory go back to the laws of chance developed from gambling in the 15th and 16th centuries. So we will deal with coins, dice, and cards. However, there are many applications of probability theory to activities other than gambling, as you will find when you take a course in statistics—insurance, politics, risks of cancer, AIDS, car accidents, for example.

If an event is absolutely certain to happen, the probability of this event is 1. If it is absolutely certain that the event will not happen, then the probability is 0. All other probabilities are expressed as fractions between 0 and 1.

When a coin is tossed 1000 times, we can be fairly certain that heads and tails will come up approximately 500 times each. If a head comes up 500 times out of 1000 tosses, we say that the experimental *probability* of a head on a single toss is $500/1000$ or $1/2$.

In tossing one penny, the outcome is a head or a tail; there are two possibilities. Out of these two possibilities, one is a head. The theoretical probability of tossing a head is $1/2$. In the long run (with many trials) the theoretical probability is the same as the experimental probability.

The desired outcome is called a *success*. In the language of probability, a success does not have to be something good; it is simply the outcome we are looking for. In probability experiments and theoretical calculations we assume that *fair coins* are used, i.e., the coins are new and not tampered with.

EXAMPLE

If a nickel is tossed twice, what is the probability of
(a) two heads?   (b) one tail?

> **Solution**   The number of possible results is four: head head, head tail, tail head, and tail tail, abbreviated HH, HT, TH, TT.
> (a) There is one success (HH). The probability of two heads is one out of four, or $1/4$, written $P(\text{HH}) = 1/4$.
> (b) This time a success is either of two events (HT, TH). The probability of one tail is two out of four, or $2/4 = 1/2$, written $P(\text{T}) = 1/2$.

The probability that an event will *not* happen is $1 - P$ (event happening). For example, the probability of not getting two heads when two coins are tossed is $1 - 1/4 = 3/4$. This can also be concluded from the outcomes

(HH, HT, TH, TT). There are three possibilities of getting something other than two heads.

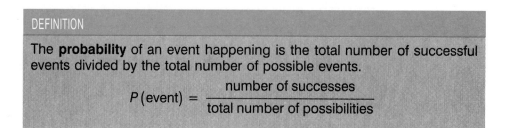

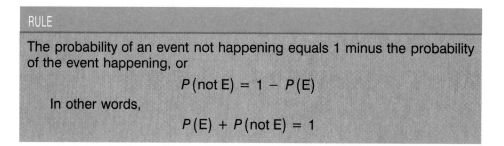

EXAMPLE

Two coins are tossed. Find the probability of getting
(a) two heads   (b) at least one head   (c) at most one head   (d) no heads.

> **Solution**   When you toss two coins, they are equally likely to come up Head-Head, Head-Tail, Tail-Head, or Tail-Tail {HH, HT, TH, TT}.
> (a) The probability of two heads is one out of four, or $1/4$.
> (b) For *at least* one head (HT, HH, TH) the probability is $3/4$.
> (c) For *at most* one head (TT, TH, HT) the probability is again $3/4$.
> (d) For *no* heads (tails only), the probability is $1/4$.

EXAMPLE

Three coins are tossed. Find the probability of
(a) three heads   (b) at least one tail.

> **Solution**   How many different ways can three coins come up?
> $$HHH, HHT, HTH, THH, TTH, THT, HTT, TTT$$
> (a) $P(HHH) = \dfrac{1}{8}$
> (b) $P(\text{at least one tail}) = \dfrac{7}{8}$

In some forms of gambling, dice are rolled. A single die has six sides, each with $1, 2, 3, 4, 5,$ and 6 spots.

EXAMPLE

A die is rolled. Find the probability of rolling (a) an odd number; (b) a number less than 6.

Solution
(a) The probability of rolling an odd number {1, 3, 5} is 3 out of 6, or 3/6.
(b) The probability of rolling less than 6 {1, 2, 3, 4, 5} is 5 out of 6, or 5/6.

With two dice, in how many ways can they come up? See Table 12.11.
There are 36 ways in all. Therefore the probability of any particular outcome is 1/36.

Table 12.11

	Possible Sums of Two Dice					
Sum	Outcomes					
2	(1, 1)					
3	(1, 2)	(2, 1)				
4	(1, 3)	(3, 1)	(2, 2)			
5	(1, 4)	(4, 1)	(3, 2)	(2, 3)		
6	(1, 5)	(5, 1)	(4, 2)	(2, 4)	(3, 3)	
7	(1, 6)	(6, 1)	(2, 5)	(5, 2)	(3, 4)	(4, 3)
8	(6, 2)	(2, 6)	(3, 5)	(5, 3)	(4, 4)	
9	(6, 3)	(3, 6)	(4, 5)	(5, 4)		
10	(6, 4)	(4, 6)	(5, 5)			
11	(6, 5)	(5, 6)				
12	(6, 6)					

EXAMPLE

Find: (a) $P(2, 1)$ (b) $P(4, 5)$

Solution   (a) $P(2, 1) = 1/36$ (b) $P(4, 5) = 1/36$

Now look at the probability of getting a particular *sum*.

EXAMPLE

What is the probability of getting a sum of 7?

Solution   There are six possible sums equal to 7. Therefore,

$$P(\text{sum } 7) = \frac{6}{36} = \frac{1}{6}$$

EXAMPLE

What is the probability of getting (a) a sum equal to 11? (b) an odd-numbered sum?

Solution

(a) There are two possible sums equal to 11.

$$P(\text{sum } 11) = \frac{2}{36} = \frac{1}{18}$$

(b) Sums 3, 5, 7, 9, and 11 are odd sums. From Table 12.11 we find the corresponding numbers of outcomes:

$$2 + 4 + 6 + 4 + 2 = 18$$

Therefore

$$P(\text{odd-numbered sum}) = \frac{18}{36} = \frac{1}{2}$$

Now we will look at examples dealing with playing cards. The box shows the composition of a pack of cards.

PLAYING CARDS

There are 52 cards in all. They can be sorted into four suits of 13 cards each.

Clubs	♣	black
Diamonds	◇	red
Hearts	♡	red
Spades	♠	black

The cards in each suit are 2, 3, 4, 5, 6, 7, 8, 9, 10, Jack, Queen, King, Ace. The ace stands for either 1 or 14.

There are 12 face cards: a jack, a queen, and a king in each suit.

Since there are four kings in a pack of cards and 52 cards altogether, the probability of drawing a king from a shuffled pack is 4/52 or 1/13.

EXAMPLE

Find the probability of drawing:
(a) a red card   (b) not a red card   (c) a face card   (d) a jack

Solution

(a) Since there are 26 red cards out of 52, $P(\text{red}) = 26/52 = 1/2$.
(b) $P(\text{not red}) = 1 - 1/2 = 1/2$
(c) There are 12 face cards out of 52, so $P(\text{face}) = 12/52 = 3/13$.
(d) There are four jacks out of 52, so $P(\text{jack}) = 4/52 = 1/13$.

EXERCISE 12.3.1

1. Five hundred tickets are sold at a church raffle.
   (a) You buy one. What probability do you have of winning?
   (b) You buy five tickets. What is your probability of winning?

2. What is the probability of tossing three heads in one toss of three coins?

3. What is the probability of tossing exactly two heads in one toss of three coins? (In how many ways can you get two heads?)

4. What is the probability of tossing at least two heads in one toss of three coins?

5. What is the probability of rolling a sum of 4 in one toss of two dice?

6. You roll two dice. What is the probability of rolling
   (a) more than a sum of 4?
   (b) less than a sum of 4?

7. What is the probability of rolling two dice so that you don't get a sum of 7, 8, or 9?

8. What is the probability of drawing a red face card from a shuffled pack of cards?

9. You remove one card from a pack of cards. Assuming that the ace is high, find the probability that your card is:
   (a) a heart
   (b) a king or a queen
   (c) a card higher than 10
   (d) not an ace
   (e) a diamond less than 5 or a spade or a club

10. You toss a coin three times. Find the probability that:
    (a) You get a head on the first toss but not on the others.
    (b) You get only one head.
    (c) You don't get a head at all.
    (d) You get at least one head.
    (e) You get two heads.

11. When you roll two dice, what are the probabilities of the following outcomes?
    (a) You get a double six.
    (b) You get exactly one six.
    (c) You don't get a six at all.
    (d) You get at least one six.

12. In a box there are four green, seven red, six blue, and three yellow marbles. Find the probability that (without looking) you will pick out:
    (a) a green marble
    (b) a blue marble
    (c) a red or a yellow marble
    (d) a green, a blue, or a yellow marble
    (e) a marble that is not blue

Odds

In each of the earlier examples we have had a number of *successful* events, and therefore in each we have had a number of *failures*.

For example, in tossing two coins, the number of possible outcomes is four. If we define a success as two heads, then one of these outcomes would be a success and the other three would be failures: $P(\text{HH}) = 1/4$ and $P(\text{not HH}) = 3/4$.

DEFINITION

Odds are the ratio of the number of possible successes to the number of possible failures:

$$\text{Odds} = \frac{\text{success}}{\text{failure}}$$

Again with two coins, $P(H) = 2/4 = 1/2$. The total possibilities equal four, of which total successes are two and total failures are two. The odds in favor of tossing a head are 2 to 2 (also written 2/2).

In a bingo game, five people win out of each 26 who play. $P(\text{winning}) = 5/26$. Of the total number of 26 a total of 5 are successes, and a total of 21 are failures. The odds of winning are 5/21.

If the odds for something happening (an *event*) are 3 to 5, then the odds *against* the event are 5 to 3.

EXAMPLE

Find the odds for and against getting: (a) one tail in the toss of one coin; (b) two tails in the toss of two coins; (c) at least one head in the toss of two coins.

Solution
(a) With one toss of one coin the set of possible events is {H, T}. Therefore there is the possibility of one success and one failure, and the odds *for* one tail are 1 to 1. The odds *against* are also 1 to 1.
(b) With one toss of two coins, success is TT, failure is HT, TH, or HH, so odds for are 1 to 3. The odds against are 3 to 1.
(c) Success is HT, TH, HH, and failure is TT. Odds for are 3 to 1 and the odds against are 1 to 3.

EXAMPLE

In rolling one die, find the odds of getting
(a) a three;   (b) an odd number;   (c) not a six.

Solution
(a) Success 3. Failure 1, 2, 4, 5, 6. Odds are 1 to 5.
(b) Success 1, 3, 5. Failure 2, 4, 6. Odds are 3 to 3.
(c) Success 1, 2, 3, 4, 5. Failure 6. Odds are 5 to 1.

EXERCISE 12.3.2

1. What are the odds for drawing an ace from a deck of 52 cards?
2. What are the odds against rolling a sum of 7 with two dice?
3. What are the odds against rolling at least a sum of 8 with two dice?
4. What are the odds for tossing four heads with four coins?
5. What are the odds for each part of Problem 9 of Exercise 12.3.1?
6. What are the odds for each part of Problem 12 of Exercise 12.3.1?

SUMMARY

Definitions

Standard deviation: The square root of the mean of the squares of the differences between each score and the mean.

The probability of an event happening is the total number of successful events divided by the total number of possible events.

$$P(\text{event}) = \frac{\text{number of successes}}{\text{total number of possibilities}}$$

Odds are the ratio of number of successes to number of failures, written as a fraction.

Rules

To construct a histogram or frequency polygon:

1.   Each score must occur in one and only one interval.
2.   There should generally be 6 to 15 intervals.
3.   All intervals must be of equal size. (Occasionally exceptions are allowed, usually for the highest or lowest interval.)

The probability of an event not happening equals 1 minus the probability of the event happening:

$$P(\text{E}) = 1 - P(\text{not E})$$

This can also be expressed as

$$P(\text{E}) + P(\text{not E}) = 1$$

VOCABULARY

Bar graph: Bars showing the frequency of data.

Central tendency: Numbers that describe where data are centered.

Circle graph: A circle showing the percents of components of the whole. Also called a pie chart.

Class: A group of data.

Descriptive statistics: A method of analyzing data based on percentages and central tendency.

Dispersion: Variation of data from the mean.

Frequency: How often something occurs.

Frequency distribution: An ordered table of frequencies.

Frequency polygon: A line graph of frequencies.

Histogram: A bar graph of grouped data.

Inferential statistics: Conclusions are made from analysis of a sample.

Interval: A range of values for grouped data.

Mean: The numerical average of a set of data.

Median: The number in the middle of a set of ordered data.

Mode: The number or scores that occur most often in a set of data.

Odds: The ratio of successful outcomes to failures.

Probability: The ratio of successful outcomes to number of possible outcomes.

Range: The difference between the highest and lowest values in a set of data.

Raw data: Unorganized data.

Standard deviation: A measure of the dispersion of data from the mean.

Statistics: Treatment of data.

Success: The outcome we are looking for.

Tally: The use of marks for counting.

Variance: A measure of how data vary from the mean within a set of data.

CHECK LIST

Check the box for each topic you feel you have mastered. If you are unsure, go back and review.

☐ Setting up frequency distributions
  Graphing:
    ☐ Histograms
    ☐ Frequency polygons
    ☐ Circle graphs
☐ Grouping data into intervals
  Finding:
    ☐ Mean
    ☐ Median
    ☐ Mode
    ☐ Standard deviation
    ☐ Variance
  Calculating:
    ☐ Probability
    ☐ Odds

REVIEW EXERCISES

1. Toss five coins a total of 36 times. Keep track of the number of tails.
   (a) Make a bar graph of the results.
   (b) Make a frequency polygon of the results.
   (c) Make a circle graph of the results.
2. The ages of students taking developmental math courses were as follows:

30	41	50	25	32	41	60	37
40	70	36	51	18	53	39	48
33	43	37	47	21	61	23	33
51	35	54	34	45	67	30	36
51	28	31	45	74	41	39	43

   (a) Choose an interval size and divide data into classes.
   (b) Make a frequency distribution table.
   (c) Construct a histogram.
   (d) Construct a frequency polygon.
3. The 1990 median prices of existing single-family homes in 66 U.S. cities, according to the National Association of Realtors, are listed in Table 12.12. (a) Round these numbers to the nearest thousand dollars, (b) group them, and (c) construct a histogram.
4. Given the following cost of a two-week vacation trip to Scotland, change to percents and construct a circle graph.

Air fare	$600
Car rental and fuel	$800
Meals	$400
Lodging	$400
Souvenirs	$200
Miscellaneous	$100

5. Find the mean, median, mode, range, and standard deviation for each set of data. (Round to whole numbers.)
   (a) 2, 9, 8, 11, 6, 5, 2, 9
   (b) 11, 3, 14, 8, 9, 19, 4
   (c) 14, 7, 9, 6, 4, 2, 8, 9
6. For each of the following, find the probability and the odds.
   (a) Getting at least two tails when you toss three coins
   (b) Getting at most two heads when you toss three coins
   (c) Getting less than two tails when you toss three coins
   (d) Getting more than three heads when you toss three coins
   (e) Getting a sum of 6 when you roll two dice
   (f) Getting a sum equal to an even number when you roll two dice
   (g) Getting a sum that is a multiple of 3 when you roll two dice

Table 12.12

67,700	89,500	60,800	142,000	76,700	63,000
69,600	352,000	153,300	105,900	74,100	75,700
63,200	212,800	70,100	79,800	89,300	71,400
157,300	81,800	259,300	60,500	53,200	174,200
64,800	84,000	86,400	72,400	137,100	81,600
59,100	183,600	116,800	78,100	80,700	68,300
108,700	150,200	86,400	174,900	80,800	93,000
63,600	242,400	74,800	127,900	77,100	88,700
63,900	76,200	82,300	55,500	63,600	62,800
84,500	64,100	67,800	64,900	75,400	69,400
77,200	70,700	79,500	80,600	84,400	62,800

(h) Getting a sum of 5 when you roll two dice
(i) Getting a 3, 4, 5, or 6 when you draw from a pack of cards
(j) Getting a black card when you draw one card from a pack
(k) Getting a heart when you draw one card from a pack
(l) Getting a red jack when you draw one card from a pack

READINESS CHECK

Solve the problems to satisfy yourself that you have mastered Chapter 12.
For Problems 1–6 use the following information.
The test scores in Mrs. Smith's algebra class were as follows:

63	75	98	86	77
85	81	68	73	83
92	86	79	81	94
73	84	94	87	83

1. Make a frequency distribution of the data. Use class intervals of 5.
2. Make a histogram of the data.
3. Make a frequency polygon of the data.
4. Find the mean.
5. Find the median.
6. Find the mode(s).
7. There were 106 jelly beans in a bowl: 41 yellow, 36 red, 11 white, and the rest orange.
   (a) What is the probability of picking an orange jelly bean?
   (b) What is the probability of picking a yellow or a red jelly bean?
   (c) What are the odds against picking a yellow jelly bean?
   (d) Make a circle graph of the percentages of jelly beans of different color in the bowl.

Chapter 1

1.1.1 **1** (a) {2, 3, 4, 5, 6, 7} (b) {6, 7, 8, ...} (c) {0, 1, 2}
(d) {1, 3, 5, 7, 9}
2 (a) 50 (b) 5 (c) 5000 (d) 5
3 (a) 405 (b) 650 (c) 3056 (d) 6400
4 (a) Five thousand two hundred thirty-six
(b) Eight thousand two hundred four
(c) Seven thousand twenty-nine
(d) One thousand two

1.2.1 **1**

+	0	1	2	3	4	5	6	7	8	9
0	0	1	2	3	4	5	6	7	8	9
1	1	2	3	4	5	6	7	8	9	10
2	2	3	4	5	6	7	8	9	10	11
3	3	4	5	6	7	8	9	10	11	12
4	4	5	6	7	8	9	10	11	12	13
5	5	6	7	8	9	10	11	12	13	14
6	6	7	8	9	10	11	12	13	14	15
7	7	8	9	10	11	12	13	14	15	16
8	8	9	10	11	12	13	14	15	16	17
9	9	10	11	12	13	14	15	16	17	18

1.2.2 **1** (a)

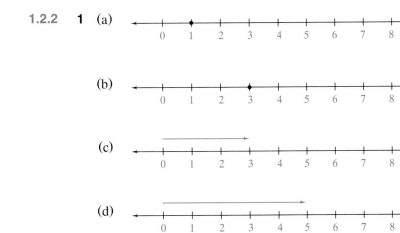

(b)

(c)

(d)

(e)

2 (a) 185 (b) 867 (c) 1329 (d) 1352 (e) 789 (f) 1272
(g) 145 (h) 9288 (i) 7511 (j) 4953
1.2.3 **1** (a) 1, 2, 3, 5, 6, 10, 15, 30 (b) 1, 2, 3, 6, 7, 14, 21, 42
(c) 1, 7, 49 (d) 1, 53

2 53, 59

3 37

4 21, 22, 24, 25, 26, 27, 28

1.2.4 1

×	0	1	2	3	4	5	6	7	8	9
0	0	0	0	0	0	0	0	0	0	0
1	0	1	2	3	4	5	6	7	8	9
2	0	2	4	6	8	10	12	14	16	18
3	0	3	6	9	12	15	18	21	24	27
4	0	4	8	12	16	20	24	28	32	36
5	0	5	10	15	20	25	30	35	40	45
6	0	6	12	18	24	30	36	42	48	54
7	0	7	14	21	28	35	42	49	56	63
8	0	8	16	24	32	40	48	56	64	72
9	0	9	18	27	36	45	54	63	72	81

1.2.5 1 234 **2** 600 **3** 1156 **4** 2905 **5** 9724 **6** 8858 **7** 197,744
8 184,212 **9** 31,150 **10** 86,580

1.2.6 1 (a) Identity element for addition   (b) Commutative
(c) Identity element for multiplication   (d) Associative
(e) Commutative   (f) Commutative and associative
(g) Commutative   (h) Commutative   (i) Distributive

1.2.7 1 279 **2** 828 **3** 1125 **4** 717 **5** 3821 **6** 2595

1.2.8 1 4 **2** 2 **3** 3 **4** 11 **5** 9 **6** 7 **7** 42 **8** 103 **9** 27 **10** 39
11 56 **12** 41 **13** (a) 0 (b) 0 **14** (a) 0 (b) undefined
15 (a) 0 (b) undefined **16** 101

1.3.1 1 9 **2** 9 **3** 27 **4** 8 **5** 30 **6** 4 **7** 19 **8** 15 **9** 50 **10** 16
11 1 **12** 4 **13** 1 **14** 36 **15** 4 **16** 8 **17** 6 **18** 10

1.3.2 1 $(7 + 4) \times 5 = 55$ **2** $2(6 - 3) = 6$ **3** $3(4 + 2)(1) = 18$
4 $18 \div (5 - 3) = 9$ **5** $(16 - 8) \div 4 = 2$ **6** $8 \times 7 \div 4 = 14$
7 $24 \div (3 \times 2) = 4$ **8** $8 \div (2 \div 2) = 8$

1.3.3 1 $(6 + 6) \div 6 + 6 = 8$ **2** $6 + 6 + 6 \div 6 = 13$
3 $4 \div 4 + 4 \div 4 = 2$ **4** $4 \times 4 + 4 \div 4 = 17$
5 $1 \times 1 + 1 \times 1 = 2$ **6** $1 + 1 + 1 \div 1 = 3$
There are other possibilities for solutions.

1.4.1 1 1785 **2** 5804 **3** 1566 **4** 7920 **5** 6699 **6** 2336 **7** 24 **8** 667
9 335 **10** 59 **11** 127 **12** 253 **13** 2860 **14** 1599 **15** 3403
16 4964 **17** 11,571 **18** 33,258 **19** 51 **20** 896 **21** 34 **22** 46
23 104 **24** 425

1.5.1 1 (a) tens (b) ones (c) millions (d) thousands (e) hundreds
(f) ten thousands (g) ten millions (h) billions
2 (a) 4,389 Four thousand three hundred eighty-nine
(b) 247,932 Two hundred forty-seven thousand nine hundred
thirty-two

{(c) 9,999,999 Nine million nine hundred ninety-nine thousand nine
hundred ninety-nine

(d) 45,592,640 Forty-five million five hundred ninety-two thousand
six hundred forty

(e) 20,681,030 Twenty million six hundred eighty-one thousand
thirty

(f) 367,006,973,800 Three hundred sixty-seven billion six million
nine hundred seventy-three thousand eight hundred

1.6.1   **1** 2450   **2** 200   **3** 1400   **4** 2600   **5** 18,000   **6** 10,800   **7** 103,000
      8 200,000   **9** 9,000,000   **10** 7,890,000

1.6.2   **1** 2400; 2187   **2** 4000; 3809   **3** 0; 42   **4** 10; 15   **5** 1200; 1050
      6 3300; 3377   **7** 4000; 3838   **8** 2; 3   **9** 300; 288   **10** 100; 117
     11 140; 140   **12** 1,000,000; 1,005,056   **13** 40,000; 36,450   **14** 50; 74
     15 2000; 2142   **16** 20,000; 23,201

1.7.1   **1** 6 apples   **2** 12¢; 20¢; 28¢; 44¢   **3** 20 miles   **4** 10 legs   **5** $35
      6 $8   **7** $60   **8** 10¢

1.7.2   **1** 3,280,000 people   **2**   (a) $51 billion   (b) $3 billion
      (c) $54 billion
      3 $55   **4** $49,385   **5** $4500 million   **6** (a) 180 cans
      (b) 12,960 cans   (c) 14,040 cans   **7**   (a) 18 gallons
      (b) 6570 gallons
      8 $13,000   **9** (a) $40   (b) $20   (c) 40,000 lire
     10   (a) 14,700,000 troy ounces   (b) 18,700,000 troy ounces

Review Exercises

1

2   (a) Commutative

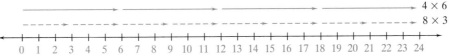

     (b) Associative

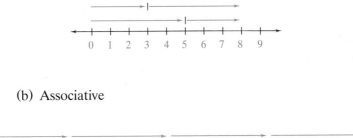

(c)

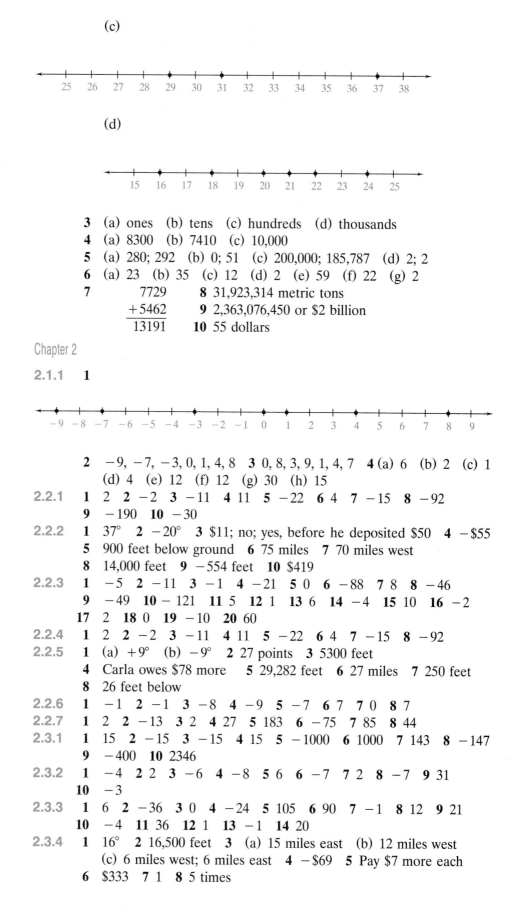

(d)

3 (a) ones   (b) tens   (c) hundreds   (d) thousands
4 (a) 8300   (b) 7410   (c) 10,000
5 (a) 280; 292   (b) 0; 51   (c) 200,000; 185,787   (d) 2; 2
6 (a) 23   (b) 35   (c) 12   (d) 2   (e) 59   (f) 22   (g) 2
7   7729   **8** 31,923,314 metric tons
          $\underline{+5462}$ **9** 2,363,076,450 or \$2 billion
           13191   **10** 55 dollars

Chapter 2

2.1.1   **1**

2 $-9, -7, -3, 0, 1, 4, 8$ **3** 0, 8, 3, 9, 1, 4, 7   **4** (a) 6   (b) 2   (c) 1
     (d) 4   (e) 12   (f) 12   (g) 30   (h) 15

2.2.1   **1** 2   **2** -2 **3** -11 **4** 11   **5** -22 **6** 4   **7** -15 **8** -92
        9 -190 **10** -30

2.2.2   **1** $37°$ **2** $-20°$ **3** \$11; no; yes, before he deposited \$50   **4** $-\$55$
        5 900 feet below ground   **6** 75 miles   **7** 70 miles west
        8 14,000 feet   **9** -554 feet   **10** \$419

2.2.3   **1** -5 **2** -11 **3** -1 **4** -21 **5** 0   **6** -88 **7** 8   **8** -46
        9 -49 **10** -121 **11** 5   **12** 1   **13** 6   **14** -4 **15** 10   **16** -2
        17 2   **18** 0   **19** -10 **20** 60

2.2.4   **1** 2   **2** -2 **3** -11 **4** 11   **5** -22 **6** 4   **7** -15 **8** -92

2.2.5   **1** (a) $+9°$ (b) $-9°$ **2** 27 points   **3** 5300 feet
        4 Carla owes \$78 more   **5** 29,282 feet   **6** 27 miles   **7** 250 feet
        8 26 feet below

2.2.6   **1** -1 **2** -1 **3** -8 **4** -9 **5** -7 **6** 7   **7** 0   **8** 7

2.2.7   **1** 2   **2** -13 **3** 2   **4** 27   **5** 183   **6** -75 **7** 85   **8** 44

2.3.1   **1** 15   **2** -15 **3** -15 **4** 15   **5** -1000 **6** 1000   **7** 143   **8** -147
        9 -400 **10** 2346

2.3.2   **1** -4 **2** 2   **3** -6 **4** -8 **5** 6   **6** -7 **7** 2   **8** -7 **9** 31
        10 -3

2.3.3   **1** 6   **2** -36 **3** 0   **4** -24 **5** 105   **6** 90   **7** -1 **8** 12   **9** 21
        10 -4 **11** 36   **12** 1   **13** -1 **14** 20

2.3.4   **1** $16°$ **2** 16,500 feet   **3** (a) 15 miles east   (b) 12 miles west
        (c) 6 miles west; 6 miles east   **4** $-\$69$ **5** Pay \$7 more each
        6 \$333   **7** 1   **8** 5 times

2.4.1 **1** 1 **2** 32 **3** 25 **4** 125 **5** 64 **6** 1 **7** 64 **8** 49 **9** −9 **10** 9
11 −8 **12** 16 **13** −125 **14** −36

2.4.2 **1** 10 **2** 12 **3** 0 **4** −3 **5** cannot be done **6** −2 **7** 8 **8** 8
9 −2 **10** cannot be done

2.5.1 **1** 18 **2** 36 **3** 25 **4** 20 **5** 400 **6** 81 **7** −21 **8** 441 **9** 150
10 108 **11** 20 **12** −48 **13** −115 **14** 26 **15** 3 **16** 10

Review Exercises

1

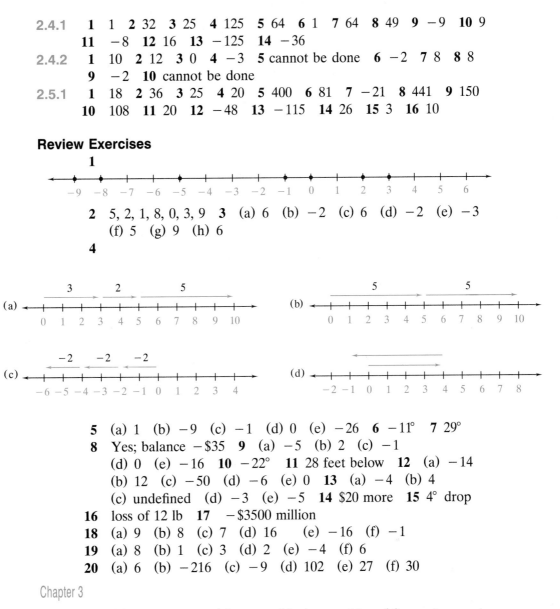

2 5, 2, 1, 8, 0, 3, 9 **3** (a) 6 (b) −2 (c) 6 (d) −2 (e) −3
(f) 5 (g) 9 (h) 6

4

5 (a) 1 (b) −9 (c) −1 (d) 0 (e) −26 **6** −11° **7** 29°
8 Yes; balance −$35 **9** (a) −5 (b) 2 (c) −1
(d) 0 (e) −16 **10** −22° **11** 28 feet below **12** (a) −14
(b) 12 (c) −50 (d) −6 (e) 0 **13** (a) −4 (b) 4
(c) undefined (d) −3 (e) −5 **14** $20 more **15** 4° drop
16 loss of 12 lb **17** −$3500 million
18 (a) 9 (b) 8 (c) 7 (d) 16 (e) −16 (f) −1
19 (a) 8 (b) 1 (c) 3 (d) 2 (e) −4 (f) 6
20 (a) 6 (b) −216 (c) −9 (d) 102 (e) 27 (f) 30

Chapter 3

3.1.1 **1** (a) hundredths (b) ones (c) thousandths (d) ten thousands
(e) tenths (f) hundred-thousandths
2 (a) Ninety-eight and six tenths (b) Forty-five and thirty-four
hundredths (c) Six hundred seventy-eight thousandths
(d) Seven and one hundred ninety thousandths
(e) Ten and six hundredths
(f) Fifteen and three thousand eight hundred twenty-nine ten-
thousandths
3 (a) 6.5 (b) 7.03 (c) 22.15 (d) 30.102 (e) 100.06 (f) 0.0036

3.1.2 **1** (a) 2 (b) 0.2 (c) 0.060 (d) 0.95 (e) 1.1 (f) 2.13
2 (a) 0.020, 0.2, 2 (b) 0.09, 0.9, 0.95 (c) 0.120, 1.2, 12
(d) 0.0005, 0.005, 0.05, 0.5 (e) 0.089, 0.091, 0.10, 0.19, 0.91
(f) 0.010, 0.014, 0.019, 0.020, 0.050

3.1.3 **1** 45.9; 45.94; 46 **2** 0.8; 0.85; 1 **3** 3.9; 3.90; 4 **4** 0.1; 0.08; 0
5 1.0; 0.97; 1 **6** 1.5; 1.53; 2 **7** 5.3; 5.28; 5 **8** 10.6; 10.63; 11
9 8.6; 8.62; 9 **10** 16.0; 15.98; 16

3.2.1 **1** 3.4 **2** 19.8 **3** 40.69 **4** 2.229 **5** 23.353 **6** 0.5 **7** 3.37
8 3.09 **9** 0.891 **10** 6.11 **11** 3.45 **12** -58.5 **13** -0.024
14 -3.0287 **15** -122.017 **16** 72.093

3.2.2 **1** 20 **2** 0.35 **3** 7.583 **4** 5.72 **5** 23.9 **6** 6,000 **7** 0.178
8 0.00950

3.2.3 **1** 25; 25.44 **2** 0.009; 0.008 **3** 0.4; 0.4095 **4** 4500; 4500
5 2000; 2205.06 **6** 0.5; 0.6435 **7** 16; 14.181 **8** 15,000; 12,689.95
9 0.24; 0.247752 **10** 0.00001; 0.00000795 The estimated answers
may vary.

3.2.4 **1** 4; 4 **2** 25; 25 **3** 100; 113.5 **4** 40; 44 **5** 0.01; 0.0125
6 0.02; 0.021 **7** 100; 105 **8** 11,000; 11,000 **9** 300; 200
10 11,000; 11,200 **11** 10; 12.8 **12** 200; 263 **13** 45; 50
14 50; 50 The estimated answers may vary.

3.2.5 **1** $0.0\overline{5}$ **2** $3.\overline{3}$ **3** $0.08\overline{3}$ **4** 6.3636... **5** 1.8181...
6 1.4285714... **7** $0.741\overline{6}$ **8** 0.1523076923...

3.2.6 **1** 2.83 **2** 3.61 **3** 4.24 **4** 5.10

3.2.7 **1** $2.07 **2** 68 acres **3** $87.33 **4** 137 bushels **5** $755.60
6 805.9 yards **7** 150 gallons **8** 1470 miles **9** 0.192 inch
10 79 gal; yes

3.3.1 **1** (a) 0.25 (b) 0.75 **2** (a) 0.025 (b) 0.0825 **3** (a) 2 (b) 7.43
4 (a) 0.003 (b) 0.005 **5** (a) 0.0001 (b) 0.0005

3.3.2 **1** (a) 25% (b) 68% **2** (a) 130% (b) 105% **3** (a) 500%
(b) 20,000% **4** (a) 250% (b) 620% **5** (a) 0.75%
(b) 0.91%

3.3.3 **1** 25% **2** 200% **3** 3000 **4** 32 **5** 5 **6** 14.43 **7** 120% **8** 500
9 40 **10** 1250 **11** 1050 liberal arts majors **12** 25% **13** 56%
14 80% **15** 6.25% **16** 7% **17** 4% **18** $12
19 18 correct answers **20** 0.00025%

3.3.4 **1** d **2** c **3** d **4** a **5** b **6** b **7** c **8** b **9** d **10** c

3.3.5 **1** 6.7% **2** $24,480 **3** $28,620 **4** 5% **5** 14% **6** 13% **7** 14%
8 57% **9** 33% **10** 96% **11** 22% **12** 588 persons
13 284 freshmen **14** The price after discount is $654.68 in either
case. The tax *before* discount is $40.50 and *after* discount it is
$37.06.

Review Exercises

1 (a) Ones (b) tenths (c) thousands (d) ten-thousandths
2 (a) Seven and two thousandths (b) one and seven hundred
eighty-five thousandths (c) one hundred five ten-thousandths
(d) two hundred and three hundred ten-thousandths
3 (a) 0.291, 0.3, 0.4002 (b) 0.973, 1.58, 2.003
4 (a) 10.487 (b) 4.929 (c) 22.13 (d) 0.799
5 (a) 300 (b) 41.6 (c) 0.032 (d) 650.8
6 (a) 56; 54.94 (b) 320; 290.852 (c) 2000; $2333.\overline{3}$ (d) 20; 27.64
(e) 0.3; $0.\overline{18}$ (f) 9; 9 (g) 4; 4.12 (h) 20,000,000; 21,530,435
7 (a) 72% (b) 80% (c) 12.3% (d) 135% (e) 2600%
8 (a) 0.31 (b) 0.07 (c) 0.455 (d) 0.00062 (e) 2.57
9 35.6 **10** 25% **11** 300 **12** 30 months **13** $48.69
14 $154.88 **15** 19% **16** 24,752 people **17** $24,829.05

Chapter 4

4.1.1   **1**   (a) $2\frac{1}{3}$ (b) $2\frac{2}{5}$ (c) $7\frac{1}{2}$ (d) $3\frac{1}{7}$ (e) $4\frac{1}{3}$ (f) $6\frac{4}{9}$ (g) $6\frac{5}{32}$

  (h) $6\frac{10}{16}$ **2** (a) $\frac{7}{3}$ (b) $\frac{11}{2}$ (c) $\frac{15}{4}$ (d) $\frac{17}{3}$ (e) $\frac{111}{11}$ (f) $\frac{76}{3}$

  (g) $\frac{39}{5}$ (h) $\frac{702}{7}$

4.2.1   **1** $\frac{32}{40}$ **2** $\frac{27}{63}$ **3** $\frac{18}{45}$ **4** $\frac{30}{90}$ **5** $\frac{150}{350}$ **6** $\frac{2}{4}$ **7** $\frac{3}{4}$ **8** $\frac{5}{15}$ **9** $\frac{1}{5}$ **10** $\frac{2}{4}$

4.2.2   **1** 2, 13, 37   **2** (a) $2 \cdot 2 \cdot 2 \cdot 3$ (b) $2 \cdot 19$ (c) $2 \cdot 2 \cdot 2 \cdot 7$

  (d) $2 \cdot 2 \cdot 2 \cdot 2 \cdot 2 \cdot 3$ (e) $2 \cdot 2 \cdot 2 \cdot 2 \cdot 3 \cdot 3$ (f) $2 \cdot 2 \cdot 5 \cdot 13$

  3 (a) 3   (b) 6   (c) 9   (d) 14   (e) 85   (f) 16

  4 (a) $\frac{3}{5}$ (b) $\frac{1}{6}$ (c) $\frac{1}{9}$ (d) $\frac{8}{9}$ (e) $\frac{3}{8}$ (f) $\frac{3}{5}$

4.2.3   **1** 36 weeks   **2** \$183.82   **3** $6\frac{1}{2}$ hours   **4** 33 gallons

4.3.1   **1** $1\frac{1}{5}$ **2** $\frac{2}{15}$ **3** $\frac{6}{77}$ **4** $\frac{4}{27}$ **5** $\frac{3}{10}$ **6** $\frac{63}{160}$ **7** 1   **8** 3   **9** $\frac{4}{9}$ **10** $-\frac{1}{8}$

  11 $-3\frac{3}{8}$ **12** 4   **13** $7\frac{9}{16}$ **14** $\frac{5}{12}$ **15** $\frac{1}{9}$ **16** -1

4.3.2   **1** $1\frac{1}{2}$ **2** $\frac{2}{3}$ **3** 2   **4** $\frac{14}{15}$ **5** $\frac{3}{8}$ **6** $\frac{1}{6}$ **7** 7   **8** $4\frac{1}{2}$ **9** $1\frac{1}{5}$ **10** $\frac{4}{5}$

  11 6   **12** 14   **13** $3\frac{1}{2}$ **14** $\frac{2}{5}$ **15** $\frac{3}{4}$ **16** $2\frac{1}{10}$

4.4.1   **1** (a) $\frac{1}{3}$ (b) 1   (c) $\frac{6}{11}$ (d) $\frac{3}{4}$ (e) $5\frac{4}{7}$ (f) $7\frac{2}{9}$

  2 (a) $\frac{2}{13}$ (b) $\frac{1}{3}$ (c) $\frac{1}{3}$ (d) $1\frac{1}{5}$ (e) $\frac{1}{2}$ (f) -3

  3 (a) $\frac{21}{100}$ (b) $\frac{9}{1000}$ (c) $\frac{37}{125}$ (d) $\frac{253}{5,000}$

  4 (a) 24   (b) 60   (c) 36   (d) 240

  5 (a) $1\frac{1}{2}$ (b) $1\frac{11}{24}$ (c) $1\frac{19}{21}$ (d) $\frac{41}{70}$ (e) $\frac{2}{5}$ (f) $\frac{2}{3}$

  6 (a) $1\frac{2}{3}$ (b) $1\frac{3}{5}$ (c) $1\frac{1}{4}$ (d) $\frac{3}{7}$ (e) $2\frac{1}{2}$ (f) $\frac{4}{5}$

  7 (a) $2\frac{11}{30}$ (b) $9\frac{37}{40}$ (c) $9\frac{7}{15}$ (d) $3\frac{5}{88}$ (e) $5\frac{11}{24}$ (f) $-2\frac{1}{5}$

4.5.1   **1** $\$368\frac{1}{2}$ **2** $2122\frac{1}{2}$ cents   **3** $\frac{1}{6}$ bushel   **4** $20\frac{7}{13}$ cents   **5** $24\frac{2}{5}$ rods

  6 $\frac{87}{125}$ ton copper and $\frac{87}{500}$ ton tin   **7** 30 fl oz; $\frac{15}{16}$ **8** \$8   **9** $\frac{1}{12}$

  10 (a) $1\frac{1}{6}$ yards   (b) $36\frac{2}{3}$ square yards   (c) $24\frac{1}{3}$ yards

  11 $61\frac{1}{2}$ hours   **12** 9 cookies

4.6.1   **1** $\frac{37}{100}$ **2** $1\frac{1}{4}$ **3** $12\frac{3}{10}$ **4** $17\frac{223}{250}$ **5** $245\frac{49}{200}$ **6** $1\frac{9}{200}$ **7** $\frac{1}{50}$ **8** $\frac{1}{200}$

  9 $\frac{3}{40,000}$ **10** $\frac{3}{10,000}$

4.6.2   **1** 0.2 **2** 0.08 **3** $0.\overline{2}$ **4** $0.\overline{142857}$ **5** $0.\overline{27}$ **6** $2.\overline{3}$ **7** $3.\overline{6}$

  8 $0.\overline{153846}$

4.6.3   **1** (a) $\frac{2}{25}$ (b) $\frac{3}{10}$ (c) $\frac{33}{100}$ (d) $1\frac{1}{20}$ **2** (a) $\frac{3}{40}$ (b) $\frac{1}{12}$

  (c) $\frac{7}{300}$ (d) $\frac{7}{80}$

4.6.4   **1** (a) $83\frac{1}{3}\%$ (b) $73\frac{1}{3}\%$ (c) $54\frac{6}{11}\%$ (d) $133\frac{1}{3}\%$

  2 (a) $4\frac{56}{111}\%$ (b) $733\frac{1}{3}\%$ (c) $1566\frac{2}{3}\%$ (d) $3636\frac{4}{11}\%$

4.7.1   **1** $\frac{4}{7} < \frac{3}{5} < \frac{2}{3}$ **2** $\frac{3}{10} < \frac{8}{15} < \frac{4}{5}$ **3** $\frac{5}{12} < \frac{7}{16} < \frac{9}{18}$ **4** $\frac{3}{14} < \frac{4}{17} < \frac{5}{19}$

Review Exercises

1   (a) $2\frac{1}{4}$ (b) $2\frac{2}{5}$ (c) $-8\frac{2}{7}$ (d) $-2\frac{1}{2}$

2   (a) $\frac{4}{3}$ (b) $\frac{23}{4}$ (c) $\frac{94}{9}$ (d) $-\frac{37}{5}$

3   (a) $\frac{24}{40}$ (b) $\frac{40}{56}$ (c) $\frac{5}{12}$ (d) $\frac{2}{12}$

4   2, 3, 19, 83, 101

5   (a) $2 \cdot 2 \cdot 3 \cdot 3$ (b) $3 \cdot 13$ (c) $2 \cdot 2 \cdot 2 \cdot 7$ (d) $2 \cdot 2 \cdot 2 \cdot 3 \cdot 5$

6   (a) 4   (b) 9   (c) 12   (d) 7   (e) 1

7   (a) $\frac{2}{3}$ (b) $\frac{2}{9}$ (c) $\frac{3}{4}$ (d) $\frac{1}{9}$

8   (a) $\frac{2}{15}$ (b) $3\frac{1}{2}$ (c) $\frac{1}{4}$ (d) $-\frac{9}{16}$ (e) $2\frac{1}{4}$ (f) $\frac{14}{15}$ (g) $-\frac{3}{4}$

  (h) 2   (i) $1\frac{3}{5}$ (j) $-7\frac{1}{2}$ (k) $-2\frac{8}{9}$ (l) $\frac{12}{37}$

9   (a) 60   (b) 126   (c) 144   (d) 1134

10   (a) $1\frac{2}{9}$ (b) $1\frac{1}{10}$ (c) $\frac{7}{24}$ (d) $1\frac{2}{3}$

11   (a) $4\frac{9}{10}$ (b) $1\frac{5}{16}$ (c) $10\frac{3}{10}$ (d) $4\frac{11}{12}$

12   (a) $\frac{1}{4}$; 25%   (b) $\frac{1}{200}$; 0.5%   (c) $1\frac{23}{100}$; 123%   (d) 2; 200%

13 (a) 0.5; 50% (b) 0.75; 75% (c) 0.625; 62.5% (d) 3; 300%
14 (a) $\frac{4}{25}$; 0.16 (b) 1; 1 (c) $\frac{3}{100}$; 0.03 (d) $\frac{13}{200}$; 0.065
15 $4\frac{1}{4}$ lb
16 $15\frac{7}{24}$ min
17 20 points
18 $3\frac{1}{12}$ ft
19 $1\frac{13}{120}$ ohms
20 (a) $397\frac{5}{6}$ sq ft (b) $82\frac{1}{3}$ ft (c) $17\frac{2}{3}$ ft

Chapter 5

5.1.1 **1** $4500 **2** (a) $52 (b) $40 (c) $75 (d) $80 (e) $90
3 (a) $15 (b) $24 (c) $35 (d) $128 (e) $225.12
4 (a) 19.4% (b) 19.4%

5.2.1 **1** (a)

15	Mouton Cadet	4.98	$74.70
2.5	Chatau Ricon	4.85	$12.125
7	Mise Joseph	5.25	$36.75
9	Volnay	16.91	$152.19
12	Valpolicella	3.35	$40.20
36	Mt Vender	13.00	$468.00
Total 81.5			$783.965

(b) Average cost $9.62 per bottle.

2 (a)

Barb	35	$354.60	$10.13
Nancy	33	$335.30	$10.16
Cis	33	$328.20	$ 9.95
Halina	30	$319.60	$10.65
Lee	30	$305.25	$10.18
Luis	34	$342.85	$10.08
Hugo	32	$324.20	$10.13
Total	227	$2310.00	$71.28

(b) $10.18

3 (a)

Bill	8	$673.95	$84.24
Hung	8	$469.15	$58.64
Bob	8	$850.00	$106.25
Mary	8.5	$784.45	$92.29
Maureen	6	$702.40	$117.07
Lisa	8.5	$861.70	$101.38
Alfred	7.5	$513.52	$68.47
Total	54.5	$4855.17	

(b) $89.76 (c) $89.09

4

	Week 1	Week 2	Week 3	Total
Dominick	$242.00	$266.20	$242.00	$750.20
Lucas	320.16	266.80	266.80	853.76
Oritz	73.44	440.64	367.20	881.28
Rodriguez	290.40	290.40	242.00	822.80
Spearman	312.00	260.00	260.00	832.00
Jiminez	313.39	285.56	242.00	840.95
Mitchell	352.58	303.84	253.20	909.62
Brown	338.24	422.80	338.24	1099.28
Azzarano	416.30	283.20	283.20	982.70
Borras	94.88	118.60	94.88	308.36
Krykes	474.24	395.20	395.20	1264.64
Manos	346.08	247.20	247.20	840.48
Tran	339.36	282.80	282.80	904.96
White	302.40	252.00	252.00	806.40
Bloomer	336.38	242.00	242.00	820.38
McKenna	440.64	440.64	367.20	1248.48
Totals	$4992.49	$4797.88	$4375.92	$14,166.29

Total payroll for 3 weeks = $14,166.29

5.2.2	**1**	$106,848.66
	2	(a) $48,681.90 (b) $327.64 (c) $57,839.12
	3	$1,439.08
5.3.1	**1**	$85\frac{5}{8}$ **2** $-\frac{1}{2}$ **3** $\frac{7}{8}$ **4** $76\frac{1}{2}$ **5** $70\frac{5}{8}$ **6** $93\frac{1}{4}$
5.3.2	**1**	$120 **2** (a) $61.60 (b) $296.80
5.4.1	**1**	$169 **2** $-$196
5.4.2	**1**	$2.36 **2** $4.46 **3** $1.95 **4** (a) zone 4; $5.75 (b) zone 3; $2.90
		(c) zone 6; $10.55 (d) Total $19.20
5.4.3		$10.92
5.4.4	**1**	yes **2** (a) same (b) $1.59 half gallon (c) 89¢/12 oz
	3	(a) yes (b) yes (c) $-$ (d) $+$ (e) $+$ (f) yes (g) $+$ (h) $-$
5.4.5		Roads, $119.79; Education, $336.58; Gen. Admin., $48.50;
		Fire, $9.35; County, $23.96; Misc., $35.07
5.4.6	**1**	$3654.50; $3375.00; $4090.00; $3375.00
	2	$7434.50; $5660.00; $7870.00; $6531.00
	3	$11,995.50; $10,140.00; $12,677.75; $11,011.00
5.5.1	**1**	$385.00 **2** 6 coins **3** yes, one
5.5.2	**1**	(a) 31 min (b) 55 min (c) 1 hr 35 min
	2	(a) 10 hr 15 min (b) 14 hr 2 min
	3	(a) 24 min (b) 4 hr 48 min (c) 3 hr 30 min (d) 5 hr 15 min
		(e) 7 hr 45 min
5.5.3	**1**	(a) 11 A.M. (b) 9 P.M.
5.5.4	**1**	(a) $7\frac{1}{2}$ hr (b) 10:20 P.M. **2** First party 3 hr 23 min; the flight
		leaving at 1320 takes 6 hr 8 min.
5.5.5	**1**	(a) 11:20 P.M. (b) 10:20 P.M. **2** 5 A.M.

Review Exercises

1 (a) $9611 (b) $480.55 (c) $10,091.55

2 $1050 **3** 7,894,408 **4** (a) $32,566 (b) 14.3 students

5 (a) grills 40%; tennis racquets 25%; beach towels 44%; cordless telephones 28%; air conditioners 28%; average 33%

6 (a) yen -8.8%; mark -16.5%; dollar $+3.7\%$; pound $+15.2\%$
(b) average 1.6% loss

7 (a) $70 (b) 0 **8** 98¢ **9** (a) 5 P.M. (b) 3 A.M. (c) 11 P.M.
(d) 10 P.M.   **10** $\frac{3}{4}$ pound for 89¢

Chapter 6

6.1.1 **1** (a) $1/10$; 0.1 (b) $1/100$; 0.01 (c) $1/1000$; 0.001 (d) $1/10,000$; 0.0001 **2** (a) 10^{-5}; 0.00001 (b) 10^{-6}; 0.000001
3 (a) 10^{-10}; $1/10,000,000,000$ (b) 10^{-11}; $1/100,000,000,000$

6.1.2 **1** 2.3 **2** 1456 **3** 249,500 **4** 24 **5** 2.3 **6** 0.003 **7** 4.5
8 0.0122 **9** 0.0059 **10** 8.3 **11** 25.6 **12** 1,247,000

6.2.1 **1** (a) 134,000 (b) 4780 (c) 820 (d) 7,352,000 (e) 0.0046
(f) 0.000018 (g) 0.00000000295 (h) 0.000000000097
2 (a) 4.68×10^3 (b) 8.39×10^2 (c) 3.4×10 (d) 7.1×10^{-4}
(e) 1.96×10^{-7} (f) 2.3×10^{-9} (g) 1.68×10^{-18}
(h) 2.7914×10^8
3 (a) 3.5×10^{-3} (b) 1.26×10^4 (c) 3×10^2 (d) 3.58×10^{-2}
(e) 1.982×10^{-3} (f) 5.75×10^{-3} (g) 6×10^5
(h) 1.9853×10^8
4 0.0000000000000000000001672 g
5 0.000000000000000000000053 g **6** 1.5×10^8 km
7 1.4×10^9 years **8** 2.5×10^8 to 3×10^8 years

6.3.1 **1** (a) 500 cm (b) 300 cg (c) 1000 cL (d) 2500 cm (e) 40 cg
(f) 5 cL (g) 0.30 mm (h) 0.53 mg **2** (a) 0.2 m (b) 0.3 g
(c) 4.5 L (d) 12 kg (e) 1.315 km (f) 2 kg (g) 3.8 L
(h) 0.8 kg

6.3.2 **1** (a) 5 dm^2 (b) 30 mm^2 (c) 1.6378 m^2 (d) 2890 mm^2
2 (a) 31.4 dm^2 (b) 40 km^2 (c) 10.3 dm^2 (d) 4900 mm^2

6.3.3 **1** (a) 0.8 dm^3 (b) 20,000 cm^3 (c) 6 dm^3 (d) 1.675 m^3
2 (a) 5 cm^3 (b) 500 mm^3 (c) 5000 mm^3 (d) 0.00305 dm^3

6.3.4 **1** (a) 4 dm^3 (b) 50 cm^3 (c) 6000 cm^3 (d) 5 L **2** (a) 50 cm^3
(b) 0.3 dL (c) 4 L (d) 5 cm^3

6.3.5 **1** 40 L **2** 3.5 dm **3** (a) 181 m (b) 1794 m^2 **4** $2857.50
5 (a) 395 cm^2 (b) 379 dm^2 **6** $89.60 **7** 5×10^{-7} cm
8 (a) 1000 (b) 10,000 Å

6.4.1 **1** 2 gal **2** (a) 720 sq ft (b) 18 outlets (c) 80 sq yd
3 (a) 1 tsp = $\frac{1}{48}$ cup, $\frac{1}{96}$ pt, $\frac{1}{192}$ qt; 1 Tbsp = $\frac{1}{64}$ qt; 1 pt = 96 tsp, 32 Tbsp (b) $4\frac{1}{2}$ (c) 21 Tbsp + 1 tsp (d) 1920 cal
4 1 oz cheese 0.48 fl oz mayo **5** 400 servings
6 (a) $12\frac{1}{2}$ lb peaches (b) $5\frac{5}{9}$ cups jam

6.5.1 **1** 165 cm **2** 6 ft 4 in **3** 2727 kg **4** $1\frac{2}{3}$ ft **5** 0.66 lb **6** 2.4 dL
7 $\frac{1}{8}$ in. **8** 9.5 mm **9** 72 mph **10** 205 lb **11** 71 cm; 86 cm
12 $\approx 16,000$ ft

6.5.2 **1** 2 L for $1.09 **2** 34¢/L **3** $4\frac{3}{8}$ oz for 85¢ **4** same

6.5.3 **1** (a) $3.98 $2.18 $4.77
(b) $1.99 $3.35
(c) $6.99 $6.33
(d) $3.72 $4.20
(e) $2.76 $2.68
(f) $2.59 $5.72
(g) $1.24 $1.07
(h) $0.59 $0.80 $0.86
(i) $.89 $0.90
(j) $2.99 $1.90
(k) $5.98 $3.57
(l) $2.40 $2.96
(m) $0.69 $0.35 $0.53
(n) $0.63 $0.53
(o) $0.32 $0.26
(p) $0.36 $0.30

2 (a) $12.77 $6.23
(b) $2.79 $5.56
(c) $8.99 $5.80
(d) $1.99 $2.37
(e) $3.99 $7.95

Review Exercises

1 (a) 1; 1 (b) 10^{-1}; 0.1 (c) 1/1000; 0.001 (d) 10^{-5}; 1/100,000
(e) 10^{-2}; 0.01 (f) 10^{2}; 100

2 (a) 2.3 (b) 170 (c) 13,840 (d) 342,000 (e) 29,000,000
(f) 380 (g) 0.000000106 (h) 1247

3 (a) 48,300 (b) 820 (c) 0.000002108 (d) 0.003256

4 (a) 4.8×10^{3} (b) 3.16×10^{2} (c) 1.6×10^{-8} (d) 3.8104×10^{7}

5 (a) 6.7 mm (b) 4.892 L **6** (a) 8 fl oz (b) 8.5 miles

7 (a) 88 km (b) 1.3 lb **8** (a) 1.806×10^{24} atoms
(b) 1.204×10^{24} atoms (c) 3.01×10^{24} atoms

9 1.673×10^{-24} g **10** 2.8 glasses **11** $559\frac{1}{6}$ sq ft

Chapter 7

7.2.1 **1** $7y$ **2** $5b$ **3** $5x$ **4** $-4a$ **5** $-x$ **6** a **7** $-5c$ **8** $-4b$ **9** 0
10 $5y$ **11** $10c + 16d$ **12** $13ab - 6bc$ **13** $3x - 11y$ **14** $7m - m^{2}$
15 $2a^{2}b - 2ab^{2}$ **16** $2z^{2} + 3z$ **17** $3a^{2}y$ **18** $-x^{3} + 2x^{2} - 5x$
19 $-3a + c$ **20** $5x^{2}y + 4xy^{2}$

7.3.1 **1** (a) 2^{3} (b) $(-2)^{3}$ (c) g^{4} (d) h^{2} (e) i^{5} (f) j^{3} (g) k^{6}
(h) $(-m)^{4}$ (i) $-t^{3}$ (j) $-(-d)^{4}$ **2** (a) 4 (b) -125 (c) 125
(d) q^{4} (e) $-r^{5}$ (f) $-p^{3}$ (g) p^{3} (h) $-x^{4}$ (i) -108
(j) $-x^{6}$

7.3.2 **1** m^{9} **2** q^{6} **3** x^{6} **4** t^{7} **5** y^{8} **6** p^{5} **7** a^{3} **8** z^{8}

7.3.3 **1** $6x^{3}$ **2** $3t^{6}$ **3** $-12p^{10}$ **4** $15a^{2}x^{4}$ **5** $30s^{6}t^{3}$ **6** $63a^{6}b^{4}c^{5}$
7 $24x^{5}y^{8}$ **8** $-100s^{12}t^{14}$ **9** $30a^{8}b^{3}$ **10** $-30x^{7}y^{4}$

7.3.4 **1** t^{8} **2** c^{6} **3** s^{24} **4** w^{10} **5** b^{12} **6** p^{20}

7.3.5 **1** $-8s^{3}$ **2** $81q^{4}$ **3** $2.25n^{2}$ **4** $a^{6}b^{2}c^{4}$ **5** $25s^{4}t^{2}$ **6** $-24x^{3}y^{6}z^{9}$

7.4.1 **1** x^{3} **2** a **3** $x^{5}y^{2}$ **4** 25 **5** $-a^{3}$ **6** x^{3} **7** $x^{3}y^{2}z$ **8** ac

7.4.2   **1** $6p^3$ **2** $5x^2$ **3** $4y$ **4** $-3z^4$ **5** $4x^2y^2$ **6** $-6q^2r$ **7** $4a^2b$
   8 $-3ac$

7.4.3   **1** 1   **2** 1   **3** 15   **4** 64   **5** 1   **6** 2   **7** 0   **8** 1

7.4.4   **1**   (a) $1/3$ (b) $1/9$ (c) $1/27$ (d) $1/x$ (e) $1/y^2$
      (f) 3   (g) 9   (h) 1   (i) p^2 (j) $-1/a^2$ (k) $-1/a^2$
   2   (a) $1/x^2$ (b) $b^4/2a^4$ (c) y^3/x (d) $-1/(5rs^4t^5)$ (e) $3x^2/y^3$

7.5.1   **1** $4x-12$ **2** xy^2-4xy **3** $15a^2+30ab$ **4** $3a^2b^2-3a^3bc$
   5 $-24x^3y^2+36x^2y$ **6** $-x+13$ **7** $-3x^2+10x-9$
   8 $3xy-4z-z^2$ **9** $5-2/x$ **10** $x+5$ **11** b^2-abc
   12 $10x+1$ **13** $2x+1+1/x$ **14** $1-27t$ **15** $1-15/s+8/s^2$
   16 $5-22/z$ **17** $x+9$ **18** $17-3x$ **19** $x+8$ **20** $10x-10$
   21 $-8x-8$ **22** $x-5y$ **23** $5x-2$ **24** $5a-3$

7.6.1   **1** $2\cdot2\cdot2\cdot3$ **2** $2\cdot2\cdot2\cdot3\cdot3$ **3** $-2\cdot3\cdot3\cdot a\cdot a\cdot b\cdot b\cdot b\cdot c$
   4 $5p(p-7q)$ **5** $4x(2-3x)$ **6** $3y(5x^2-x+10)$
   7 $2ab(-2a+4-3b)$ **8** $3p^2r^2(5r-9p)$
   9 $(1/5)tuv(t^2u+3v)$ **10** $4c(2c-1)$

7.7.1   **1** 7   **2** -3 **3** 12   **4** 7   **5** -18 **6** 5   **7** 13   **8** 39

7.7.2   **1** -216 **2** -16 **3** $25/3$ **4** -1 **5** 21   **6** -17 **7** -66 **8** 18

7.8.1   **1** 32 sq ft   **2** 24 ft   **3** 85 cm^2 **4** 9.2 ft   **5** 1.13 sq ft   **6** 21.4 m
   7 217 miles   **8** 14 miles   **9** 32° F   **10** 100° C   **11** $480   **12** $65

Review Exercises

   1   (a) $5t$ (b) $3c$ (c) $3s$ (d) $-10w$ (e) $3a$ (f) 0
   2   (a) $4a+11b$ (b) $5z+4$ (c) $3x^2-2x+5$
      (d) $3ps+2p+8s$ (e) $3cd-2cd^2$ (f) $c-9c^2$
   3   (a) t^3 (b) v^5 (c) m^4 (d) $-p^4$ (e) x^5 (f) $-y^3$
   4   (a) 4   (b) $-x^3$ (c) y^3 (d) 81   (e) -1 (f) 1
   5   (a) c^4 (b) d^5 (c) $6x^7$ (d) $-6a^6$ (e) x^8 (f) $-30t^9$
   6   (a) p^3 (b) $-s$ (c) -1 (d) $9p^3$ (e) $2a$ (f) 1
   7   (a) 1   (b) 1   (c) -1 (d) 2   (e) a^3 (f) 1
   8   (a) $1/5$ (b) $1/8$ (c) $1/x^2$ (d) 3   (e) 16   (f) 8   (g) $-ab^2c^3$
      (h) $-s^8/r$
   9   (a) $3t-15$ (b) $2x^2+3x$ (c) $-3r^3+3rs^2$ (d) $3ab^3-3a^3b$
      (e) $-17c+49$ (f) $-6x^2y+2xy^2+6x^2-24$
   10   (a) $6x+9$ (b) $3tv^2-v$ (c) $-2/r-3s/r^2+4$
      (d) $21-7/s+35/s^2$ (e) $5a-17-35/a$ (f) $x^2/3+4x-2$
   11   (a) $2\cdot2\cdot3\cdot3$ (b) $2\cdot2\cdot2\cdot3\cdot5$ (c) $-3\cdot5\cdot a\cdot a\cdot a\cdot b\cdot b$
      (d) $23\cdot a\cdot a\cdot c$ (e) $3t(t-5s)$ (f) $2y(8-9y)$ (g) $xy(x-2)$
      (h) $(1/2)(pq)(-p+q/3)$ (i) $5abc(a+2b-20c)$
      (j) $0.01x^2y^2(y+10x-2)$
   12   (a) 12   (b) 7   (c) 6   (d) 5   (e) -22 (f) 12
   13   (a) 30 m^2 **14** 42 ft   **15** 176.6 cm^2 **16** 245 miles   **17** 212° F
   18   0° C

Chapter 8

8.1.1   **1**   (a) $x=10$ (b) $x=3$ (c) $x=-3$ (d) $x=8$ (e) $x=-18$
      (f) $x=10$ (g) $x=9$ (h) $x=64$
   2   (a) $x=1$ (b) $m=6$ (c) $x=6$ (d) $x=3.5$ (e) $y=-3$
      (f) $y=-2$ (g) $x=27$ (h) $x=16$

8.1.2   **1** $x=2$ **2** $x=-1/2$ **3** $a=-4.5$ **4** $x=-13$ **5** $t=2/3$
   6 $a=-2/3$ **7** $x=2$ **8** $x=-2$

8.2.1   **1**   (a) $x = 30$ (b) $t = 32$ (c) $a = 20$ (d) $c = 6/5$
              (e) $x = -3$ (f) $x = 3/4$ (g) $x = 11/5$ (h) $x = -3$
              (i) $x = 15$ (j) $x = 42$ (k) $x = 3/2$ (l) $x = -12$
       2   (a) $x = 2$ (b) $x = 2$ (c) $x = 3$ (d) $x = 6{,}500$

8.3.1   **1**   (a) 9 in.   (b) 8 cm   (c) 9 cm   (d) 4 cm   **2**   (a) 4 years
              (b) \$4000   **3** (a) 2 years   (b) 6%   **4** (a) 2   (b) 4
       5   (a) 32° F   (b) $-40°$ F   **6** (a) 10 g   (b) 8 cm^3 **7** 4 hr 48 min
       8   46 mph   **9** 135 km   **10** 1 hr 25 min

8.4.1   **1**   (a) $x < 9$ (b) $x > -9$ (c) $x > 5$ (d) $x < 8$ (e) $x < 9$
              (f) $x < -8$ (g) $x < 2$ (h) $x > 16$
       2   (a) $x < 4$ (b) $x > 6$ (c) $x < 2$ (d) $x > -4/5$
       3   (a) $x < 5/2$ (b) $x > 2$ (c) $x > 7$ (d) $x < 1$ (e) $x < 8$
              (f) $x < 4$

8.5.1   **1** $x = 5; y = 1$ **2** $x = 1; y = 1$ **3** $x = 1; y = 1$ **4** $x = 2; y = 1$
       5 $x = 0; y = 2$ **6** $x = 2; y = 3$

8.5.2   **1** $x = 3; y = 1$ **2** $x = 1; y = 4$ **3** $x = 3; y = 1$ **4** $x = -1;$
            $y = -1$ **5** $x = 2; y = 2$ **6** $x = 2; y = 0$ **7** $x = 1; y = 1$
       8 $x = 1; y = 2$

Review Exercises

       1   (a) Subtract; $x = 9$ (b) Add; $x = 15$ (c) Divide; $x = 4$
              (d) Multiply; $x = 36$
       2   (a) $n = 4$ (b) $n = 21$ (c) $n = 13$ (d) $n = 2$ (e) $n = 6$
              (f) $n = 1$ (g) $n = 18$ (h) $n = 11$ (i) $n = -20$ (j) $q = 66$
              (k) $p = 36$ (l) $n = 12$ (m) $c = -10/3$
       3   (a) $x = 3$ (b) $x = 20/3$ (c) $d = 6$ (d) $s = -1/5$ (e) $p = 2$
       4   (a) $x = 66$ (b) $x = 180$ (c) $x = 1$ (d) $x = -0.3$
       5   (a) $w = 5$ cm   (b) $b = 10$ in.   (c) $r = 10$ cm   (d) $b = 12$ cm
              (e) $t = 4$ years   (f) $r = 6\%$ (g) $C = -40°$ C   (h) $F = -40°$ F
       6   (a) $x < 32$ (b) $x < 23$ (c) $x > 4$ (d) $x > 10$ (e) $x > -5$
              (f) $x < -2$ (g) $x < -2$ (h) $x > 1/3$
       7   (a) $x = 3; y = 2$ (b) $x = 1; y = 1$ (c) $x = 1; y = 1$
              (d) $x = 2, y = 2$
       8   (a) $x = 3; y = 1$ (b) $x = 1; y = 1$ (c) $x = 3; y = 2$
              (d) $x = 4; y = -3$

Chapter 9

9.1.1   **1**   **2**

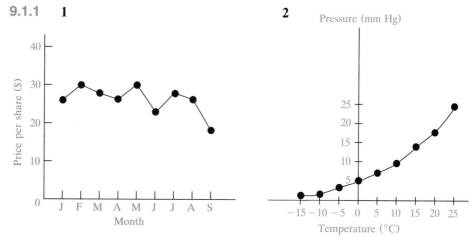

3

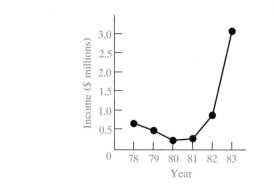

9.2.1 **1**

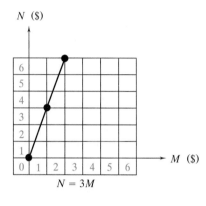

$$N = 3M$$

2

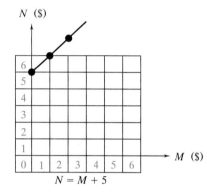

$$N = M + 5$$

3

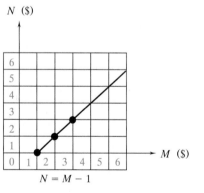

$$N = M - 1$$

4

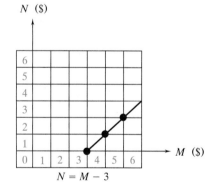

$$N = M - 3$$

5

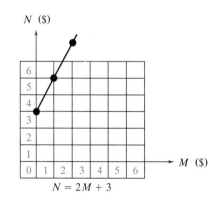

$$N = 2M + 3$$

9.2.2 1

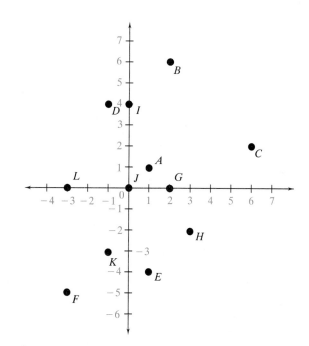

2

A $(-6, 6)$	H $(-4, -2)$
B $(0, 7)$	I $(4, -2)$
C $(6, 7)$	J $(-6, -5)$
D $(-4, 3)$	K $(0, -5)$
E $(4, 4)$	L $(6, -4)$
F $(-6, 0)$	
G $(6, 0)$	

9.3.1 1 $y = 2x$

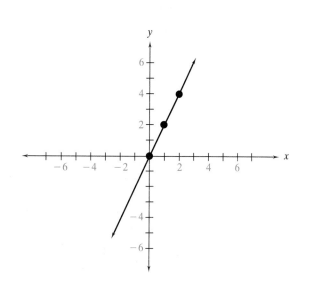

2 $y = x + 4$

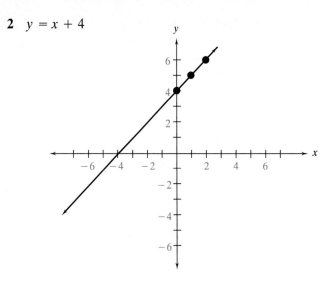

3 $y = -\frac{1}{3}x$

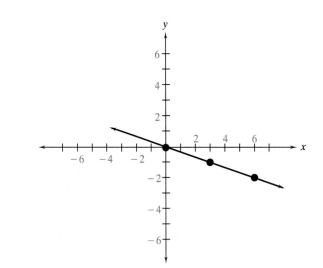

4 $y = x - 4$

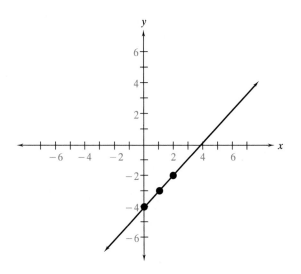

5 $y = 2x + 1$

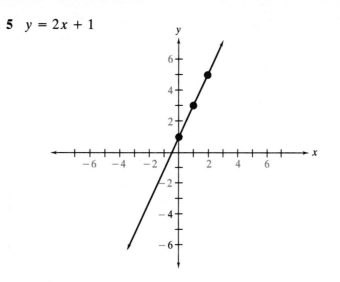

6 $y = 3x - 4$

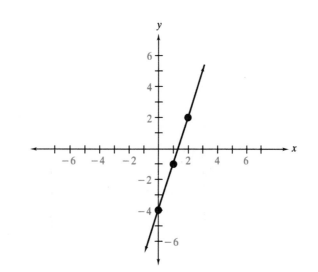

7 $y = \frac{2}{3}x + 1$

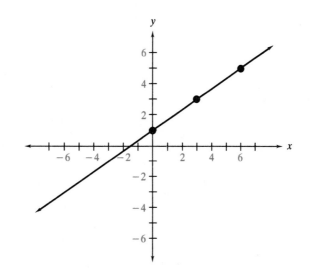

8 $y = -x + 5$

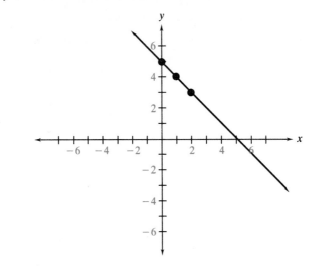

9 $y = -3x$

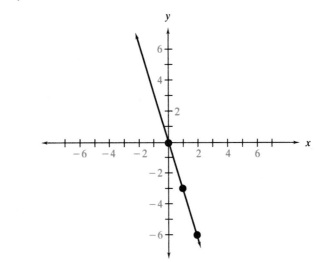

10 $y = -3x + 1$

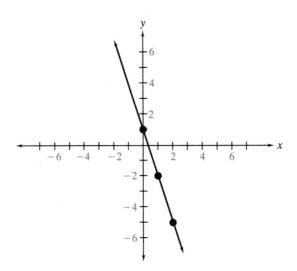

11 $y = -x - 2$

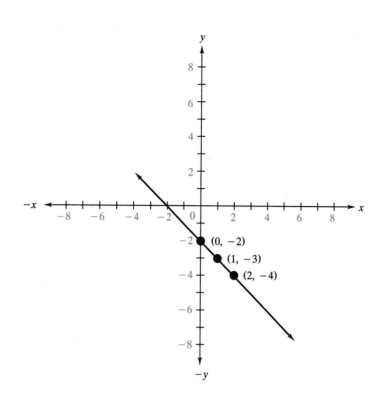

12 $y = \frac{1}{2}x$

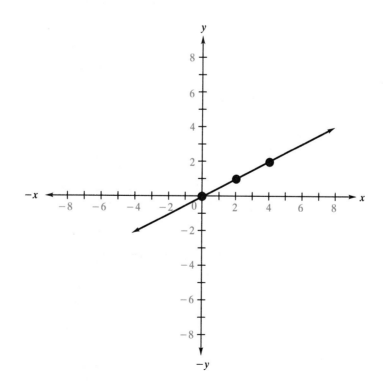

13 $y = -2x + 5$

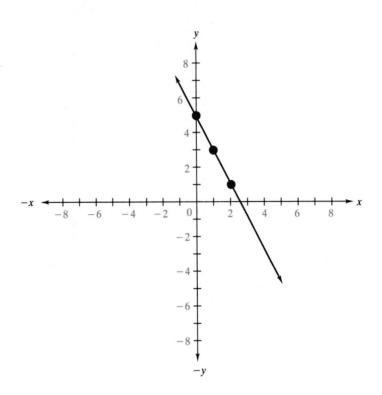

14 $y = -2x + 3$

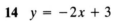

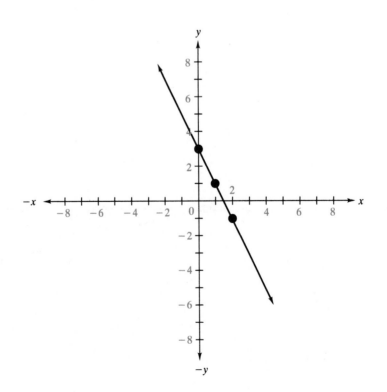

15 $y = -2x - 1$

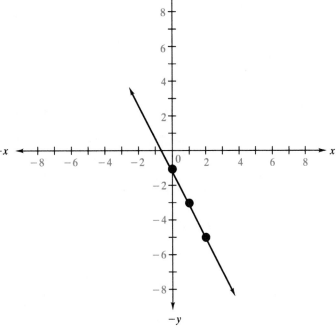

16 $y = -\frac{1}{4}x - 2$

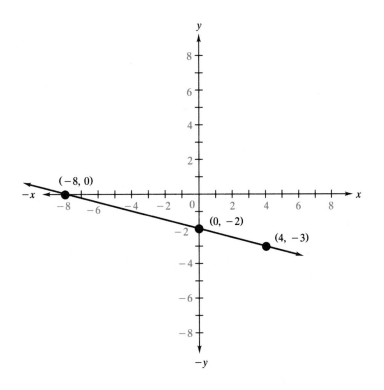

17 $y = 2$

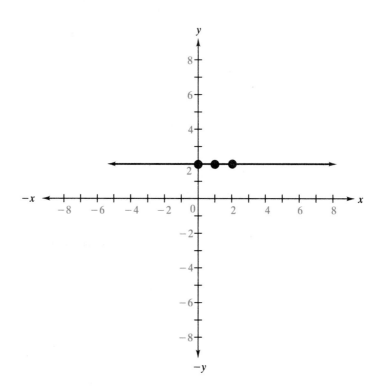

18 $x = -3$

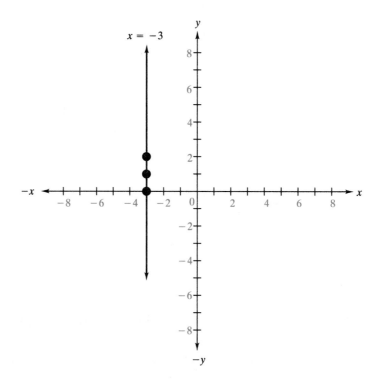

19 $y = -3$

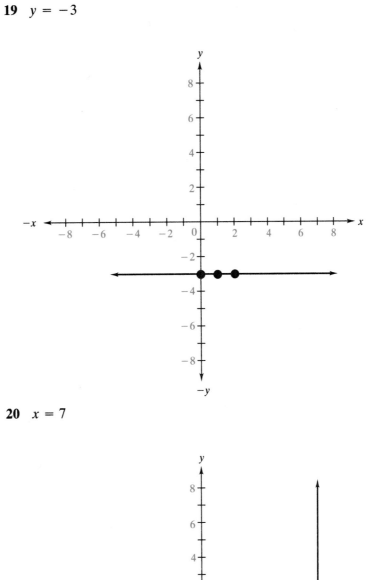

20 $x = 7$

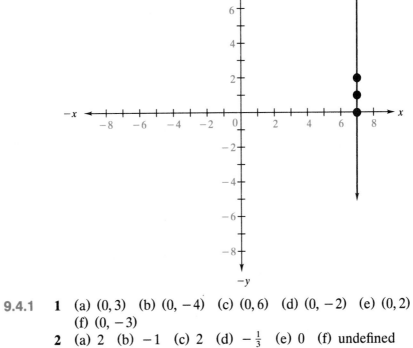

9.4.1 **1** (a) $(0, 3)$ (b) $(0, -4)$ (c) $(0, 6)$ (d) $(0, -2)$ (e) $(0, 2)$
(f) $(0, -3)$

2 (a) 2 (b) -1 (c) 2 (d) $-\frac{1}{3}$ (e) 0 (f) undefined

9.5.1   **1** $m = 1; b = -2$

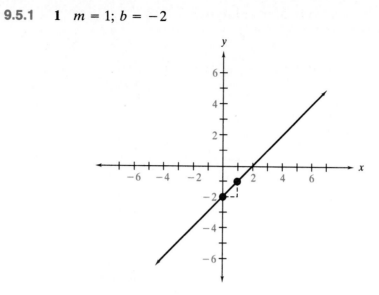

2 $m = 5; b = 0$

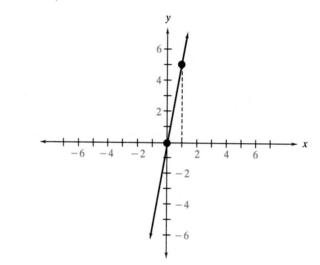

3 $m = 1/2; b = -5$

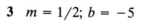

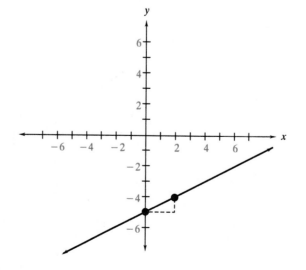

4 $m = -3; b = -1$

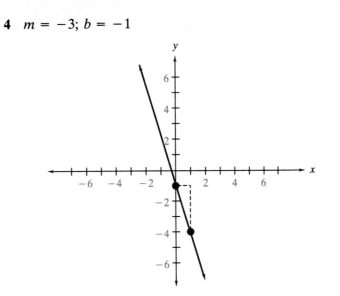

5 $m = 3; b = 2$

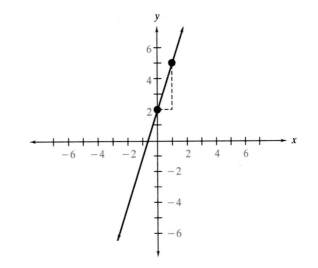

6 $m = -2/3; b = 1$

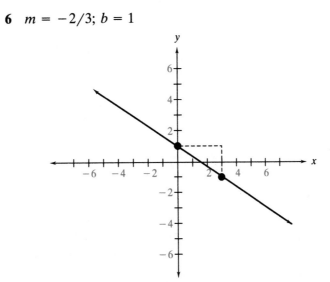

7 $m = 3/2; b = -2$

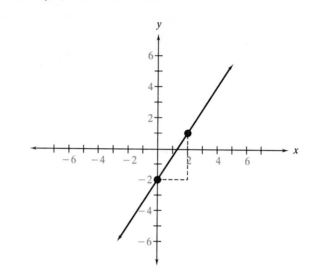

8 $m = 3/4; b = -6$

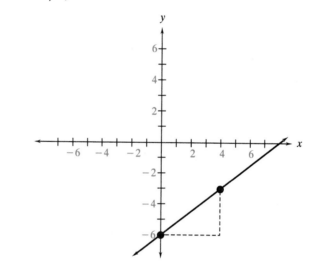

9 $m = -3/2; b = 3$

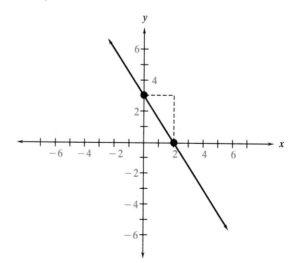

10 $m = 2/5; b = -2$

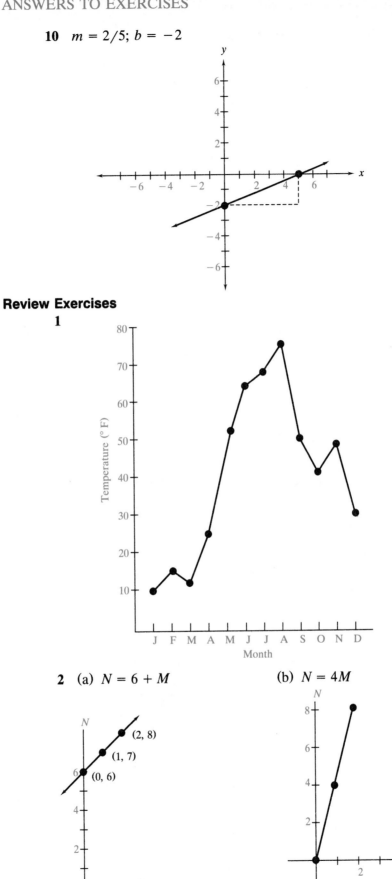

Review Exercises

1

2 (a) $N = 6 + M$ (b) $N = 4M$

(c) $N = M - 3$

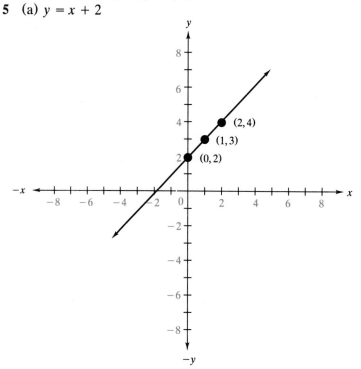

(d) $N = 1 + 2M$

3

4 A $(0, 0)$, B $(2, 2)$, C $(-1, 6)$ D $(0, 4)$, E $(-7, 1)$, F $(-2, -2)$, G $(0, -6)$, H $(5, -1)$, I $(8, 0)$

5 (a) $y = x + 2$

(b) $y = x - 3$

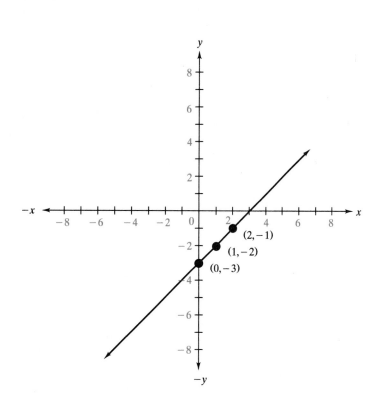

(c) $y = 3x$

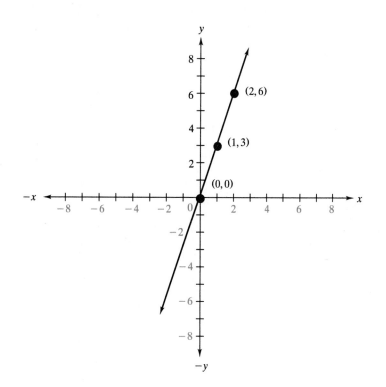

(d) $y = 3$

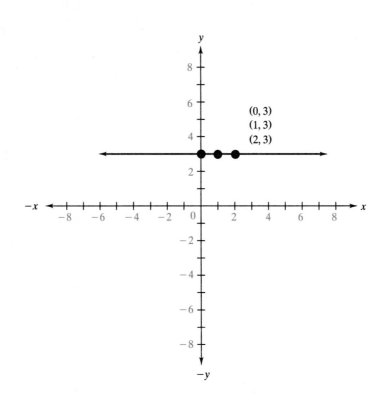

(e) $y = 2x + 1$

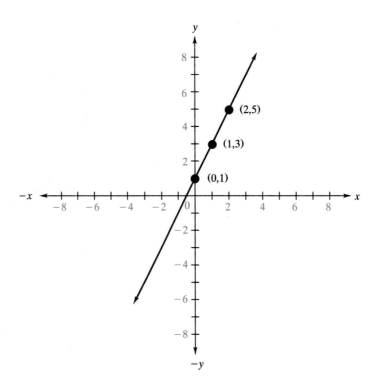

(f) $x = 4$

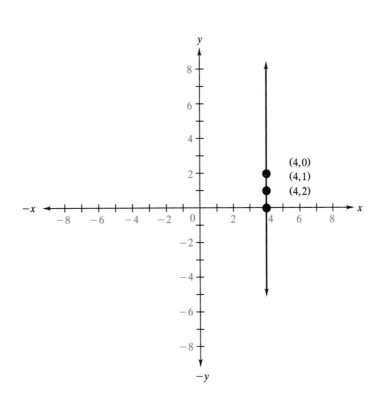

(4,0)
(4,1)
(4,2)

(g) $y = -x + 2$

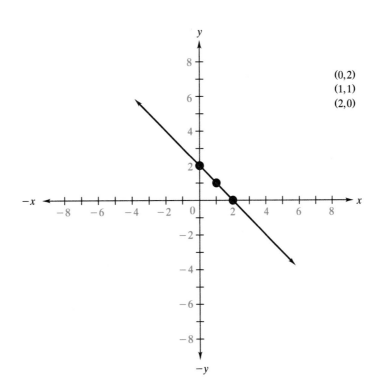

(0,2)
(1,1)
(2,0)

(h) $y = 3x - 2$

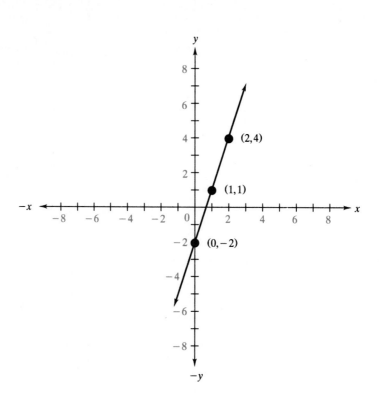

6   (a) $m = \frac{3}{2}$

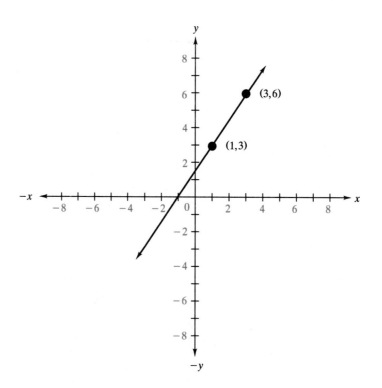

(b) $m = 1$

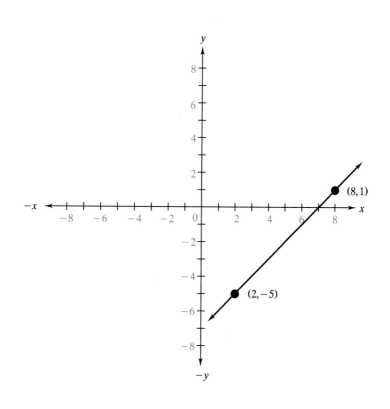

(c) $m = -\frac{7}{3}$

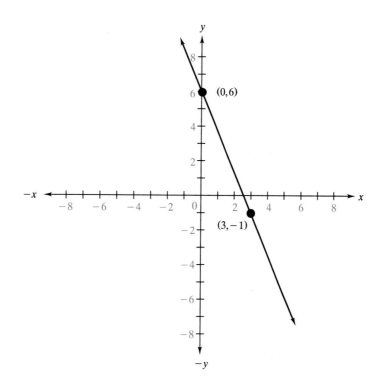

(d) $m = -\frac{1}{2}$

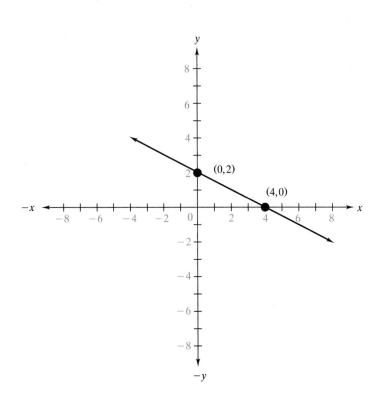

(e) $m = 0$

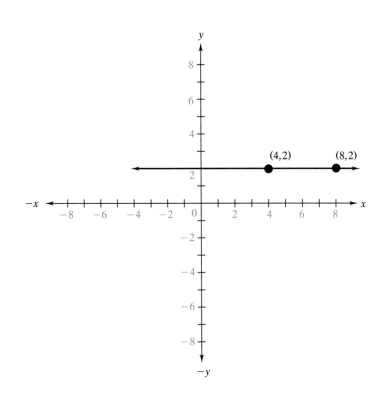

(f) $m = -1$

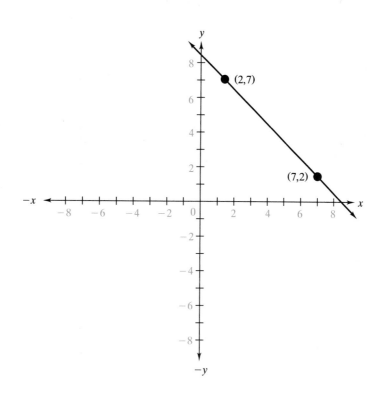

7   (a) $m = 1$; $b = 2$

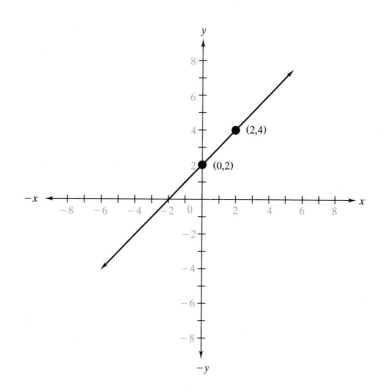

(b) $m = 1$; $b = -5$

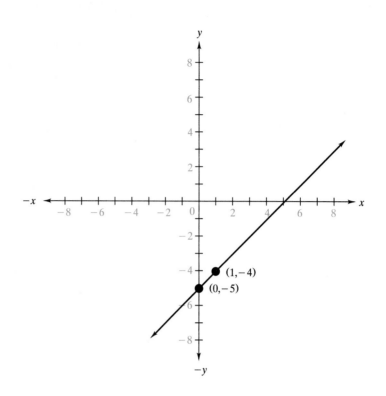

(c) $m = 2$; $b = 0$

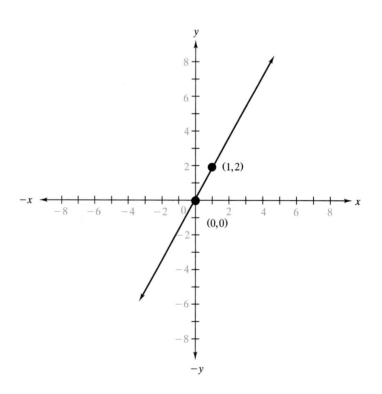

(d) $m = -2$; $b = 4$

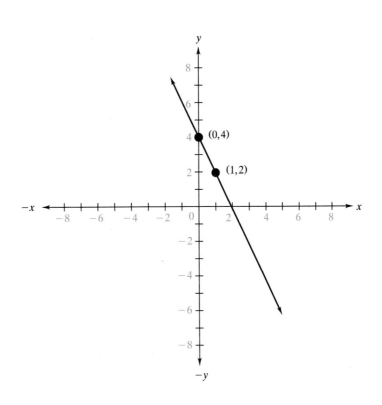

(e) $m = 1/2$; $b = 0$

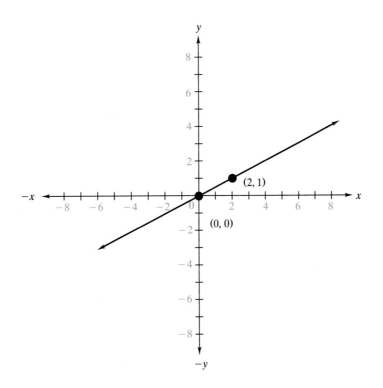

(f) $m = 0$; $b = -3$

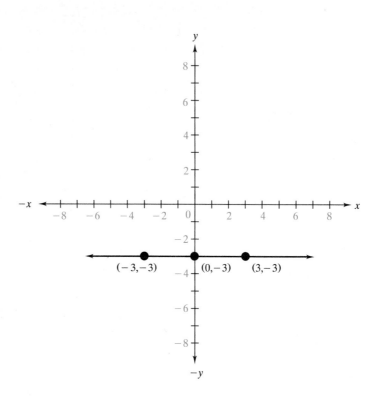

(g) $m = -1$; $b = 6$

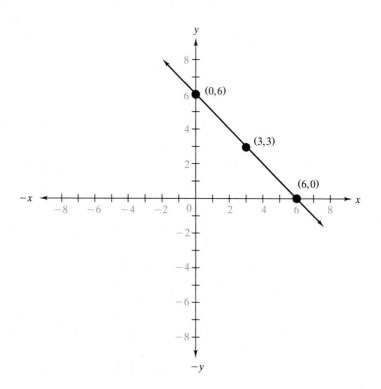

(h) $m = 2; b = -2$

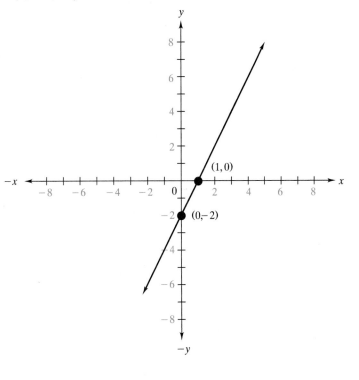

Chapter 10

10.1.1 **1** $3 + 25$ **2** $12 + 5$ **3** $11 + 5$ **4** $15 - 3$ **5** $10 - 3$ **6** $5(4 + 7)$
7 $2 - 9$ **8** $5 \div 6$ **9** $12 - x$ **10** $x - 5$ **11** $5(6 + y)$
12 $(3p + 5q)4$ **13** $20 \div x$ **14** $3t \div x$ **15** $4x \div (a + b)$
16 $5c(4x - 9y)$

10.2.1 **1** $x + 4 = 10; x = 6$ **2** $x + 25 = 36; x = 11$ **3** $x - 3 = 20$;
$x = 23$ **4** $7 - x = 11; x = -4$ **5** $2x = 18; x = 9$
6 $5x = 40; x = 8$ **7** $2x + 3 = 31; x = 14$ **8** $3x + 25 = 40$;
$x = 5$ **9** $5x - 8 = 20; x = 5.6$ **10** $8x - 9 = 7; x = 2$
11 $2x + 13 = 25; x = 6$ **12** $25 + 3x = 34; x = 3$
13 $5x - 8 = 22; x = 6$ **14** $6x + 7 = 37; x = 5$
15 $2x = 5 + x; x = 5$ **16** $7 + x = 2x; x = 7$

10.2.2 **1** $\frac{1}{5}x < 4; x < 20$ **2** $2(5 + x) \geq 26; x \geq 8$ **3** $\frac{2}{3}x \geq 10; x \geq 15$
4 $15 - 2x \leq 41; x \geq -13$ **5** $60 - 6x \leq 132; x \geq -12$
6 lunch ≤ 400; breakfast ≤ 200; dinner ≤ 600
7 width ≤ 10.5 cm, length ≤ 31.5 cm **8** at least 91

10.3.1 **1** 7, 8, 9 **2** 17, 18, 19 **3** 13, 15, 17 **4** 20, 22, 24 **5** 16 **6** 14, 18
7 10, 12 **8** 94

10.3.2 **1** 20% **2** 16% **3** 40% **4** 30% **5** 3.75% **6** 14% **7** 600 **8** 80
9 16,000 **10** 45,000 **11** 400 **12** $120 **13** $25.95 **14** $4500
15 $126 **16** $7000 at 10%; $5000 at 14% **17** $800 at 6%;
$400 at 5% **18** $401.50

10.3.3 **1** 45 ft **2** 33 gal **3** 24¢ **4** 7.5 gal **5** width 72 ft; length 108 ft
6 24 in.; 36 in. **7** 10 acres corn; 5 acres potatoes **8** $420

10.3.4 **1** $l = 18$ in.; $w = 9$ in. **2** $w = 4$ ft; $l = 8$ ft **3** $17\frac{1}{3}$ cm; $17\frac{1}{3}$ cm;
$25\frac{1}{3}$ cm **4** $w = 6$ units; $l = 20$ units **5** 6 ft; 6 ft; 2 ft
6 7 m; 8 m; 12 m **7** 55 sq ft **8** 14.9 sq ft

10.3.5 **1** 3 hr **2** 46 mph; 56 mph **3** 2 mph **4** (a) 1 hr (b) 4 miles
5 192.5 miles **6** 17 mph **7** Jim 200 mph; John 600 mph
8 1.54 mph

10.3.6 **1** Jane 9 yr; Mary 24 yr **2** Liz 30 yr; John 36 yr **3** Betsey 5 yr;
Father 36 yr **4** Casey 8 yr; Aaron 4 yr **5** Fiori 11 yr; Omn 8 yr
6 Deborah 34 yr; Daughter 9 yr **7** Aaron 4 yr; Cassandra 8 yr
8 David 35 yr; Jonathan 42 yr

10.3.7 **1** 18 min **2** $37\frac{7}{9}$ days **3** Assad 18 days and Abduhl 36 days
4 $2\frac{11}{12}$ days **5** 2 hr 24 min **6** 12 hr **7** 2:43 P.M.

10.3.8 **1** 36 dimes, 72 pennies **2** 14 dimes, 33 nickels **3** 27 qt @ 28¢,
18 qts @ 33¢ **4** 1756 adults, 744 children **5** 100 @ 23¢,
50 @ 29¢ **6** 66 dimes, 80 quarters **7** 23.3% **8** 20 fl oz

10.4.1 **1** Lars 165 cm; Hans 147 cm **2** wine 85 gal; cider 35 gal
3 25 yr **4** 84 **5** In-state 6250; out-of-state 2500 **6** 23 days
7 Dave 30 min; Deb 20 min; Judy 60 min

10.4.2 **1** $14\frac{1}{6}$, $15\frac{1}{6}$, $16\frac{1}{6}$, $17\frac{1}{6}$, $18\frac{1}{6}$, $19\frac{1}{6}$ **2** 60 years **3** (a) 3600 stades
(b) 2400 stades **4** $14\frac{28}{97}$ **5** 60 drachmas **6** 336 walnuts
7 18 mangos **8** mother $218\frac{2}{11}$; son $363\frac{7}{11}$; daughter, $218\frac{2}{11}$.

Review Exercises

1 (a) ab (b) $a + b$ (c) $a - b$ (d) a/b (e) a/b (f) $a - b$
(g) $a - b$ (h) $a - b$ **2**(a) $x - 4 = 11$; $x = 15$ (b) $x/8 = 2$;
$x = 16$ (c) $8 \div x = 2$; $x = 4$ (d) $2x - 5 = 11$; $x = 8$
(e) $(x + 6) \div 7 = 3$; $x = 15$ (f) $3x + 3 = 24$; $x = 7$

3 21, 22, 23 **4** 11, 13, 15, 17
5 22% **6** 106.25 **7** 36,000 **8** $3\frac{3}{4}$ days **9** $3000
10 2 yr **11** 18 yr **12** Dick is 5 yr old; Tom is 10 yr old;
Harry is 16 yr old **13** Bob 17 yr; Pete 7 yr **14** 1 hr 20 min
15 9 days **16** 5 quarters, 12 dimes

Chapter 11

Note: Answers may vary slightly based on the number of decimals used.

11.1.1 **1**

Month	Principal	Interest	Loan Repayment
1	6000.00	80.00	464.39
2	5535.61	73.80	470.58
3	5065.03	67.53	476.86
4	4588.17	61.18	483.21
5	4104.96	54.73	489.66
6	3615.30	48.20	496.19
7	3119.11	41.59	502.80
8	2616.31	34.88	509.51
9	2106.81	28.09	516.30
10	1590.51	21.21	523.18
11	1067.32	14.23	530.16
12	537.17	7.16	537.23

Finance charge = $532.60

11.1.2 **1**

Month	Principal	Interest	Loan Repayment
1	574.79	3.83	46.17
2	528.62	3.52	46.48
3	482.15	3.21	46.79
4	435.36	2.90	47.10
5	388.26	2.59	47.41
6	340.85	2.27	47.73
7	293.12	1.95	48.05
8	245.08	1.63	48.37
9	196.72	1.31	48.69
10	148.02	0.99	49.01
11	99.01	0.66	49.34
12	49.67	0.33	49.67

11.1.3 **1** $\frac{36}{78}$ **2** $\frac{10}{78}$
11.1.4 **1** $49,722.01 **2** $3590.44
11.2.1 **1**

Principal	Time	New Principal
1000.00	1	1008.30
1008.30	2	1016.67
1016.67	3	1025.11
1025.11	4	1033.62
1033.62	5	1042.19
1042.19	6	1050.84
1050.84	7	1059.57
1059.57	8	1068.36
1068.36	9	1077.23
1077.23	10	1086.17
1086.17	11	1095.18
1095.18	12	1104.27

2 Compound interest $16.60; simple interest $16.00
3 $382.81; $370
11.2.2 **1** (a) $1088.39 (b) $1087.75 (c) $1088.71
11.2.3 **1** OK
11.2.4 **1** $1999.89 **2** $1973.59 **3** $2025.82 **4** $1979.93 **5** $2012.20
        6 $1967.15 **7** $1999.00 **8** $1992.56 **9** $1948.72
11.2.5 **1** (a) 6 years (b) 8 years **2** 14%

Review Exercises

1

Year	Principal		New Principal
1	$5000		
2	$5416.39	+ $5000 =	$10,416.39
3	$11,283.84	+ $5000 =	$16,283.84
4	$17,639.92	+ $5000 =	$22,639.92
5	$24,525.32	+ $5000 =	$29,525.32
6	$31,984.12	$\approx$	$32,000

2

Year	Principal	Interest	Repayment
1	2873.97	19.16	230.84
2	2643.13	17.62	232.38
3	2410.75	16.07	233.93
4	2176.82	14.51	235.49
5	1941.33	12.94	237.06
6	1704.28	11.36	238.64
7	1465.64	9.77	240.23
8	1225.41	8.17	241.83
9	983.58	6.56	243.44
10	740.14	4.93	245.07
11	495.07	3.30	246.70
12	248.37	1.66	248.34

3

Year	Principal	Interest	Repayment
1	60,000.00	450.00	89.84
2	59,910.20	449.33	90.51
3	59,819.60	448.65	91.19
4	59,728.50	447.96	91.88
5	59,636.60	447.27	92.57
6	59,544.00	446.58	93.26
7	59,450.80	445.88	93.96
8	59,356.80	445.18	94.66
9	59,262.10	444.47	95.37
10	59,166.80	443.75	96.09
11	59,070.70	443.03	96.81
12	58,973.90	442.30	97.54

4 $88.87   **5** $23.01   **6** 4.6%   **7** 15 years   **8** $5.81 in 1992
(They actually cost about $1/lb.)

Chapter 12

12.1.1   **1**   Use your own results.
              2

Score	Frequency
97-99	4
94-96	1
90-93	10
87-89	4
84-86	5
80-83	3
77-79	5
74-76	5
70-73	7
67-69	2
64-66	1
60-63	5
0-59	2

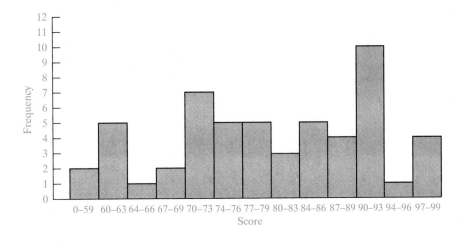

3 Graphs will differ depending on intervals.

4

Score	Frequency
0	1
1	0
2	3
3	5
4	7
5	8
6	6
7	4
8	2
9	3
10	1

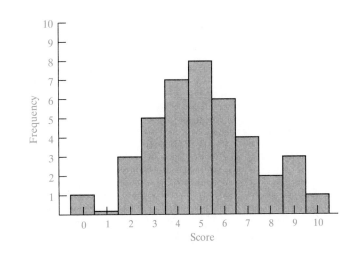

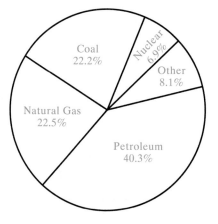

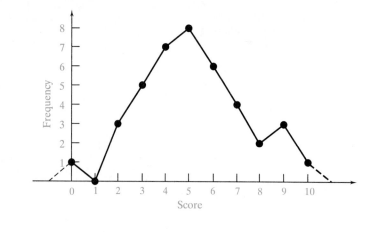

5

Coal	22.2%	80°
Natural gas	22.5%	81
Petroleum	40.3%	145
Nuclear	6.9%	25
Other	8.1%	29
Total	100%	360°

12.2.1 **1** Mean 5; median 4; mode 4; range 8; SD = 2.6

2 Mean 5; median 6; mode 1; 7 (bimodal); range 8; SD = 3

3 Mean 9; median 5; mode 5; range 37; SD = 10

12.3.1 **1** (a) $\frac{1}{500}$ (b) $\frac{5}{500}$ **2** $\frac{1}{8}$ **3** $\frac{3}{8}$ **4** $\frac{4}{8}$ **5** $\frac{3}{36}$ **6** (a) $\frac{30}{36}$ (b) $\frac{3}{36}$ **7** $\frac{21}{36}$

8 $\frac{6}{52}$ **9** (a) $\frac{13}{52}$ (b) $\frac{8}{52}$ (c) $\frac{16}{52}$ (d) $\frac{48}{52}$ (e) $\frac{29}{52}$

10 (a) $\frac{1}{8}$ (b) $\frac{3}{8}$ (c) $\frac{1}{8}$ (d) $\frac{7}{8}$ (e) $\frac{3}{8}$ **11** (a) $\frac{1}{36}$ (b) $\frac{10}{36}$ (c) $\frac{25}{36}$

(d) $\frac{11}{36}$ **12** (a) $\frac{4}{20}$ (b) $\frac{6}{20}$ (c) $\frac{10}{20}$ (d) $\frac{13}{20}$ (e) $\frac{14}{20}$

12.3.2 **1** 4 to 48 **2** 30 to 6 **3** 21 to 15 **4** 1 to 15

5 (a) 13 to 39 (b) 8 to 44 (c) 16 to 36 (d) 48 to 4 (e) 29 to 23

6 (a) 4 to 16 (b) 6 to 14 (c) 10 to 10 (d) 13 to 7 (e) 14 to 6

Review Exercises

1 Use your own results.

2 Example of one solution:

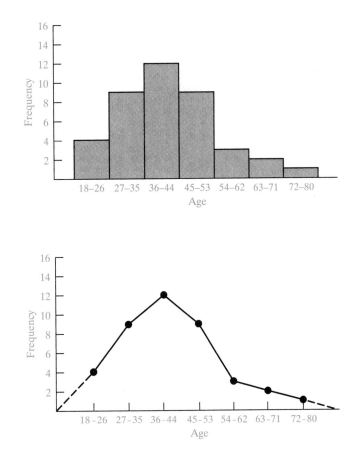

Interval	Frequency
18–26	4
27–35	9
36–44	12
45–53	9
54–62	3
63–71	2
72–80	1

3 (a)

68	85	242	86	106	65	81	174
70	77	76	117	80	81	77	82
63	90	64	86	61	77	64	68
157	352	71	75	72	74	75	93
65	213	61	82	78	89	84	89
59	82	153	68	175	53	63	63
109	84	70	80	128	137	76	69
64	184	259	142	56	81	71	63
64	150						

(b) Range 53–352. Ten intervals (There are other possibilities depending on what intervals you choose.)

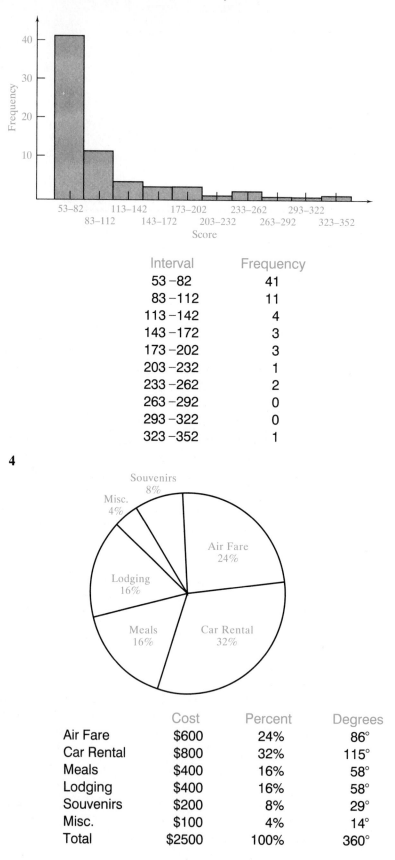

Interval	Frequency
53–82	41
83–112	11
113–142	4
143–172	3
173–202	3
203–232	1
233–262	2
263–292	0
293–322	0
323–352	1

4

	Cost	Percent	Degrees
Air Fare	$600	24%	86°
Car Rental	$800	32%	115°
Meals	$400	16%	58°
Lodging	$400	16%	58°
Souvenirs	$200	8%	29°
Misc.	$100	4%	14°
Total	$2500	100%	360°

5

	(a)	(b)	(c)
Mean	7	10	7
Median	7	9	8
Mode	2, 9	None	9
Range	9	16	12
SD	3	5	3

6 (a) $\frac{4}{8}$; 4 to 4 (b) $\frac{7}{8}$; 7 to 1 (c) $\frac{4}{8}$; 4 to 4 (d) 0; no odds
(e) $\frac{5}{36}$; 5 to 31 (f) $\frac{18}{36}$; 18 to 18 (g) $\frac{12}{36}$; 12 to 24 (h) $\frac{4}{36}$; 4 to 32
(i) $\frac{16}{52}$; 16 to 36 (j) $\frac{26}{52}$; 26 to 26 (k) $\frac{13}{52}$; 13 to 39 (l) $\frac{2}{52}$; 2 to 50

READINESS CHECKS
ANSWERS AND SOLUTIONS TO SELECTED PROBLEMS

Chapter 1, first part

1 The counting numbers are 1, 2, 3.... Multiples of 3 are 3, 6, 9, ...
Answer: {3, 6, 9, ...}
2 Undefined **3** 5020 **4** 1, 2, 3, 4, 6, 9, 12, 18, 36 **5** Composite, because 51
is divisible by 3 and 17 **6** 84, 91, 98 **7** 75 > 31 **8** $10 - 2(2 - 1 + 2 \times 2) =$
$10 - 2(2 - 1 + 4) = 10 - 2(5) = 10 - 10 = 0$ **9** Associative and commutative **10** $4080 \div 4 = \$1020$

Chapter 1, second part

1 Twenty million **2** Three million fifty-six thousand seventy **3** 376,307
4 101, 103, 107, 109, 113 **5** 40,000 **6** 9 is in the thousands place. Add one to
49 and get 50. Three zeros must be added to show place value.
Answer: 50,000
7 Not prime. It has 17 and 23 as factors. **8** 349,687,349
9 $5,000,000 - 350,000 = 4,650,000$ **10** 63,000 pesos

Chapter 2

1 $8 + 5 = 13$ **2** $19 + (-11) = 19 - 11 = 8$ **3** $-11 + 17 = 6$ **4** 36
5 -5 **6** $-5 + 7 - 3 + 12 - 4 = 19 - 12 = 7$ **7** $-9 + 9 = 0$
8 $\sqrt{25 + 144} = \sqrt{169} = 13$ **9** $3(-3) = -9$ **10** $-11 - (-2) = -11 +$
$2 = -9$ The drop was 9°.

Chapter 3

1 9 **2** Fifteen and seventy-eight thousandths **3** $0.9 > 0.0956$ **4** 4.60 (add 1
to 59; keep the zero.) **5** Three places to the right **6** 0.0005 (move the
decimal point two places to the left) **7** 123% (multiply by 100%)
8 $0.0825 \times 12 = 0.99$ **9** $270 \div 1.08 = 250$
10 Decrease = 30¢; $30 \div 125 = 0.24 = 24\%$

Chapter 4

1 $1\frac{7}{8}$ **2** $\frac{18}{7}$ **3** $45 \div 5 = 9$; $9 \times 4 = 35$; $\frac{36}{45}$ **4** $62 = 2 \times 31$; $93 = 3 \times 31$;
$\frac{62}{93} = \frac{2}{3}$ **5** $5\frac{2}{24} + 4\frac{9}{24} = 9\frac{11}{24}$ **6** $1\frac{1}{5}$ **7** $\frac{15}{4} \times \frac{12}{5} = 9$ **8** $\frac{8}{3} \times \frac{1}{6} = \frac{4}{9}$ **9** $\frac{125}{1000} = \frac{1}{8}$
10 $\frac{7}{13} = 0.538$; $\frac{5}{11} = 0.454$; $\frac{7}{13} > \frac{5}{11}$

Chapter 5

1 $23.00 \div 1.15 = 20$; $20.00   **2** $\dfrac{325.00 - 25.00}{5} = 60$; $60.00   **3** $892.45
4 29¢ + 4 × 23¢ = 121¢; $1.21   **5** $120 \div 0.8 = 150$; $150   **6** 38 min + 17 min = 55 min (It is 5 minutes less than one hour.)   **7** Price per coin = $\frac{1}{4}$ × $340 = $85; $600 \div 85 = 7.06$; 7 coins   **8** $4630 - 334 = 4296$; $4296 \div 334 = 12.86 = 1286\%$
9 $3(-2) + 2(0.5) + 5(-2.5) = -17.5$; $17.5 \div 10 = 1.75$; drop of 1.75 lb
10 Single: $3052.50 + 28\%($15,650) = $7434.50 Twice single: $14,869; Married: Income $72,000; Tax $5100.00 + 28\%($38,000) = $15,740; Difference: $871; they must pay $871 more.

Chapter 6

1 0.0001   **2** 5.260×10^3 **3** 0.40 m   **4** 5.380 kg   **5** $0.4 \text{ dm}^3 = 0.4 \text{ L} = 4 \text{ dL}$
6 $\frac{1}{3}$ × 12 in. = 4 in.   **7** $\frac{47}{16}$ lb = $2\frac{15}{16}$ lb   **8** $165/2.54 = 65$ in. = 5 ft 5 in.
9 99¢/2 L = 49.5¢/L or 49.5¢ × 0.946 = 46.8¢/qt
199¢/(12 × 12) = 1.38¢/fl oz = 1.38¢ × 32 = 44.16¢/qt;
The second option is slightly cheaper.
10 $8.5 \times 2.54 \times 4.5 \times 2.54 \times 2.54 \times 2.54 \text{ cm}^3 = 1567 \text{ cm}^3 = 1.6 \text{ L}$, and 1 qt = 0.946 L. There is plenty of room.

Chapter 7

1 $-2x + 5y$ **2** $5x - 4x + 20 = x + 20$ **3** $6x^9$ **4** $5x^2$ **5** $3^4 x^8 = 81x^8$
6 $1/x^5$ **7** $1 + 1 = 2$ **8** $5x(x + 2)$ **9** $(1) - (-1) = 1 + 1 = 2$
10 $C = (5/9)(5 - 32) = (5/9)(-27) = 5(-3) = -15$; $-15°$ C

Chapter 8

1 $x = 4$ **2** $x = 49$ **3** $x = 6\frac{1}{2}$ **4** $x = 12$ **5** $x = 3$ **6** $x = 0.85$
7 The base is 4 cm   **8** $x > -11$ **9** $x = 2$; $y = 1$ **10** $x = 1$; $y = 5$

Chapter 9

1 $A = (-5, 0)$
2

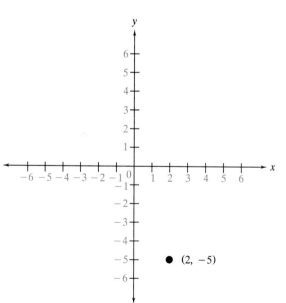

• (2, −5)

3

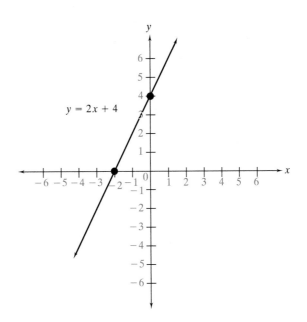

$y = 2x + 4$

4

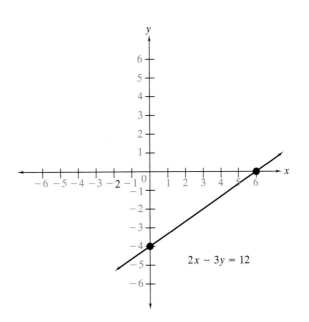

$2x - 3y = 12$

5

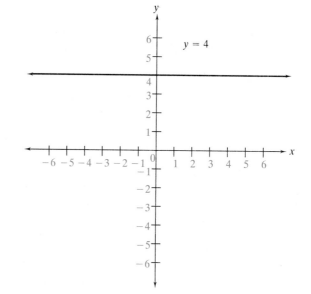

$x = -6$

6

$y = 4$

7 $m = 6$ **8** $b = 4$ **9** $m = \dfrac{9 - 3}{4 - 2} = 3$

10

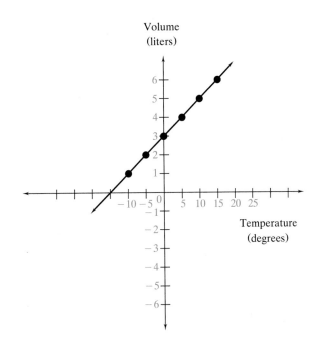

Volume
(liters)

Temperature
(degrees)

Chapter 10

1 $8 - x = 3$; $x = 5$ **2** $13 < 4x$; $x > 3\frac{1}{4}$ **3** $x + x + 2 + x + 4 = 39$; $x = 11$
The numbers are 11, 13, and 15. **4** $6\%x = 36$; $x = 600$ **5** $\frac{2}{3} = x/1500$;
$x = 1000$
6 $x + x + 5 + x + x + 5 = 22$; $x = 3$; width = 3 ft, length = 8 ft
7

	Rate	Time	Distance	
Downstream	$x + y$ km/hr	3 hr	51 km	$3(x + y) = 51$
Upstream	$x - y$ km/hr	3 hr	39 km	$3(x - y) = 39$

$x = 15$ 15 km/hr in calm water
$y = 2$ 2 km/hr rate of current

8 $x + x + 2 = 68$; $x = 33$; $x + 2 = 35$ Answer: Ellen is 35 yr old.
9 $x/15 + x/30 = 1$; $x = 10$; It takes 10 min.
10

	Number of Coins	Value per Coin	Total Value
	x	5¢	$5x$¢
	y	10¢	$10y$¢
Total	$\overline{32}$		$\overline{245}$¢

$x + y = 32$; $5x + 10y = 245$; $x = 15$; $y = 17$; 15 nickels

Chapter 11

1 6%   **2** $(10\%/12)(\$30,000) = \$250$ **3** $146.45   **4** $20,000(1 + 5\%)^{20} = $53,066   **5** $20,000(1 + 10\%)^{20} = \$134,550$ **6** $\frac{6}{78} = \frac{1}{13}$
7 $1000(1 + 4.5\%/12)^{12} = 1045.94$; $1045.94 - 1000 = 45.94$; Interest is $45.94.
8 $45.94/1000 \times 100\% = 4.59\%$ **9** $70 \div 6.2 = 11.3$; After 11 yr
10 It doubles every 15 yr (approx.); $70 \div 15 \approx 5$
It is more accurate to say that it doubles every 12 yr. Then the increase is 5.8%.

Chapter 12

1

Scores	Frequency
63–67	1
68–72	1
73–77	4
78–82	3
83–87	7
88–92	1
93–98	3

2

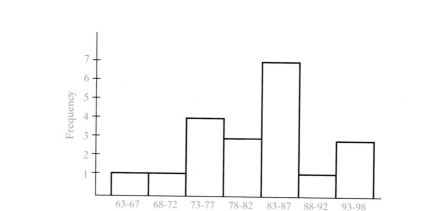

3

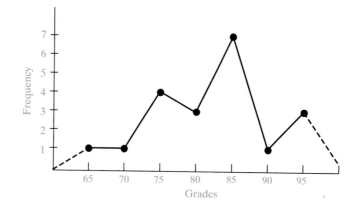

4 Mean 82

5 Median 83

6 Modes 73, 81, 83, 86, 94

7 (a) 18 orange jelly beans; probability $\frac{18}{106}$

(b) Yellow or red: $41 + 36 = 77$; probability $\frac{77}{106}$

(c) Odds $\frac{65}{41}$

(d)

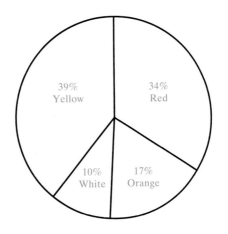

Yellow	39%	140°
Red	34%	122°
White	10%	36°
Orange	17%	61°
	100%	359°

Page numbers in boldface type refer to highlighted rules in the text